Distribution Atlas
of Grasshoppers and Praying Mantises
in Germany

Dedicated to **Prof. Dr. Michael J. Samways** with deep gratitude. He is one of the world's most renowned orthopterists and played a decisive role in putting insect conservation on the global agenda of nature conservation.

Thomas Fartmann | Dominik Poniatowski
(Ed.)

Distribution Atlas of Grasshoppers and Praying Mantises in Germany

201 Pictures
83 Distribution Maps

Contents

Species Profiles of Grasshoppers and Praying Mantises in Germany

Service

Supported by: Observation.org

Supported by:

Federal Ministry
for the Environment, Nature Conservation,
Nuclear Safety and Consumer Protection

Federal Agency for
Nature Conservation

based on a decision of
the German Bundestag

leben.natur.vielfalt
das Bundesprogramm

Preface

We associate the song of grasshoppers with a summerly walk. It can be heard from field margins, grasslands, shrubberies or along Alpine rivers. What we hear is the buzzing, chirping or creaking that the animals create by means of their wings or legs. These sounds, together with their morphological characteristics, are the basis for the determination of the grasshopper species in their different habitats.

The threat status of each of the grasshopper species in Germany has been documented since 1977 through national red lists by the Federal Agency for Nature Conservation. The recent red-data book from 2024 includes 82 grasshopper and one praying mantis species. Altogether, 31,7 % of the species are considered threatened and four have even become extinct.

More than 20 years after the publication of the first atlas of German grasshopper species, this book gives a summary of the current distribution of all native species. It served as the basis for the threat analysis within the national red list in 2024. It was intensively prepared for many years, and extensive voluntary surveys preceded it. The authors gathered large data sets and, therefore, also used observation portals such as OrthopteraWeb. OrthopteraWeb was developed within the project 'Tools for the Collection of Biological Observation Data in Germany' (Werkzeuge zur Erfassung biologischer Beobachtungsdaten in Deutschland (WerBeo)) within the Federal Programme Biological Diversity, which was funded by the Federal Agency for Nature Conservation. It is now maintained by Osnabrück University with the help of the University of Rostock and is available under the URL https://www.heuschrecken-monitoring.de.

Overall, 35 experts have intensively contributed to this atlas and have successfully worked together with voluntary researchers, natural history associations, non-governmental and governmental nature conservation organisations as well as scientific institutions. At this point, I would like to thank them – especially all the voluntary researchers – for all their tireless work!

The atlas brings together an extensive set of available grasshopper and praying mantis data. It shows the current distribution patterns of all species native to Germany. To document possible range shifts in the future, the distribution maps in the book are linked by a QR code with the online maps of https://www.heuschrecken-monitoring.de.

Besides information on the distribution of each species, the present book includes a detailed description of their preferred habitats and main threats. Moreover, it deals with the biogeography, habitat selection and monitoring of grasshoppers. Ecological indices are also presented. Thus, for nature conservationists, it is an important basis for the planning and conduction of conservation measures. I hope that this atlas will contribute significantly to the continued commitment to grasshopper conservation.

Grasshoppers have become increasingly important due to their response to land-use and climate change and within the Insect Conservation Action Programme of the federal government. They are key organisms in grassland ecosystems since they are essential primary consumers and a crucial food source for species of higher trophic levels. Accordingly, they are an integral part of the German Insect Monitoring Scheme, which started in 2020/21 with the pilot phase.

The book at hand is a reference work that sums up our current knowledge on many interesting topics. Both the text and the illustrations are vivid. The information on the distribution of the species invites us to continue recording regularly together and, thus, to contribute to the conservation of biological diversity in the future.

Dr. Sandra Balzer
Head of the Department of Animal Conservation,
Federal Agency for Nature Conservation

Acknowledgements

THOMAS FARTMANN & DOMINIK PONIATOWSKI

An atlas project's success hinges on the availability of historical and current distribution data. Thus, our special thanks go out to the people who have either reported their observations to the grasshopper specialist groups and the state agencies in recent years or who have been active on observation platforms. Without their efforts, this atlas would not have been possible!

However, due to the decentralised recording possibilities, it was a great challenge to bring together the available data. The project 'Tools for recording biological observation data in Germany' (WerBeo), which was developed in cooperation with the University of Rostock and the 'Institute for Enduring and Environmentally Friendly Development of Natural Spaces' (DUENE e.V.) in Greifswald and funded by the Federal Agency for Nature Conservation (BfN) with funds from the Federal Ministry for the Environment, Nature Conservation, Nuclear Safety and Consumer Protection (BMUV) as part of the Federal Biological Diversity Programme (Grant no.: 3517685C08), helped in this matter.

The data from the WerBeo project formed the basis for the revision of the new red list of the grasshoppers and praying mantises in Germany, which was recently published (Poniatowski et al. 2024a). In this context, we would like to thank the team of the German Red List Centre (RLZ) for the always very pleasant and constructive cooperation. The new red list was funded by the German Aerospace Center (DLR) on behalf of the Federal Agency for Nature Conservation (BfN) with funds from the Federal Ministry for the Environment, Nature Conservation, Nuclear Safety and Consumer Protection (BMUV) (Grant no.: D/396/67279024).

The two projects 'Nationwide insect monitoring: pilot phase' (Bundesweites Insektenmonitoring: Pilotphase) (Grant no.: 3519810100; duration: 11/2019 to 4/2023) and 'Retrospective data analyses and re-surveys of local grasshopper populations as a supplement to insect monitoring' (Retrospektive Datenanalysen und Wiederholungskartierungen lokaler Heuschreckenbestände als Ergänzung zum Insektenmonitoring) (Grant no.: 3522810700; duration: 10/2022 to 12/2025) should also be highlighted at this point. As an outcome, several thousand new grasshopper records were generated. These observations were also incorporated into the distribution maps. Both projects were funded by the Federal Agency for Nature Conservation (BfN) with funds from the Federal Ministry for the Environment, Nature Conservation, Nuclear Safety and Consumer Protection (BMUV).

In addition, many thanks are due to the numerous grasshopper specialist groups and institutions that have supported our work and/or provided data:

Grasshopper specialist groups

- Faunistic-ecological Working Group (FÖAG), Grasshopper specialist group – Andreas Klinge and Christian Winkler,
- Hesse Grasshopper Network – Inga Hundertmark, Leo Meier, Stefan Stübing and Rosina Weber,
- Society for Nature Conservation and Ornithology Rhineland-Palatinate (GNOR), Grasshopper specialist group – Manfred Alban Pfeifer,
- Specialist group grasshoppers Baden-Württemberg – Prof. Dr. Peter Detzel, Dr. Hubert Neugebauer, Maren Niehues as well as Juliane and Peter Zimmermann,
- Specialist group grasshoppers Mecklenburg-Western Pomerania – Dr. Volker Meitzner (†) and Dr. Wolfgang Wranik,
- Specialist group grasshoppers North Rhine-Westphalia – Dr. Andreas Kronshage and
- Specialist group grasshoppers Saxony – Tommy Kästner, Raik Moritz and Dr. Sabine Walter.

State agencies and institutes

- Baden-Württemberg State Institute for the Environment (LUBW) – Barbara Grünes and Dr. Florian Theves,
- Bavarian Environment Agency (LfU) – Johannes Voith,
- Hamburg Ministry of Environment, Climate, Energy and Agriculture (BUKEA) – Dr. Hannes Hoffmann and Dr. Nina Klar,
- Hessian Agency for Nature Conservation, Environment and Geology (HLNUG) – Christian Geske and Sabine Ludwig,

- Lower Saxony Water Management, Coastal and Nature Protection Agency (NLWKN) – Dr. Jakob Fahr, Thomas Herrmann, Dr. Alexander Pelzer and Dr. Annemarie Schacherer,
- Saxony-Anhalt State Office for Environmental Protection (LAU) – Stefan Ellermann, Dr. Daniel Rolke, Dr. Peer Schnitter and Marcel Seyring,
- Saxony State Office for Environment, Agriculture and Geology (LfULG) – Holger Lueg,
- State Office for Agriculture, Environment and Rural Areas of Schleswig-Holstein (LfU) – Arne Drews and
- Thuringian State Office for the Environment, Mining and Nature Conservation (TLUBN) – Dr. Frank Fritzlar, Cornelia Genßler and Dr. Heiko Korsch.

The cooperation with numerous observation platforms was also very helpful. In particular, we would like to thank the Observation International Foundation (Dr. Jan Ole Kriegs, LWL Museum of Natural History, Münster), which has provided us with several thousand verified observations every year since 2020. We would also like to thank Sofie Beckerbauer, Christopher Mollmann and Lars Poorthuis for processing this data. Large amounts of data were also provided by the observation platforms ArtenFinder Rhineland-Palatinate (Manfred Alban Pfeifer), Fauna and Flora Information Portal of Saarland and the Saar-Moselle Region (FFIpS) of Delattinia (Dr. Steffen Caspari, German Red List Centre, Bonn), Insects of Saxony (Dr. Matthias Nuß, Senckenberg Natural History Collections – Museum of Zoology Dresden) and the German grasshopper portal OrthopteraWeb operated by us. Many thanks to all involved!

Moreover, we would like to thank all those who have reported their observations in recent years. Due to the large number of active observers, this is unfortunately not possible in full here. The following list is therefore limited to people with 100 or more reported observations. Unfortunately, in some cases, we were not able to find out the full names of the persons reporting. We do not list these people here. We have also refrained from writing titles since they were not always mentioned. The following people reported a particularly large number of grasshoppers (> 500 observations): Rolf Angersbach, Kurt-Werner Augenstein, Thomas Bamann, Hans-Jürgen Beck, Armin Beckmann, Christian Beckmann, Ronald Bellstedt, Stefan Beyer, Ralf Bolz, Hermann Borsutzki, Jaap Bouwman, Markus Bräu, Manfred Braun, Ursula Braun, Karl Breinl, Ralf Brettfeld, Wolfgang Brose, Frank Brozowski, Klaus Brünner-Garten, Matthias Buchweitz, Klaus Burbach, Steffen Caspari, Armin Coray, Lutz Dalbeck, Anja Danielczak, Urte Delft, Peter Detzel, Jürgen Deuschle, Heinrich Distler, Daniel Doer, Matthias Dolek, Hans-Peter Döler, Gerd Döring, Arne Drews, Marco Drung, Thomas Dürst, Heidrun Düssel-Siebert, Marko Eigner, Klaus Eisenreich, Richard Engelschall, Maria Engl, Ilse Englmaier, Thomas Fartmann, Thomas Fechtler, Rob Felix, Thomas Findeis, Elmar Fischer, Jan A. Fischer, Uwe Fischer, Othmar Fischer-Leipold, Hans-Joachim Flügel, Alexander Franzen, Anja Freese-Hager, Felix Maximilian Freienstein, Frank Fuchs, Kay Fuhrmann, Florian Fumy, Werner Gebhardt, Bahram Gharadjedaghi, Frank Gnoth-Austen, Martin Goldschalt, Ursula Gönner, Günter Grein, Josline Griese, Martin Grimm, Norbert Grosser, Sarah Grünfelder, Robert Güsten, Ronny Gutzeit, Manfred Haacks, Joachim Hable, Stefan Häcker, Angelika Hafner, Jörg Hahnenbruch, Michael Hamann, Walter Hanschitz-Jandl, Peter Hartmann, Mario Harzheim, Karin-Simone Hauth, Stefan Heitz, Christoph Hercher, Dietmar Herold, Corinna Herr, Gerd Heusinger, Benjamin Hill, Georg Hiller, Stefan Hintsche, Axel Hochkirch, Gunnar Höpstein, Angela Höse, Gerhard Hübner, Inga Hundertmark, Markus Hundsdorfer, Kathrin Jäckel, Stefan Jansen, Ralf Joest, Silvia Johna, Jan Juffa, Caspar Jung, Jana Kaiser, Anne Kästner, Tommy Kästner, Marcel Kettermann, Josef Kiechle, Andreas Klinge, Georg Knipfer, Günter Köhler, Claude Kolwelter, Luisa König, Sebastian König, Martin Königsdorfer, Heiko Korsch, Sascha Koslowski, Mathias Kramer, Winfried Krämer, Helmut Kriegbaum, Jan Ole Kriegs, Almut Kroehling, Klaus-Bernhard Kühnapfel, Michael Kunkel, Bernd Kunz, Peter Kunze, Andreas Lange, Sebastian Lehmeier, Henry Lemke, Harry Lipsky, Wolfgang Lissak, Dieter Lode, Wilfried Löderbusch, Franz Löffler, Götz Heinrich Loos, Horst Lukas, Stephan Maas, Walter Mandl, Leo Meier, Thomas Meineke, Volker Meitzner, Gisela Merkel-Wallner, Dirk Mertens, Winrich Mertens, Angelika Meschede, Ulrich Meßlinger, Martin Miethke, Erwin Möhrlein, Sören Möller, Bernhard Moos, Christoph Morgen, Raik Moritz, Thorsten Münsch, Martin Muth, Hubert Neugebauer,

Andreas Nunner, Ernst Obermeier, Julian Overweg, Udo Pankratius, Jürgen Patzer, Christian Paulus, Marko Peltzer, Dominik Poniatowski, Herbert Pröse, Stefan Radlmair, Manfred Rauh, Tobias Rautenberg, Peter Reger, Bernhard Remme, Carsten Renker, Jörg Rietze, Frank Röbbelen, David Roderus, Oliver Röller, Konrad Roth, Wolfgang Rowold, Bernd-Ulrich Rudolph, Torsten Ruf, Jens Sachteleben, Jörg Samietz, Sebastian Sändig, Matthias Sanetra, Julian Sattler, Martin Schädler, Björn Schäfer, Norbert Scheydt, Jens Schirmel, Jürgen Schleef, Helmut Schlumprecht, Martin Schlüpmann, Klaus Schmidt, Annalena Schotthöfer, Erich Schraml, Ralf Schreiber, Sebastian Schröder-Esch, Axel Markus Schulte, Tom Schulte, Sabine Schulz, Dietmar Schuster, Hans Schwaiger, Katrin Schwerdtfeger, Markus Schwibinger, Malte Seehausen, Helmut Siebert, Klaus Siedle, Falk Siegenthaler, Holger Siemers, Maximilian Simmnacher, Heiko Sparmberg, Hartwig Stadelmaier, Herbert Stadelmann, Thomas Stalling, Aloysius Staudt, Rosmarie Steglich, Lilli Steiger, Holger Sternad, Christian Strätz, Stefan Stübing, Jürgen Thein, Reinhold Treiber, Josef Tumbrinck, Sönke Twietmeyer, Justus Vogel, Johannes Voith, Thomas von Peter, Jörn Vorwald, Georg Waeber, Norman Wagner, Wolfgang Wagner, Michael Wallaschek, Sabine Walter, Rosina Weber, Torsten Weidner, Andreas Weigel, Ulrich Weiland, Jörg Weipert, Anne Wiese, Michael Wimbauer, Christian Winkler, Frank Wolf, Kathrin Worschech, Wolfgang Wranik, Julia Zehlius, Wolfgang Zehlius-Eckert, Ute Zengerling-Salge, Elke Zimmermann and Peter Zimmermann.

Many thanks also go to the following people (100 to 499 observations): Tilmann Adler, Cinzia Angenendt, Hubert Anwander, Claus Albrecht, Heike Albrecht, Friedrich Alkemeier, Thomas Alpers, Nils Anthes, Friedemann Arndt, Ursula Arnold-Reich, Roman Aßhoff, Reinhard Baar, Wim Bakker, Jeremy Barker, Anita Bastian, Tino Bastian, Robert Bauer, Volkhard Bauer, Michael Bäumler, Burkhard Beinlich, Jochen Bellebaum, Ulrich Bense, Ricarda Bergen, Leon Berghaus, Christoph Bernd, Anja Betzin, Axel Beutler, Oliver Böck, Sven Bodingbauer, Jörg Böhringer, Andre Bönsel, Hermann Borsutzki, Felix Bossong, Nina Bramm, Jeroen Brandjes, Daniel Brandt, Oliver Brauner, Floris Brekelmans, Walter Bretz, Robert Brinkmann, Jonas Brüggeshemke, Florian Buchkremer, Marc Bulte, Michael Bußmann, Dietrich Cerff, Rainer Cezanne, Torsten Cloos, Bernd Cogel, Dierk Conrady, Klaus-Jürgen Conze, Armin Dahl, Tom Damm, Gottfried Dechert, Klaus Demuth, Rainer Deschle, Alexa Dettinger-Klemm, Michael Deussen, Armin Deutsch, Axel Didion, Til Dieterich, Bernhardt Disch, Claudia Distler, Klaus Dühr, Hannes Egle, Oliver Eller, Peter Endl, Svea Endler, Elias Engel, Richard Engelschall, Julian Enß, Dirk Esplör, Markus Faas, Jakob Fahr, Stefan Feer, Dirk Felzmann, Martin Fichtler, Siegfried Filus, Wolfgang Fluck, Hans-Georg Folz, Philipp Foquet, Birgit Förderreuther, Kai Frank, Anja Freese, Frank Fritzlar, Gerald Fuchs, Markus Fuhrmann, Peter Gabler, Jörg Gelbrecht, Christian Gelpke, Malte Gemeinhardt, Joachim Genser, Karsten Gerland, Adam Geyer, Nicola Glaser, Ronald Glaser, Frank Gnoth-Austen, Klaus Gollmer, Alyna Görlitz, Astrid Grauel, Herbert Grimm, Thomas Guggemoos, Pitschi Günther, Matthias Haag, Lennart Haak, Sybille Hacker, Herwig Hadatsch, Ulrich Haese, Leonie Haferkemper, Jan Hähnel, Melitta Haller-Probst, Frank Hänsel, Mareike Hansen, Heiko Haupt, Oliver Hawlitschek, Doris Heimbucher, Heiko Heise, Felix Helbing, Andreas Hemp, Claudia Hemp, Udo Herkommer, Gabriel Hermann, Mike Herrmann, Philipp Herrmann, Ralph Hildenbrand, Chiara Hille, Matthias Hippke, Florian Hochkirch, Sylvain Hodvina, Lisa Holtmann, Georg Homann, Sebastian Hopfenmüller, Henri Hoppe, Peter Hoppenbrouwers, Rainer Hozak, Lennart Hudel, Jörg Huth, Sigfrid Ingrisch, Slobodan Ivkovic, Harald Jacubeit, Guus Jenniskens, Mike Jessat, Wolfgang Kaiser, Steffen Kämpfer, Inken Karst, Fabian Karwinkel, Doreen Kasper, Andre Keßler, Robert Ketelaar, Jan Erik Kikkert, Jörg Kinkele, Jürgen Kirchner, Ralf Kirchner-Heßner, Gert Klages, Dietmar Klaus, Michel Klemann, Matthias Klemm, Roy Kleukers, Ralf Klusmeyer, Andreas Knipping, Daniela Köhler, Günter Köhler, Melanie Köhler, Ronny Köhler, Helmut J. G. Kolbeck, Andreas Kopetz, Ingo Koslowski, Jens Kramer, Dieter Kraus, Tobias Krause, Detlef Krebs, Mathias Krech, Markus Krieger, Alfons Krismann, Peter Kröger, Ralf Krohm, Betina Küchenhoff, Katrin Kunkel, Jan Kuper, Volkmar Kuschka, Andreas Lang, Gerold Lang, Helmut Laußmann, Henning Lege, Peter Leupold, Catharina Lorenz, Hannah Lorenzen, Urs Rainer Lüders, Holger Lueg, Johannes Maas, Tobias Maier, Andreas Maletzky,

Niko Mandl, Christof Manhart, Til Mann, Andreas Manz, Max Marquart, Andreas Marten, Angela Martin, Tim Mattern, Jörg Maurer, Wolfgang Mayer, Verena Medinger, Michael Meier, Stefan Meisberger, Lars Meitner, Kerstin Menge, Robert Mertl, Konstantin Meßmer, Frank Meyer, René Micksch, Jonas Mittemeyer, Helge Mock, Christopher Mollmann, Gudrun Mühlhofer, Klaus Mühlhofer, Christian Müller, Katrin Müller, Martin Muth, Martin Neub, Johann Neumayer, Herbert Nickel, Christian Niederbichler, Johannes Niederstrasser, Olaf Niehus, Frank Nowotne, Hans-Markus Oelerich, Matthias Olthoff, Olaf Op den Kamp, Stefan Opitz, Jürgen Otto, Arjan Ovaa, Bert Oving, Markus Paßlick, Rolf Peschel, Falk Petzold, Eberhard Pfeuffer, Stephan Pflume, Mathis Pfreundt, Wolfgang Piepers, Amelie Pier, Holger Pieren, Ulrich Pittius, Detlef Pohlmann, Gabriele Potabgy, Anne Puchta, Oswald Rank, Gerd Rausch, Michael Reich, Klaus Reinhardt, Lisa Reiss, Claudia Reiter, Erwin Rennwald, Felix Riegel, Peter Rode, Thomas Roedl, Hanns-Jürgen Roland, Johanna Römer, Konrad Roth, Dietmar Rothmund, Kurt Rudnick, Wolf Rühle, Lukas Rühlke, Marion Rumpf, Florin Rutschmann, Thomas Sacher, Nicole Sachs, Karin Sager, Martin Salcher, Birgit Salzbrunn, Hartmut Sänger, Wolfgang Schäfer, Bernhard Schaffner, Fritz Schätzle, Konstanze Scheffler, Gwydion Scherer, Philipp Schiefenhövel, Franz-Josef Schiel, Karen Schindler, Richard Schleicher, Susann Schleip, Josef Schlögel, Gesine Schmidt, Katharina Schmidtmann, Michael Schmitz, Maximilian Schmucker, Antje Schnellbächer-Bühler, Paul Schnitzler, Harald Schnöde, Michael Schödl, Alexander Scholz, Manfred Schön, Dorian Schöter, Michael Schott, Melanie Schubert, Annette Schulte, Björn Schulz, Martin Schulze, Monika Schumacher, Thomas Schütz, Ulrich Schwab, Cinja Schwarz, Markus Schweighöfer, Malte Seehausen, Erwin Sefrin, Steffen Seidel, Carola Seifert, Klaus Selle, Ina Siebert, Klaus Siedle, Martina Sievers, Ludwig Simon, Armin Six, Martina Sommer, Frank Sonnenburg, Angelika Stacke, Michael Stadler, Rainer Stawikowski, Claude Steck, Johanna Stegherr, Manja Stempin, Helena Stoltz, Stefan Straube, Frank Stühmer, Nathalie Tent, Steffen Thoß, Johanna Trappe, Kerstin Tschiedel, Jörg Turk, Yannick van der Veen, Erik van Dijk, Christine Vanky, Christian Venne, Hanno Voigt, Robert Völkl, Wolfgang Völkl, Martin Volpers, Benno von Blanckenhagen, Felix von Blanckenhagen, Annelies Vriens, Michael Wagner, Reinhard Waldert, Julie Walter, Ralf Wancura, Siegfried Weid, Florian Weihrauch, Linda Weiler, Jörg Weipert, Heinz Weishaupt, Kilian Weixler, Udo Weller, Andreas Welsch, Karla Wenner, Paul Westrich, Siegfried Willig, Joachim Wimmer, Michael Winterholler, Karola Winzer, Elmar Witting, Anton Wolf, Andreas Wolf, Armin Woll, Jann Wübbenhorst, Friedrich Wulf, Andreas Wurm, Andreas Zahn, Sabine Zebli, Andreas Zehm, Stephan Zöller and Casper Zuyderduyn.

Marco Drung, Lennart Haak, Thorsten Münsch and Dr. Dominik Poniatowski (all Osnabrück University) were responsible for reviewing almost 300 references and processing the data. Felix Helbing prepared the majority of the figures and maps. Dr. Gregor Stuhldreher helped to develop the boxes for the ecological indices. Prof. Dr. Thomas Fartmann, Hannah Kalthoff, Dr. Marcel Kettermann, Dr. Franz Löffler, Dr. Dominik Poniatowski and Dr. Merle Streitberger (all Osnabrück University) provided individual figures and maps. We would like to thank the Umbrella Organisation for German Avifaunists (DDA) for providing a shapefile, which served as the basis for a map of the major geographical features in Germany (see inside of front and back cover). Many thanks to Jürgen Fischer, Dr. Zoltán Kenyeres, Arnold Kleiner, Dr. Petr Kočárek, Christian Roesti, Dr. Martin Schädler, Prof. Dr. Sebastian Seibold, Jennifer Theobald and Carsten Wagner for providing photos. Moreover, we are grateful to the Federal Agency for Nature Conservation (BfN), which partly funded the publication of this book. Linguistic revision was done by Benjamin Morris. Finally, we would also like to thank Ulf Müller (editor) and Birgit Heyny (book production) from Eugen Ulmer Publishers. Their trusting and constructive cooperation has contributed significantly to the success of this book!

Introduction

THOMAS FARTMANN & DOMINIK PONIATOWSKI

Scientific studies on grasshoppers have a long tradition in Germany. Here and hereinafter, the term grasshoppers is usually used as a pars pro toto for bush-crickets, crickets, grasshoppers, groundhoppers, locusts and praying mantises. Regional species lists have been published since the end of the 18th century (Poniatowski & Fartmann 2023). However, until the mid-20th century, only a few naturalists and scientists dealt with grasshoppers. The first identification key was published by Kurt Harz (Harz 1957), the initiator of the German Orthopterists' Society. However, his epochal, nearly 500-page-long book was very scientific and hence unsuitable for a wide audience. The situation changed with the field guide of Heiko Bellmann, an instructively illustrated and compact book (Bellmann 1985). For the first time, it enabled the identification of all German grasshoppers in the field. It triggered a wave of surveys and scientific studies on grasshoppers across large parts of Germany. As a result of the rapidly increasing knowledge, Sigfrid Ingrisch and Günter Köhler published a comprehensive textbook on the biology and ecology of Central European grasshoppers (Ingrisch & Köhler 1998). At roughly the same time, the first extensive distribution atlases for some federal states were released. Among them, those for Baden-Württemberg (Detzel 1998) and Bavaria (Schlumprecht & Waeber 2003) should be particularly emphasised since they included in addition to very extensive species profiles an analysis of most of the relevant Central European literature. The first complete German distribution atlas of grasshoppers was published in the early 2000s (Maas et al. 2002). Since then, the populations and distribution patterns of grasshoppers have been highly dynamic. Progressing land-use change and increasing climatic alterations have contributed to this in particular (e.g. Poniatowski et al. 2018a, 2020; see also chapter Land-use and climate change). In recent times, for example, many thermophilic species have expanded their range. Simultaneously, a lot of populations have become extinct, mainly those of habitat specialists; however, some relatively widespread species have also declined. Moreover, recording of grasshopper occurrence has strongly increased owing to the establishment of some online platforms such as OrthopteraWeb or Observation (Poniatowski & Fartmann 2023). The strong changes in distribution and significant increase in data during the past 20 years as well as many new scientific findings called for a recent overview of the distribution of grasshoppers in Germany.

The main component of the distribution atlas are the species profiles, which concisely present current knowledge concerning distribution, habitat, and population trend and threats. The text on the distribution of the species is supplemented by visually appealing grid maps. The different topics of the species profiles are also part of the introductory chapters. However, their focus is on the explanation of general distribution patterns (see chapters Distribution, Habitats as well as Land-use and climate change). In the chapter 'Monitoring', grasshopper monitoring as part of the German Insect Monitoring Scheme is introduced and the importance of re-surveys for detecting changes in grasshopper assemblages is mentioned. The chapter 'Ecological indices' presents several categorical and metric parameters of ecological preferences and functional traits of species, which are highly relevant for the analysis of grasshopper data. For example, the Species Temperature Index (STI) and the Species Farmland Index (SFI) have distinctly proven their value (Löffler et al. 2019, Fumy et al. 2020, Poniatowski et al. 2020, Riggi & Berggren 2020, Fartmann et al. 2021b, 2022b, Thorn et al. 2022; see chapter Land-use and climate change). For the first time, for all grasshopper species of open habitats, specific temperature and moisture indicator values are introduced. The indicator values are inspired by those known for plants by Ellenberg et al. (2001) and can be used to characterise the ecological conditions of a habitat. The following chapter on the data basis mentions all data sources that were used to create the grid maps. The chapter 'Range of species, nomenclature and taxonomy' explains which species were considered and why this was the case. The last of the introductory chapters deals with the structure of the species profiles. It outlines the structure and content of the text. Moreover, it provides information for the interpretation of the distribution maps.

Distribution

DOMINIK PONIATOWSKI, FELIX HELBING & THOMAS FARTMANN

The distribution range of a species is determined by various environmental factors. The macroclimate is of overriding importance. As ectothermic organisms whose development and survival depend on ambient temperatures, many species generally require high temperatures. This connection can be seen in Europe. The diversity hotspots are in southern Europe (Fig. 1). For example, over 300 species occur on the Iberian Peninsula (Heller et al. 1998) (Fig. 2). Greece and Italy have the greatest diversity with 378 and 382 species, respectively (Willemse et al. 2018, Iorio et al. 2019). As the geographical latitude increases, the number of species decreases significantly. The Central European fauna only contains around 160 species (Fischer et al. 2020). Of these, 82 species are considered native in Germany (Poniatowski et al. 2024a). In Northern Europe, the grasshopper fauna is even more impoverished. There, fewer than 40 species can cope with the low temperatures (Karjalainen 2009, Hansen & Olsen 2010, Kindvall 2018).

At the regional level, however, other factors influence species richness. Within areas with a uniform macroclimate, species richness is determined in particular by their habitat diversity. The diversity of habitats is in turn an expression of land use. Heavily cleared landscapes with large arable fields and few grassland-like habitats (Fig. 3) are significantly poorer in species than historical cul-

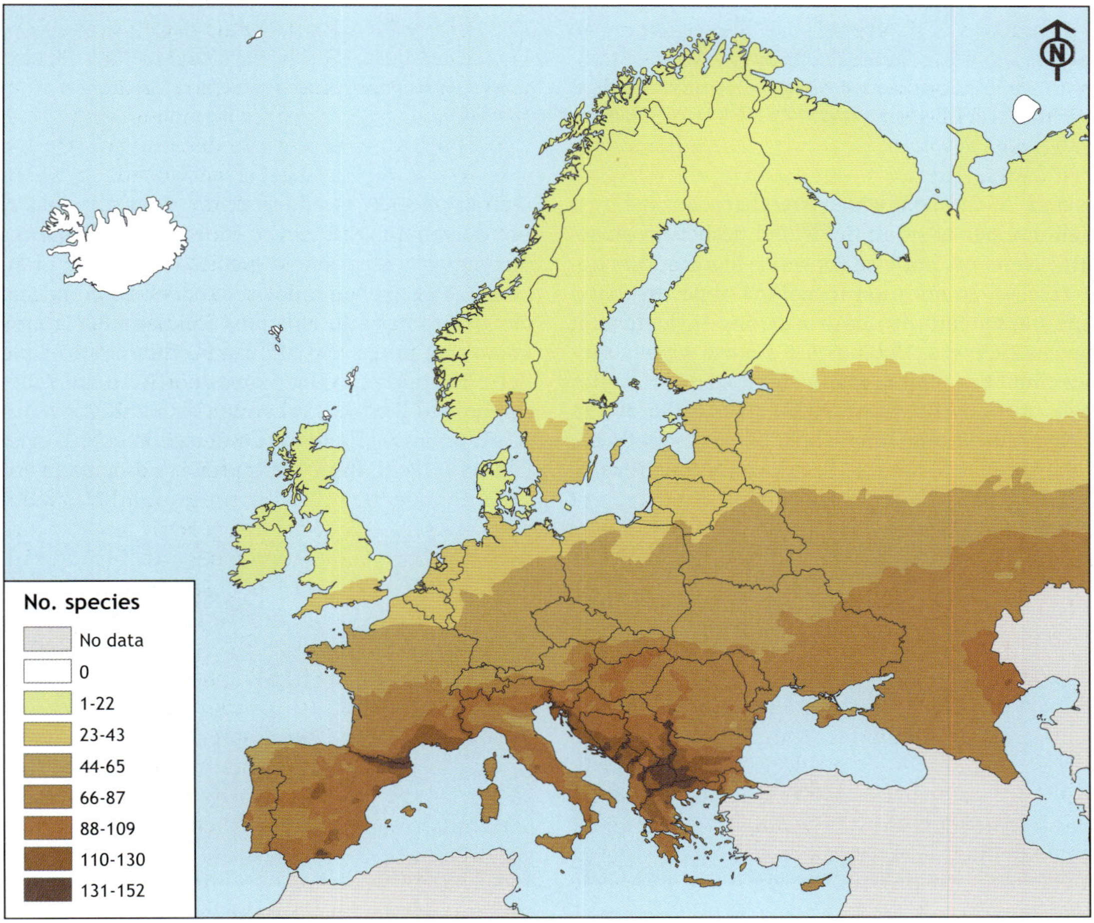

Fig. 1: Generalised illustration of grasshopper diversity in Europe. (Source: Hochkirch et al. 2016, modified)

Fig. 2: *Eumigus cucullatus* is endemic to the Iberian Peninsula. As in the whole of southern Europe, a particularly large number of grasshopper species can be found there. (Photo credit: Dominik Poniatowski)

tivated landscapes, which are characterised by a high number of small fields and numerous structural elements such as copses, fringes and hedges (Fig. 4). In addition to habitat type diversity, topography can also have a strong influence on species richness. Uplands often have higher species numbers than areas in the lowlands (Pauler et al. 2019, Löffler et al. 2023). This is due to the strong elevational and thus climatic gradient, which provides suitable living conditions for numerous species. In addition, the intensity of land use decreases with elevation in many mountainous regions (MacDonald et al. 2000, Hülber et al. 2020). The uneven relief, a harsh climate and the fact that some of the locations are difficult to access mean that intensive land use is often not possible here. Therefore, in the montane elevational zone, threatened biotope types such as heathlands, nutrient-poor grasslands and wet meadows still exist in many places; these biotopes serve as habitats for

Fig. 3: Intensively used agricultural landscape north of Warburg (North Rhine-Westphalia). (Photo credit: Felix Helbing)

Fig. 4: Heterogeneous cultural landscape in the Lower Diemel Valley (Hesse). (Photo credit: Dominik Poniatowski)

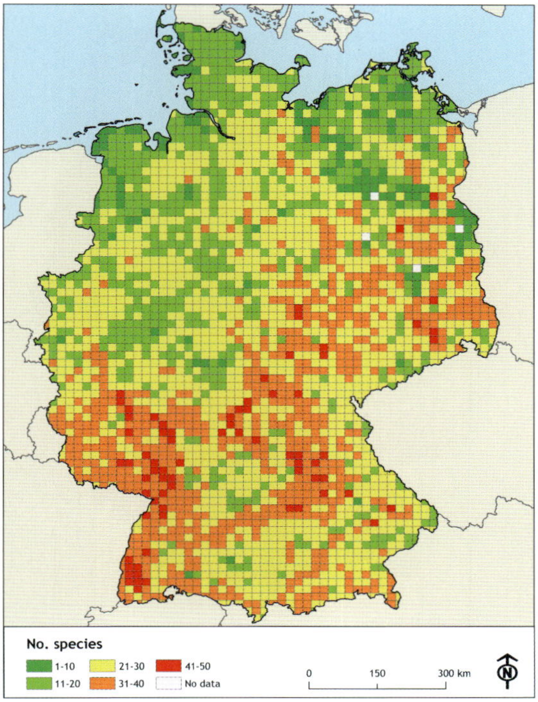

Fig. 5: Number of grasshopper species in Germany per 10 × 6 geographic minute grid cells (period: 1990 to 2023). (Source: author's own diagram)

numerous grasshopper species (Fischer et al. 2020, Fartmann et al. 2021a).

The diversity factors that apply at the European level, such as climate, landscape structure and topography, can also be witnessed in Germany. The warm summer regions of southern and eastern Germany, for example, are characterised by a particularly high species richness (Fig. 5). Among the centres of species diversity are the regional heat islands – in particular Kaiserstuhl, the Upper Rhine Valley, the Middle Rhine Valley, Main-Franconia, the Franconian Basin, Grabfeld and Kyffhäuser (Fig. 6). However, some heterogeneous low mountain ranges in southern Germany also have high species numbers. The southern part of the Black Forest as well as Swabian and Franconian Jura are particularly noteworthy. These areas are characterised by a high proportion of grassland of high nature conservation value (HNV farmland) (Fig. 7), from which numerous grasshopper species benefit.

The diversity of grasshopper species in Germany must also be seen in the context of the last ice age more than 10,000 years ago. At that time, the Fennoscandian ice sheet extended as far as the North German Lowlands. To the south of this was large-scale permafrost – i. e. permanently frozen ground. Its boundary ran roughly through southern France, along the Alps and from there eastwards through the Pannonian Basin to the northern edge of the Pontic region on the Black Sea (French 2018). The current distribution of species in Germany is therefore the result of immigration processes from various glacial refugial areas such as Anatolia and the Balkans as well as the coastal and southern areas of the Iberian and Italian peninsulas (Ingrisch & Köhler 1998, Hewitt 1999). In the course of recolonisation in Germany, various geographical barriers such as the Vosges in the west, the Alps in the south and the Bavarian and Bohemian Forests in the east had to be crossed, depending on the species. Detzel (1995) names five post-glacial immigration routes (Fig. 8):

1. Lorraine, the Moselle, Nahe and Main valleys,
2. the Burgundian Gate and the Rhine Valley,
3. the North German and Polish Plain,
4. the Moravian Gate and
5. the Danube Valley.

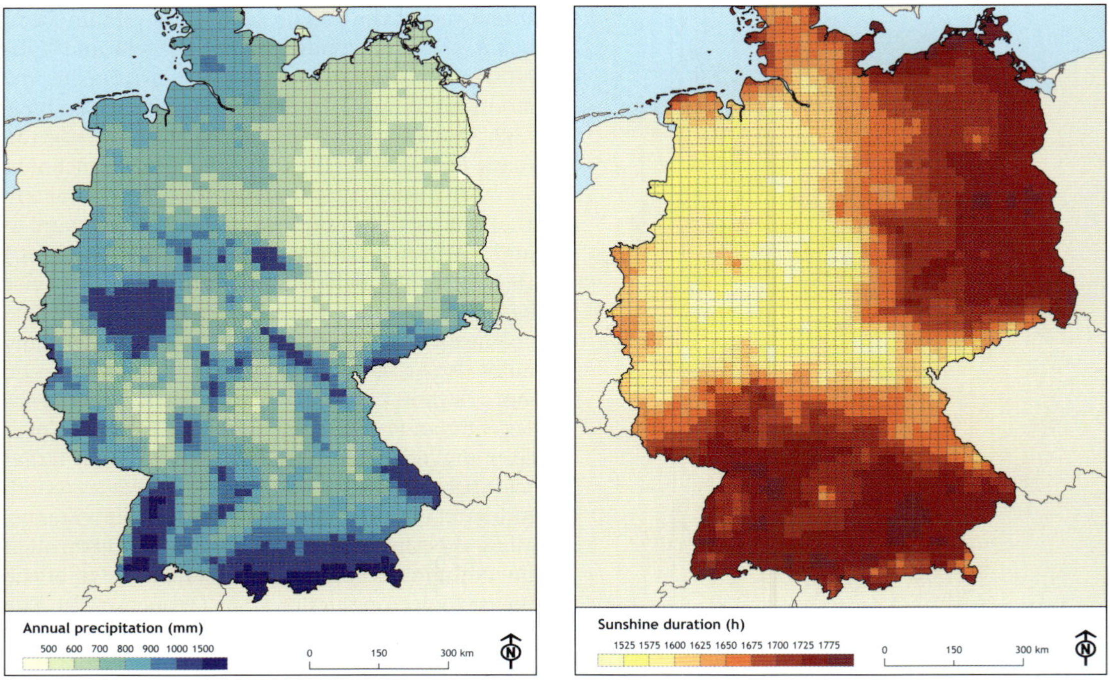

Fig. 6: Mean annual precipitation (left), hours of sunshine per year (right) and summer temperatures (April to September, top of page 17) in Germany (period: 1991 to 2020). (Data source: DWD 2024)

The natural landscape after the last ice age probably consisted of heterogeneous and sometimes very light woodlands, which were interspersed in large parts with open habitats of varying sizes (Feurdean et al. 2018). Various factors such as drought, fire, snowfall, windthrow and large wild ungulates such as Aurochs (*Bos primigenius*) and Wisent (*Bison bonasus*) probably contributed to keeping the landscape open (e. g. Svenning 2002). Locally, Beavers (*Castor fiber*) may also have been influential, regularly creating new open habitats through their activities in floodplains (Harthun 1999, Wells et al. 2000). In addition, various special sites such as bogs, coastal dunes, gravel banks, rocky knolls and screes were also free of wood (Ellenberg & Leuschner 2010). Some grasshopper species were able to recolonise Central Europe comparatively quickly. Hewitt (1990) assumes that the Meadow Grasshopper (*Pseudochorthippus parallelus*) reached Germany and subsequently the British Isles more than 8,000 years ago, i. e. at a time when there was still a land connection with mainland Europe (see also Hewitt 1999).

With the beginning of agriculture and livestock farming in Central Europe, the so-called Neolithic Revolution, around 6,000 to 5,000 years BC, the natural landscape was gradually transformed into a cultivated landscape (Poschlod 2017). This process promoted the dispersal of numerous species as it created new grasshopper habitats. Another important factor was migrating herds of domestic animals (Fischer et al. 1995, Bugla & Poschlod 2006), which already existed in the Neolithic (Poschlod 2017). The study by Fischer et al. (1996) shows that not only the diaspores of plants but also grasshopper individuals can be spread by sheep.

In the 14th and 15th centuries, transhumance experienced a strong upswing as large areas of fallow fields were available for grazing due to the late medieval deserted period. Sheep farming reached its peak towards the mid-19th century. At this time, there were over 25 million sheep in Germany (Poschlod 2017). The herds sometimes covered large distances on their migrations (Dolek & Geyer 2002). Transhumance can therefore be regarded as the most effective mobile biotope network in our cultural landscape (Fischer et al. 1995, Poschlod et

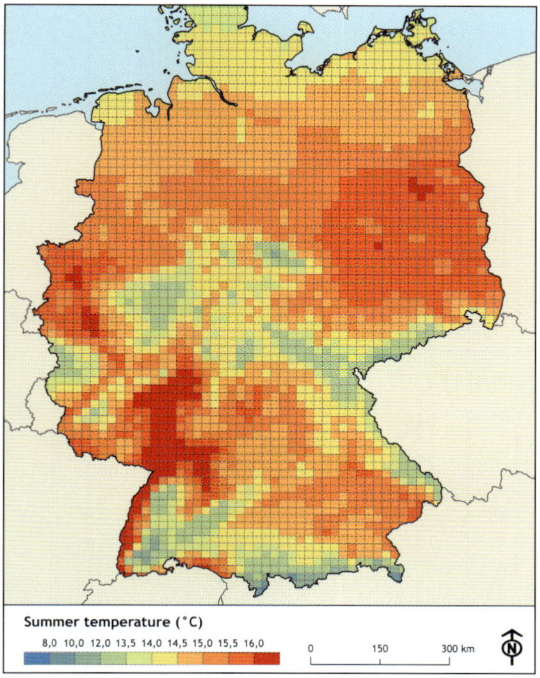

al. 1996). However, cheap wool from overseas led to a trend reversal in sheep farming towards the end of the 19th century, which caused the population to decrease to less than 10 million animals towards the beginning of the 20th century (Poschlod 2017). Around 1.5 million sheep are currently kept in Germany (Federal Statistical Office 2023).

The decline in sheep populations has not only led to a deterioration in habitat connectivity, but also to the loss of large areas of typical grasshopper habitats such as heathlands and nutrient-poor grasslands as a result of afforestation or succession (Poschlod 2017, Fartmann et al. 2021a). On nutrient-rich sites, on the other hand, land use has been heavily intensified – especially since the mid-20th century (Dierschke & Briemle 2002, Sturm et al. 2018). Both developments have led to the loss of many local grasshopper populations (Poniatowski et al. 2024a). Low-mobility habitat specialists in particular, whose populations are restricted to landscapes with a high proportion of grassland of high conservation value (high species-specific SFI, see also chapter Ecological indices, p. 33), have suffered severe population losses (Poniatowski et al. 2020). This applies, for example, to the Woodland Grasshopper (*Omocestus rufipes*), the Rattle Grasshopper (*Psophus stridulus*) and the Black-spotted Toothed Grasshopper (*Stenobothrus nigromaculatus*). All three species can now only be found scattered in Germany (pp. 154, 180 and 186).

Despite the deterioration in habitat availability, numerous grasshopper species have expanded their range northwards in Germany over the last 20 to 30 years – in line with the sharp rise in mean annual temperatures (IPCC 2023). Conspicuous new colonisations have been documented in particular for formerly summer-cool regions such as the Northwest German Lowland and the higher elevations of some low mountain ranges (e. g. Brose & Peschel 1998, Grein 2007, Löffler et al. 2019, Fartmann et al. 2021b). However, the extent of range expansion varies greatly from species to species and depends on various factors such as habitat specificity and species-specific mobility. In any case, it is significant that the expanding species generally have a high or even a very high STI (p. 33) and therefore prefer a warm macroclimate.

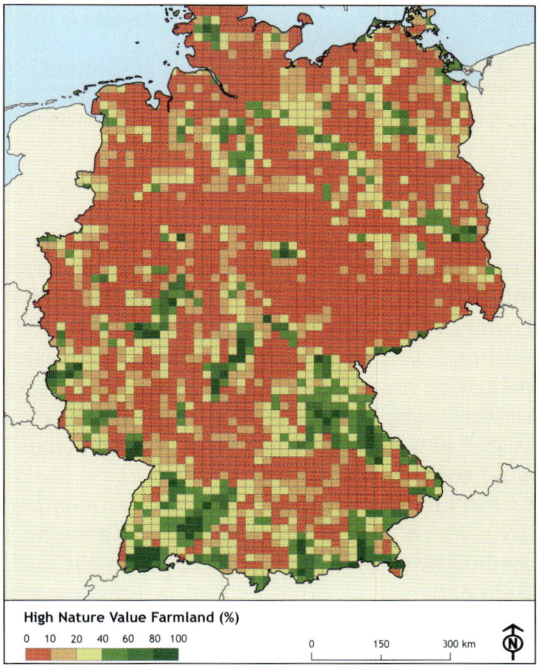

Fig. 7: Current proportion of High Nature Value farmland (HNV farmland = grassland of high nature conservation value in a broader sense) in the open landscapes of Germany. (Data source: EEA 2024)

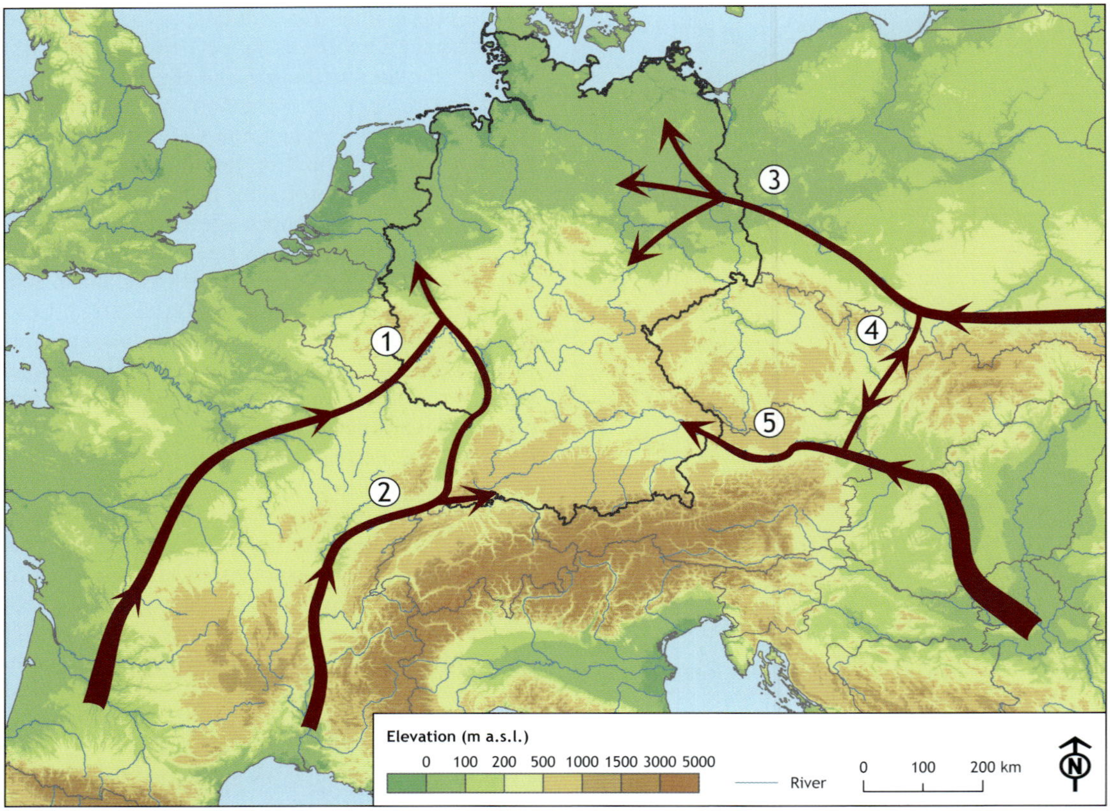

Fig. 8: Potential post-glacial migration routes of grasshopper species currently occurring in Germany (author's own diagram, based on Detzel [1995] and Hewitt [1999]): 1 – Moselle Valley, 2 – Burgundian Gate, 3 – North German and Polish Plain, 4 – Moravian Gate, 5 – Danube Valley.

Climate change is therefore largely responsible for this development (e. g. Trautner & Hermann 2008, Pfeifer 2012, Kettermann & Fartmann 2018). According to Poniatowski et al. (2018a), over 30 % of the native species currently benefit from climate change. Many of these are habitat generalists such as the Large Gold Grasshopper (*Chrysochraon dispar*), the Long-winged Conehead (*Conocephalus fuscus*) and Roesel's Bush-cricket (*Roeseliana roeselii*), which do comparatively well in intensively used cultivated landscapes (cf. Poniatowski et al. 2024a). Compared to earlier studies, however, it is now clear that both generalists and specialists have been able to disperse in recent years, provided they have a high degree of mobility (Poniatowski et al. 2018a, 2020). For example, the Blue-winged Sand Grasshopper (*Sphingonotus caerulans*), the Italian Locust (*Calliptamus italicus*) and the Large Marsh Grasshopper (*Stethophyma grossum*) are among the dispersing habitat specialists.

In some cases, passive dispersal of eggs, nymphs or adults has also contributed to the spread. Transportation by vehicles is well documented, for example, for the Southern Oak Bush-cricket (*Meconema meridionale*) (e. g. Köhler et al. 2022b, c). Other species are regularly transported with plants or potting soil – as is assumed, for example, for the Praying Mantis (*Mantis religiosa*) and the Southern Sickle Bush-cricket (*Phaneroptera nana*) (Aßmann & Ritt 2021, Beckerbauer et al. 2024, Hochkirch et al. 2023). There are also indications of anthropogenically induced occurrences for the Italian Cricket (*Oecanthus pellucens*) and the two species of Bush-crickets *Leptophyes albovittata* and *Leptophyes punctatissima* (Samu & Wischhof 2000, Sczepanski 2005, Pfeifer 2012, p. 118). In addition, species such as the Blue-winged Grasshopper (*Oedipoda caerulescens*), the Blue-winged Sand Grasshopper (*Sphingonotus caerulans*) or the Bordeaux Cricket (*Eumodicogryllus bordigalensis*) can occasionally be dispersed by gravel or sand transportation (Köhler et al. 2001, Pfeuffer 2018, Brauner & Ristow 2022). Hydrochory – i. e. the drifting of plant material with corresponding developmental stages of grasshoppers along rivers or even the seashore – has been documented for the Striped Bush-cricket (*Leptophyes albovittata*), the Short-winged Conehead (*Conocephalus dorsalis*) and the Long-winged Conehead (*Conocephalus fuscus*), among others, and is also suspected for the Italian Cricket (Fartmann 2004, Wallaschek et al. 2004, Lehmann & Haacks 2006, Pfeifer et al. 2012).

Although many species are currently expanding their range, it should not be forgotten that the ongoing intensification of land use in open areas is leading to the loss of further grasshopper habitats (Finck et al. 2017). If habitat loss continues, at some point even the mobile species will no longer be able to cover the large distances between the remaining habitat islands (Hill et al. 1999, Buse & Griebeler 2011).

Population declines in the course of climate change are difficult to detect as they are less obvious than conspicuous new colonisations in well-studied areas at the edge of the range (Trautner & Hermann 2008, Pfeifer 2012). Nevertheless, regressive population and range developments are likely for some species. Poniatowski et al. (2018a) list six species whose populations may be negatively affected by climate change. One of these is the Green Mountain Grasshopper (*Miramella alpina*) – a glacial relict. Population losses of this species have been documented for the Black Forest and the neighbouring Vosges Mountains, among other sites (e. g. Hafner & Zimmermann 2019, Fumy et al. 2020, d'Agostino & Vacher 2021, p. 148). It is currently unclear whether other species belong to the group of 'climate change losers.' In general, however, it can be stated that climate change poses a particular threat to grasshopper species in humid-wet or boreal-montane habitats (Rabitsch et al. 2011, Streitberger et al. 2016a, b, Poniatowski et al. 2018a).

Habitats

DOMINIK PONIATOWSKI & THOMAS FARTMANN

Grasshoppers occur in almost all terrestrial habitats (Fischer et al. 2020). By far the most important habitat type for grasshoppers are nutrient-poor dry grasslands (Dolek 1994, Zehm 1997, Löffler et al. 2019, Köhler et al. 2023a). Across Germany, they provide a habitat for almost 70 % of all native grasshopper species (Fig. 9). Dry grasslands include, among others, calcareous, sandy and acidic grasslands as well as steppe grasslands (Fig. 10–14). They are located on marginal sites – i. e. areas that are very unproductive due to their steep slopes and/or nutrient-poor soils, for example, and are therefore only used with low intensity. Such habitats are often characterised by a mosaic of low and tall-growing vegetation as well as patches of bare ground, stones and shrubs (Fig. 15). Occasionally, rocks are also part of the habitat. This high structural diversity results in a high number of species (Zehm 1997, Hemp & Hemp 2000, Löffler & Fartmann 2017). Other important habitats are mesic and wet grasslands as well as heathlands and tall herbaceous fringes (Fig. 16).

To a certain extent, this also applies to moorlands (Fig. 9). Grazed fens and transition mires in particu-

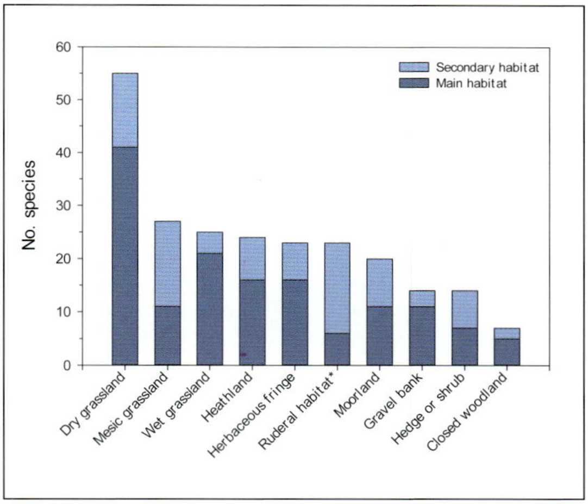

Fig. 9: Total species numbers for the most important grasshopper habitats in Germany. (Data source: Fischer et al. 2020). Secondary habitat = species occurs occasionally in this habitat type; however, it does not have its main focus here; main habitat = species occurs mainly in this habitat type; *collective term for extraction sites, industrial wasteland, post-mining landscapes, railway tracks and spoil heaps.

Fig. 10: Dry grassland on gypsum rock in the Schloßberg-Solwiesen Nature Reserve (Thuringia: Kyffhäuser; 180 m a. s. l.). Occurring species: e. g. *Mantis religiosa*, *Oecanthus pellucens*, *Stenobothrus crassipes* and *S. nigromaculatus*. (Photo credit: Dominik Poniatowski)

Fig. 11: Subcontinental sandy dry grassland in the Müllerberge Nature Reserve (Brandenburg: Uckermark; 60 m a. s. l.). Occurring species: e. g. *Bicolorana bicolor*, *Chorthippus apricarius*, *C. mollis*, *Myrmeleotettix maculatus*, *Oedipoda caerulescens* and *Omocestus haemorrhoidalis*. (Photo credit: Thomas Fartmann)

Fig. 12: Calcareous grasslands on the Stiegelesfelsen (Baden-Württemberg: Upper Danube Valley; 640 m a. s. l.). Occurring species: e. g. *Euthystira brachyptera*, *Gomphocerippus rufus*, *Platycleis albopunctata* and *Tetrix bipunctata*. (Photo credit: Dominik Poniatowski)

Fig. 13: Acidic grassland with Common Broom (*Cytisus scoparius*) near Medelon (North Rhine-Westphalia: Medebach Bay; 420 m a. s. l.). Occurring species: e. g. *Chorthippus biguttulus*, *Omocestus viridulus*, *Pseudochorthippus parallelus* and *Stenobothrus stigmaticus*. (Photo credit: Dominik Poniatowski)

Fig. 14: Heterogeneous pasture with Winged Broom (*Genista sagittalis*) in the Upper Hotzenwald (Baden-Württemberg: Black Forest; 875 m a. s. l.). Occurring species: e. g. *Decticus verrucivorus*, *Omocestus rufipes*, *Psophus stridulus* and *Stauroderus scalaris*. (Photo credit: Thomas Fartmann)

Fig. 15: Heterogeneous sandy dry grassland in the Tennenloher Forst Nature Reserve (Bavaria: Franconian Basin; 314 m a. s. l.). The area is grazed by Przewalski's horses and provides a habitat for numerous grasshopper species. (Photo credit: Dominik Poniatowski)

lar can harbour species-rich grasshopper assemblages (Dolek et al. 1999, Bamann 2018, Schwarz & Fartmann 2022) (Fig. 17). Gravel banks are also considered valuable grasshopper habitats (Lemke et al. 2010, Pfeuffer 2015, 2020) (Fig. 18). Although species numbers are low in these sparsely vegetated sites, some highly specialised and therefore highly endangered species such as the Speckled Buzzing Grasshopper (*Bryodemella tuberculata*), the Gravel Grasshopper (*Chorthippus pullus*) and the Alpine Groundhopper (*Tetrix tuerki*) occur there (Poniatowski et al. 2024a). Scrubs and closed forests, on the other hand, are less important for grasshoppers as they only serve as the main habitat for a few species. The situation is different with light woodlands such as coppice woodland and wood pastures, which have a lush herb layer or are characterised by numerous grassland-like clearings. Many typical species of open habitats can also occur in such forests (Helbing et al. 2014a, Rösch et al. 2019, Liebelt 2021) (Fig. 19 and 20). Ruderalised or heavily anthropogenically modified habitats such as extraction sites, industrial wasteland, post-mining landscapes, railway tracks and spoil heaps have a special status (Fig. 21). They can be valuable secondary habitats, especially in urban areas or in regions used intensively for agriculture (Hamann & Weber 2012, Eckert et al. 2017, Kettermann & Fartmann 2018, Pfeuffer 2018).

Fig. 16: Mesic to moist mountain grassland in the Long Rhön (Bavaria: Rhön Mountains; 750 m a. s. l.). Occurring species: e. g. *Chorthippus biguttulus*, *Omocestus viridulus*, *Pseudochorthippus parallelus* and *Tettigonia cantans*. (Photo credit: Thomas Fartmann)

Fig. 17: Litter meadow with Siberian Iris (*Iris sibirica*) in the Murnauer Moos (Bavaria: Pre-Alpine Hills and Moorland; 650 m a. s. l.). Occurring species: e. g. *Metrioptera brachyptera*, *Pseudochorthippus montanus*, *Pteronemobius heydenii* and *Stethophyma grossum*. (Photo credit: Thomas Fartmann)

Fig. 18: Dolomite alluvial fan of the Friedergries (Bavaria: Ammergau Alps; 920 m a. s. l.). Occurring species: e. g. *Bryodemella tuberculata*, *Chorthippus pullus* and *Psophus stridulus*. (Photo credit: Thomas Fartmann)

Fig. 19: Historic cattle pasture with flowing transitions between open woodland, sandy dry grassland and heathland in the Borkener Paradies Nature Reserve (Lower Saxony: Emsland; 10 m a. s. l.). Occurring species: e. g. *Chorthippus biguttulus* and *Stenobothrus stigmaticus*. (Photo credit: Thomas Fartmann)

Fig. 20: Coppice with standards with a lush herb layer near Bad Kissingen (Bavaria: Main-Franconia; 350 m a. s. l.). Occurring species: e. g. *Gomphocerippus rufus*, *Leptophyes punctatissima*, *Meconema thalassinum*, *Nemobius sylvestris* and *Pholidoptera griseoaptera*. (Photo credit: Dominik Poniatowski)

Fig. 21: Spoil heap of Mottbruch in Gladbeck-Brauck (North Rhine-Westphalia: Ruhr Area; 115 m a. s. l.). Occurring species: e. g. *Chorthippus brunneus*, *Oedipoda caerulescens*, *Sphingonotus caerulans* and *Tettigonia viridissima*. (Photo credit: Felix Helbing)

Grasshoppers reach high population densities (abundances) especially in grassland-like habitats (Fartmann et al. 2024). At the top of the list is wet grassland with an average of 25.8 individuals per 10 m^2 (Fig. 22). However, other grassland types such as acidic, calcareous and mesic grassland have similarly high abundances (Ø 20.3 to 22.3 individuals per 10 m^2). Tussock grasslands occupy an intermediary position with 12.1 individuals per 10 m^2. In other habitat types, the population densities are significantly lower. Heathlands, moorland and light woodland can be listed here in decreasing order.

Habitat requirements can vary greatly from species to species. Two types can be roughly distinguished: habitat generalists (eurytopic species) and habitat specialists (stenotopic species) (see also chapter Ecological indices). While the generalists are relatively undemanding in their choice of habitat, the specialists are restricted to a few habitats. Whether a species can develop and reproduce in a habitat is determined by several environmental factors. A distinction can be made between two spatial levels, which, taken together, determine a species' choice of habitat:

Landscape level: grasshoppers are cold-blooded organisms. Their occurrence at the landscape level is therefore strongly determined by the macroclimate (see also chapter Distribution). In addition to the macroclimate, the habitat size and connectivity of the habitat can also influence the occurrence of a species (Appelt & Poethke 1997, Carlsson & Kindvall 2001). Habitat specialists are particularly affected by this, as their habitats are often very isolated in a landscape that is otherwise unsuitable for them. The Speckled Buzzing Grasshopper (*Bryodemella tuberculata*), for example, depends on a well-connected network of these so-called habitat islands as it can only exist in the long term through a regular exchange of individuals with neighbouring populations (Stelter et al. 1997, Reich 2006, p. 160). By contrast, there are a number of less mobile species such as the Heath Grasshopper (*Chorthippus vagans*) or the Bog Bush-cricket (*Metrioptera brachyptera*) that can survive for long periods in highly isolated habitats without exchange. However, this is only possible if the requirements at the habitat level are continuously fulfilled and the habitat sizes are large enough to provide sufficient space for stable populations in the long term.

Habitat level: habitat specificity at the habitat level depends on a complex of factors that is dominated by the influence of the microclimate (Gardiner & Dover 2008, König et al. 2024). For successful development and reproduction, the temperature and moisture requirements of all developmental stages (eggs, nymphs and adults) must be achieved (Ingrisch 1978a, b, 1979a, 1983a, b, 1988). The egg is of particular importance here as it is the most sensitive stage. In contrast to the nymphs and adults, it cannot react to changing environmental conditions by changing location. The females therefore lay their eggs in places that meet the climatic requirements of the eggs – even under fluctuating environmental conditions (Ingrisch & Boekholt 1983). The degree of specialisation depends primarily on the drought resistance and/or the temperature requirements of the eggs (Ingrisch 1979a, 1983a, 1988). For example, the eggs of the Marsh Grasshopper (*Pseudochorthippus montanus*) are very sensitive to desiccation (Ingrisch 1983a). The species can

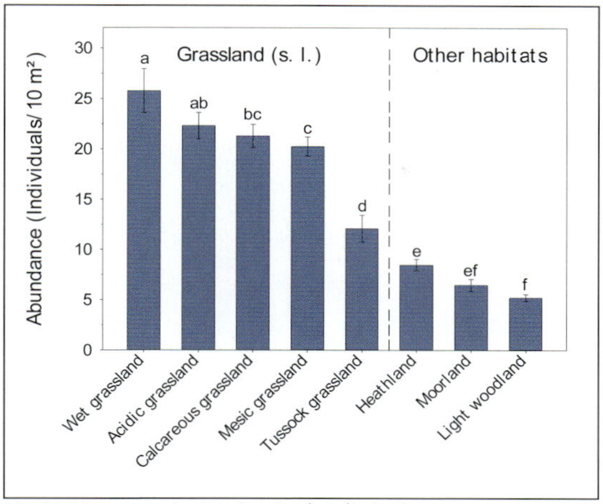

Fig. 22: Mean population density (± SE) of all grasshopper species in July/August differentiated in eight intensively sampled habitat types in Germany (N = 2,419 sample plots) (Fartmann et al. 2024, modified). Tussock grassland = initial psammophitic communities (*Koelerion glaucae* and *Spergulo-Corynephoretum*), moorland = transition mires and raised bogs. Different letters indicate significant differences between the habitat types ($P < 0.05$).

therefore only occur in moist and wet grasslands with an intact water balance (p. 220).

The vegetation structure is also of great importance for the occurrence or absence of species (Hemp & Hemp 2000, Jerrentrup et al. 2014, Kenyeres et al. 2024). In addition to its influence on the microclimate, it represents the structured space to which the species must be adapted in terms of shape, coloration and behaviour (Sänger 1977, Fartmann 1997a, Zehm 1997) (Fig. 23). According to Ingrisch & Köhler (1998), two types of behaviour can be roughly distinguished: the ground-dwelling species generally colonise open, sparsely vegetated habitats. Accordingly, they are mostly grey-brown in colour. Examples of this group would be the Blue-winged Grasshopper (*Oedipoda caerulescens*) (Fig. 24), the Mottled Grasshopper (*Myrmeleotettix maculatus*) or the Long-horned Groundhopper (*Tetrix tenuicornis*). The species living in the vegetation, on the other hand, often have a green or green-brown base colour. Three subtypes can be distinguished in this group. While the horizontally oriented species such as the Common Dark Bush-cricket (*Pholidoptera griseoaptera*) (Fig. 25) like to stay on broad leaves, the so-called long grass dwellers such as the Long-winged Conehead (*Conocephalus fuscus*) (Fig. 26) are vertically oriented species. There is also an intermediate behavioural type – i. e. species that live both on leaves and on stems. This group includes the Green Mountain Grasshopper (*Miramella alpina*) (Fig. 27) and the Rufous Grasshopper (*Gomphocerippus rufus*).

The microclimatic and structural requirements of the species are mostly linked to the influence of the habitat management (Fartmann & Mattes 1997). For example, land use has an effect on the vegetation structure and thus indirectly on the microclimate. In long-term fallows, for example, numerous species fail because the dense vegetation and litter accumulation only allow the topsoil to warm up slowly. Most species are therefore dependent on at least sporadic mowing or grazing of their habitats (Fumy & Fartmann 2023). However, if the land-use intensity is too high, the negative effects predominate (Poniatowski et al. 2024a, b). Meadows that

Fig. 23: Species-rich salt marsh (left) on the island Usedom (Mecklenburg-Western Pomerania) as an example of a dense and tall-growing habitat and tussock grassland (right) in a sand pit near Bremen (Lower Saxony) as an example of a sparsely vegetated habitat. (Photo credits: Dominik Poniatowski)

Fig. 24: The Blue-winged Grasshopper (*Oedipoda caerulescens*) is a geophilic species whose coloration is perfectly adapted to the nature of stony or sandy habitats.
(Photo credit: Dominik Poniatowski)

Fig. 25: The Common Dark Bush-cricket (*Pholidoptera griseoaptera*) – here a reddish coloured male – likes to stay on large leaves in fringes, hedges and shrubs.
(Photo credit: Dominik Poniatowski)

Fig. 26: The Long-winged Conehead (*Conocephalus fuscus*) is considered a vertically oriented species. It can therefore mainly be found in tall-growing habitats such as stands of rushes (*Juncus* spp.), sedges (*Carex* spp.) and Wood Small-reed (*Calamagrostis epigejos*). (Photo credit: Thomas Fartmann)

Fig. 27: The Green Mountain Grasshopper (*Miramella alpina*) is a phytophilous species. It requires a heterogeneous herb layer.
(Photo credit: Dominik Poniatowski)

are mowed early and often as well as heavily grazed areas lack tall-growing structures that are essential for some species – for example as a shelter, stridulation platform or oviposition substrate (Kruess & Tscharntke 2002, Fumy et al. 2023). In addition, intensive meadow use causes very high losses due to mowing and the discharge of eggs from plant-laying species (Schwarz et al. 2023, Schoof et al. 2024). On pastures with high stocking densities, the consumption of egg clutches and cattle trampling – especially under cold and wet weather – can lead to high losses (Fartmann & Mattes 1997). In combination with the influences on the vegetation structure, intensively used meadows and pastures are therefore not a suitable habitat for the majority of species living in open land (Poniatowski et al. 2024a). There are also a number of other factors that can influence the occurrence or absence of species. For example, the food supply can become a limiting factor in open, sparsely vegetated habitats.

Land-use and climate change

THOMAS FARTMANN, FELIX HELBING, FRANZ LÖFFLER & DOMINIK PONIATOWSKI

In this chapter, we introduce the general drivers of biodiversity changes. Subsequently, we explain how land-use and climate change affect the grasshopper fauna in Germany. In detail, we do this by presenting case studies.

Drivers of biodiversity change

The recent biodiversity crisis is among the most serious problems of our age (Rockström et al. 2009, Butchart et al. 2010, Naeem et al. 2012). Current species extinction rates are 10 to 100 times higher than the average of the last ten million years (Díaz et al. 2019). For terrestrial habitats, land-use change is considered the main driver of species decline (Newbold et al. 2015, Díaz et al. 2019, Cardoso et al. 2020). For a long time, natural dynamics and traditional land use have created and maintained species-rich grasshopper habitats (Bonari et al. 2017, Feurdean et al. 2018, Fartmann 2024; see also chapter Habitats). However, with increasing industrialisation of land use since the beginning of the 19th century and particularly after World War II, the area of grasshopper habitats has dramatically decreased and hence the remaining patches are today often highly fragmented (Poniatowski et al. 2024a). Mainly two large-scale processes have been responsible for this, namely: (i) intensification of land use on the one hand and (ii) abandonment and afforestation on the other hand (Squires et al. 2018, Prangel et al. 2023, Fartmann 2024). Land-use intensification has mainly occurred in the lowlands (Löffler et al. 2023), but is now also becoming increasingly relevant in the uplands (Fumy et al. 2021). By contrast, marginal soils have strongly been affected by abandonment and afforestation (MacDonald et al. 2000, Fumy et al. 2021, Prangel et al. 2023). This has especially been true for wet soils that can hardly be drained or mountain ranges with a varied relief that do not allow for intensive agriculture. Today, the remaining habitat patches are mostly embedded in a hostile matrix, which often consists of intensively used agricultural land (e. g. arable fields or improved grassland) (Öckinger et al. 2012, Poniatowski et al. 2018a). Additionally, many habitat remnants only exhibit a low habitat quality due to inappropriate land-use, eutrophication as a result of airborne nitrogen deposition or climate change (Marini et al. 2009, Fumy et al. 2021, Harvey et al. 2023, Vogels et al. 2023).

Key drivers of survival in fragmented landscapes

Habitat quality, size and connectivity are considered the main drivers of patch occupancy of insects in general and grasshoppers in particular in fragmented landscapes (Fartmann 2024; Fig. 28). Usually, habitat quality is the most decisive one among these predictors. However, species traits (e. g. habitat specificity or mobility) and landscape configuration (i. e. the spatial distribution, size and abundance of habitat patches) affect the relative importance of each of the three key drivers (Öckinger et al. 2010, Löffler & Fartmann 2017, Poniatowski et al. 2018a, Fartmann 2024). The majority of the native grasshopper species has a low mobility (see chapter Ecological indices) and builds closed populations (Poniatowski & Fartmann 2010, Poniatowski et al. 2018a). These species are able to survive in isolated habitat patches, as long as the habitat quality is permanently sufficient and the patch is large enough to harbour a long-term viable population of the species (Fartmann 2024). For an expansion in response to climate change and the colonisation of vacant or newly created habitat patches, however, these species also depend on a high habitat connectivity. By contrast, there are also some mobile grasshopper species that form metapopulations (see chapter Habitats) and necessarily require a network of high-quality habitats of sufficient size in spatial proximity.

Among the parameters that define habitat quality, the heterogeneity of a habitat has the strongest impact on species richness of grasshoppers (Löffler & Fartmann 2017). A high habitat heterogeneity fosters species richness in three different ways (Fartmann 2024):

For many grasshopper species, structurally diverse habitats are mandatory since they undergo microhabitat shifts during their ontogenetic development. A typical example is the Common Wart-biter (*Decticus verrucivorus*) (Fig. 29) (Cherrill & Brown 1992, Schuhmacher & Fartmann 2003a, p. 90): the species requires bare soil for

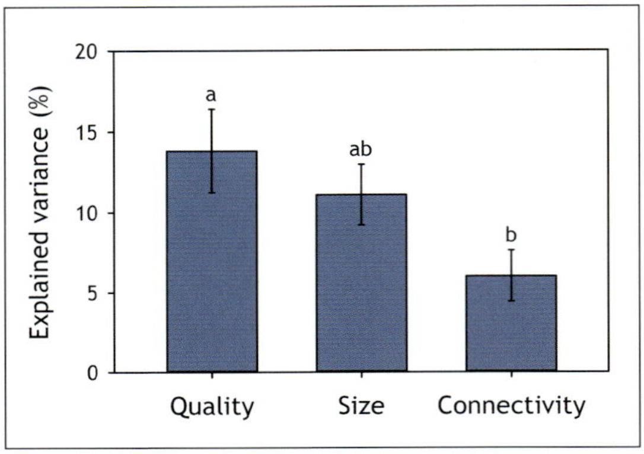

Fig. 28: Explanatory power of habitat quality, size and connectivity with respect to the patch occupancy of 13 habitat specialists (butterflies, grasshoppers, leafhoppers) in calcareous grasslands (Hesse: Diemel Valley), expressed as the mean (± SE) proportion of variance explained exclusively by the respective parameter. Different letters indicate significant differences between the parameters ($P < 0.05$). (Source: Poniatowski et al. 2018b)

egg-laying, and the young nymphs also stay there due to the warm microclimate. Larger, older nymphs, and later especially adults, however, also depend on taller vegetation that provides shelter against predators and too high temperatures. Similar but less pronounced microhabitat shifts have for example been documented for the Mottled Grasshopper (*Myrmeleotettix maculatus*) and the Common Grey Bush-Cricket (*Platycleis albopunctata*) (Lensink 1963, Gottschalk 1997, Wünsch et al. 2012, pp. 94 and 196; Fig. 30). The native species that undergoes the strongest habitat shift is the Great Green Bush-cricket (*Tettigonia viridissima*) (p. 84). The spectrum of used microhabitats during the complete life cycle ranges in this species from bare soil (oviposition) to grassy-herbaceous vegetation (habitats of nymphs and adults) up to treetops (song posts of adult males).

Heterogeneous habitats exhibit a higher variety of microhabitats and therefore enable the coexistence of many grasshopper species with sometimes very different habitat requirements (Joern 2005, Jerrentrup et al. 2014, Löffler & Fartmann 2017; Fig. 31).

Moreover, heterogeneous habitats are significantly more resistant and resilient against the impacts of climate change and generally against strong environmental variation than homogeneous ones (Fartmann 2024). This is also true for grasshoppers (Kindvall 1996). Overall, habitat heterogeneity decisively contributes to the long-term viability of grasshopper populations, even under strongly fluctuating environmental conditions, which in times of climate change is increasingly the case.

Within heterogeneous habitats, early-successional stages that are rich in sunlit bare soil play a key role for grasshoppers, particularly threatened species (Maes & Bonte 2006, Rada et al. 2017). More than 70 % of the native grasshopper species rely on open soil for oviposition (Fartmann 2017). Both habitat heterogeneity and the availability of early-successional stages rich in bare ground depend on the productivity of the habitat and the disturbance regime, which is usually equated with the land-use regime in Central Europe today (Gibson 2009, Joern & Laws 2013, Fartmann 2024). Low to intermediate intensities and frequencies of disturbance favour habitat heterogeneity and thus species richness of grasshoppers (Fumy & Fartmann 2023, Helbing et al. 2023; Fig. 32). In principle, on more fertile soils, a higher disturbance level is necessary to foster species richness than on less productive ones. Mostly, abundance and species richness of grasshoppers are highest in mid-successional stages (Schirmel et al. 2011, Fartmann et al. 2012, Helbing et al. 2014a; Fig. 33). They simultaneously provide sufficient food, an ample warm microclimate, at least some patches of bare ground for soil-breeding species

Fig. 29: A male Common Wart-biter (*Decticus verrucivorus*) that has optimised radiation absorption due to orientation of all important body parts to the sun under a cloudy sky. (Photo credit: Thomas Fartmann)

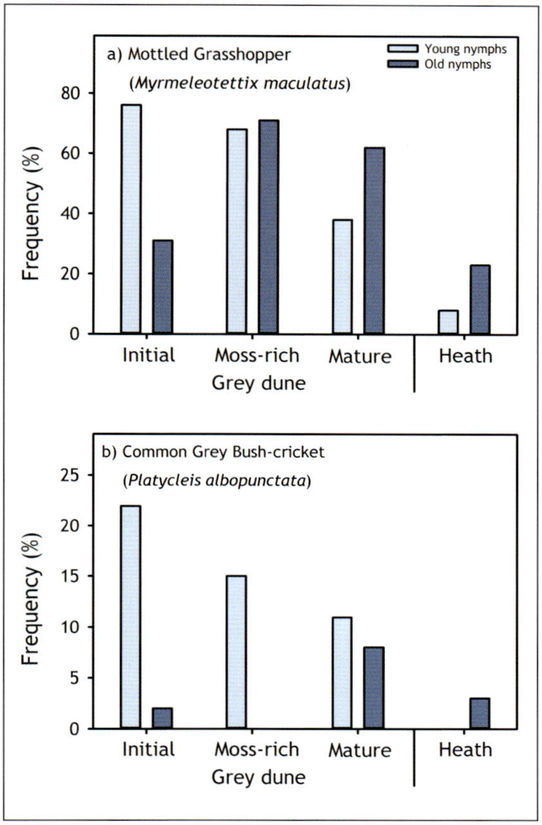

Fig. 30: Frequency (%) of young and old nymphs of the Mottled Grasshopper (*Myrmeleotettix maculatus*; a) and Common Grey Bush-cricket (*Platycleis albopunctata*; b) across a vegetation-structure gradient from initial grey dunes (open, short turf) up to heathlands (dense, high-growing) (Mecklenburg-Western Pomerania: Hiddensee). In both species, the distribution of young and old nymphs significantly differed ($P < 0.001$). (Source: Wünsch et al. (2012)

and also dense vegetation as shelter against predators or unfavourable weather conditions (Schirmel et al. 2011, Wünsch et al. 2012, Helbing et al. 2014a). Although early-successional stages normally have a lower species richness than mid-successional ones, they often exhibit more species than late-successional stages (Schirmel et al. 2011, Helbing et al. 2014a). This is mainly due to the high temperature requirements of many autochthonous species (see chapter Ecological indices) and the great importance of sunlit bare soil for egg-laying, particularly in many threatened species (see above).

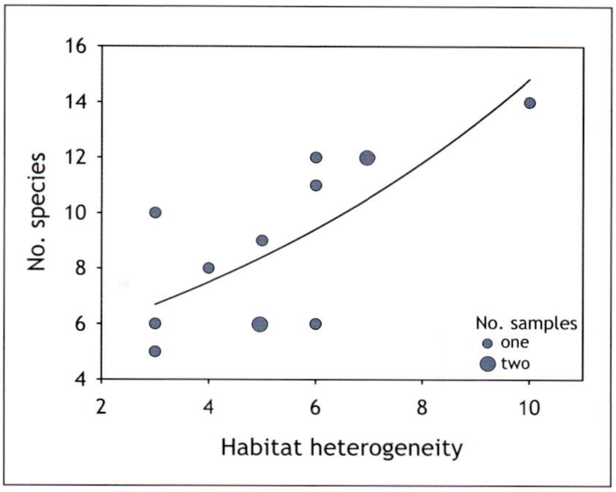

Fig. 31: Relationship between habitat heterogeneity and species richness of grasshoppers in pre-Alpine calcareous grasslands (Bavaria: Niederwerdenfelser Land). The regression curve illustrates a significant relationship ($P < 0.05$). (Source: Löffler & Fartmann 2017)

Climate change meets habitat fragmentation

Since the late 1980s, climate change has become increasingly important for biodiversity changes (Díaz et al. 2019, Cardoso et al. 2020, Harvey et al. 2023). Global warming has resulted in an increase in temperatures and extreme weather events (Streitberger et al. 2016a, IPCC 2021). By contrast, annual precipitation has only slightly changed in Central Europe. Nevertheless, periods of summer drought occur more frequently due to a precipitation shift from summer to winter and a stronger evapotranspiration as a result of higher summer temperatures. Consequently, climate change already has strong effects on native grasshoppers (Poniatowski et al. 2018a, 2020, 2024a, Engelhardt et al. 2022, Fartmann et al. 2022a, b, Fumy et al. 2020, Löffler et al. 2019, Ogan et al. 2022, Thorn et al. 2022). With ongoing warming, the effects of climate change most likely will become even more visible (Díaz et al. 2019, Stuhldreher & Fartmann 2018, Fartmann et al. 2022a, Harvey et al. 2023). Gradual climatic changes as well as the increasing frequency and intensity of extreme weather events such as droughts, fires or floods contribute to this. However, the impacts of the latter are much more severe, particularly in highly isolated habitat patches. Extreme weather events are both spatially and

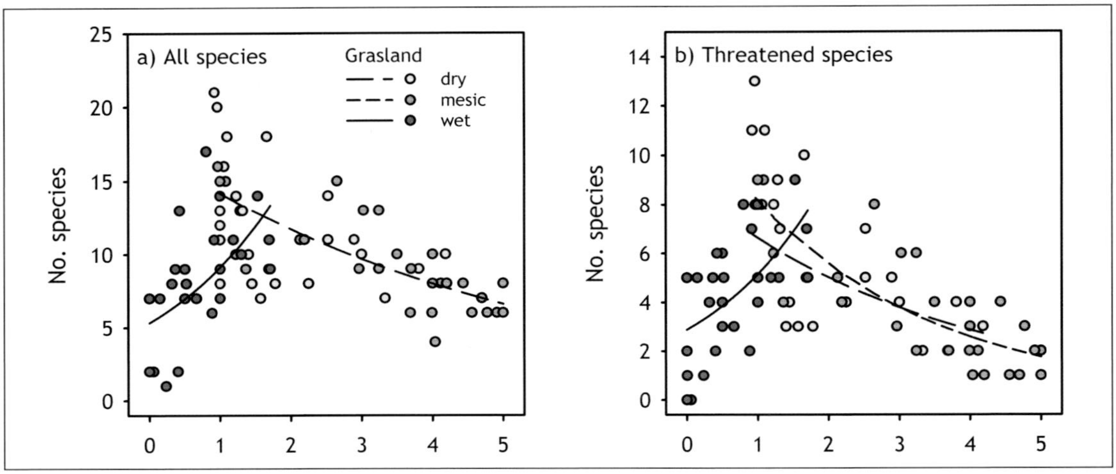

Fig. 32: Relationship between species richness of grasshoppers (a: all and b: threatened species) and land-use intensity in three different grassland types (Baden-Württemberg: southern Black Forest). Land-use intensity: 0 = fallow, 5 = very high. A regression curve illustrates a significant relationship ($P < 0.05$). (Source: Fumy et al. 2021)

temporally hardly predictable, hence, the species have nearly no adaptabilities (Harvey et al. 2023). Overall, the response to climate change is often complex and partly even differs among related species or is depending on the landscape configuration (Fartmann et al. 2021a).

These complex relationships are illustrated by the Bog Bush-cricket (*Metrioptera brachyptera*) and Roesel's Bush-cricket (*Roeseliana roeselii*): both bush-crickets are normally short-winged and flightless (Poniatowski & Fartmann 2011a, Poniatowski et al. 2012). However, the habitats of the two closely related species are quite different: *Metrioptera brachyptera* is a habitat specialist that mainly occurs in nutrient-poor habitats such as calcareous grasslands or the edges of mires in low population densities (Fig. 34). By contrast, *R. roeselii* is a habitat generalist that inhabits different types of open grassy-herbaceous habitats. Moreover, its abundance is usually much higher than that of the sibling species. In both bush-crickets, density stress in the early nymphal instars leads to the release of a stress hormone that triggers the development of long-winged (macropterous) morphs that are capable of flight (Poniatowski & Fartmann 2009). During the past 20 years, springs were often disproportionally warm and dry, which increased the survival of nymphs, leading to increases in population densities (Poniatowski & Fartmann 2011a; Fig. 34).

Since the natural abundance of *M. brachyptera* is usually low, reduced nymphal mortality due to favourable climatic conditions has mostly not been sufficient to cause density stress (Fig. 34). Even in the rare case of macropters emerging (Poniatowski & Fartmann 2011b), the high degree of habitat fragmentation in Central Europe strongly limits the dispersal of *M. brachyptera* and its ability to keep pace with climate change (Poniatowski & Fartmann 2011b). Additionally, the eggs of the species are sensitive to drought (Ingrisch 1979a). Consequently, *M. brachyptera* is increasingly threatened by drought events due to climate change, which may lead to population losses (Poniatowski et al. 2018a, 2024a; Fig. 34). By contrast, due to the naturally higher abundance of *R. roeselii*, the warmer and drier springs during the last two decades have regularly led to density stress among young nymphs. Nowadays, macropters of *R. roeselii* can be observed frequently and proportions of up to 20 % long-winged individuals within populations are quite common (Poniatowski & Fartmann 2011a). Since they are capable of flight, these morphs exhibit a higher mobility than the short-winged ones (Poniatowski et al. 2012). Due to the high availability of suitable habitats, this species can keep pace with global warming and has recently expanded its range in Western, Central and Northern Europe (Poniatowski et al. 2012; Fig. 34).

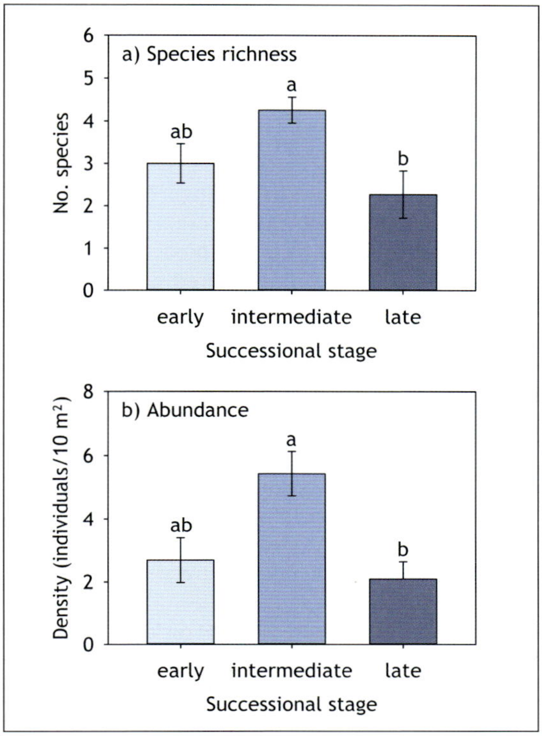

Fig. 33: Mean (± SE) species richness (a) and abundance (b) of grasshoppers across a gradient from early- to late-successional stages in light pine forests (Bavaria: Upper Isar Valley). Different letters indicate significant differences between the parameters ($P < 0.05$). (Source: Helbing et al. 2014a)

Range shifts in Germany

During past decades, the distributions of grasshoppers in Germany have strongly changed due to land-use and climate change. This was true for the historic period (pre-1990 vs. 1990–1999) as well as for the recent period (1990–1999 vs. 2000–2017) (Poniatowski et al. 2020; Fig. 35 and 36). However, the relative impacts of each of the drivers has notably changed over time. The historic period was characterised by strong range retractions due to habitat loss as a result of land-use change. In particular, this was the case for habitat specialists with low mobility (Fig. 37 and 39) and species that depend on HNV farmland (high SFI) (see also chapter Distribution). Among these, the Gravel Grasshopper (*Chorthippus pullus*), the Rattle Grasshopper (*Psophus stridulus*) and the European Mole Cricket (*Gryllotalpa gryllotalpa*) were the species that suffered most. Solely a few highly mobile and thermophilic (high STI) habitat generalists were able to expand their range already in the 1990s due to the increasing effects of climate change.

By contrast, the recent period was more strongly characterised by global warming and less by large-scale habitat loss. Many mobile, thermophilic generalists, but also some specialists, have expanded their range during that period (Fig. 38). Among the thermophilic species experiencing the strongest range expansions are the Common Sickle Bush-cricket (*Phaneroptera falcata*), the Large Conehead (*Ruspolia nitidula*) and the Southern Sickle Bush-cricket (*Phaneroptera nana*). The trend was negative only for a few species such as the Marsh Grasshopper (*Pseudochorthippus montanus*, Fig. 39), the Common Groundhopper (*Tetrix undulata*) and the Common Green Grasshopper (*Omocestus viridulus*). Due to their high moisture requirements, they can be considered losers of climate change. Moreover, the range of the Mottled Grasshopper (*Myrmeleotettix maculatus*) has contracted due to the loss of early-successional stages (Fig. 40).

The new red list of German grasshoppers confirms the patterns for the recent period mentioned before (Poniatowski et al. 2024a; Fig. 41). Currently increasing species have a lower SFI than those that are decreasing. Species that exhibit stable populations have an intermediary position between these two groups. By contrast, the STI is higher in

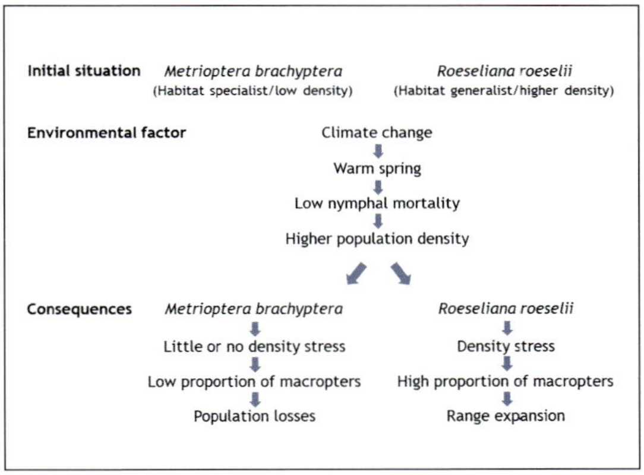

Fig. 34: Effects of climate change on two related bush-cricket species: the habitat specialist Bog Bush-cricket (*Metrioptera brachyptera*) and the habitat generalist Roesel's Bush-cricket (*Roeseliana roeselii*). (Source: Poniatowski & Fartmann 2009, 2011a, b, Poniatowski et al. 2012)

increasing species than in those that have stable populations or are decreasing.

Case studies from uplands and lowlands

In grasslands in the West German low mountain ranges the Eifel Mountains (Löffler et al. 2019) and the Black Forest (Fumy et al. 2020), parallel changes in climate and grasshopper assemblages were observed from the mid-1990s to the mid-2010s. By contrast, land use did not change during that period: the wet grasslands were still abandoned, while mesic and dry grasslands (calcareous and acidic grasslands, respectively) were characterised by a continuous low-intensity management. All grassland patches studied in the 1990s still existed in the 2010s. In the mainly abandoned wet grasslands, mean species richness did not change during that period (Fig. 42). By contrast, in the other grasslands still characterised by low-intensity management, mean species richness increased by 19–45 %. This increase in species richness was driven by an increase in temperatures due to climate change, which favoured the expansion of thermophilic and mobile habitat generalists. Among these species were the Large Gold Grasshopper (*Chrysochraon dispar*) and Roesel's Bush-cricket (*Roeseliana roeselii*).

The northern Münsterland (Westphalian Bay) is characterised by small-scale, but nevertheless intensive agriculture. Due to the small plots it is called the Münsterland park landscape and exhibits a high density of verges. In 1995, a representative sample of grassy verges and grassland patches, thereof 22 % in nature reserves, was studied (Fartmann et al. 2021b, 2022b). Re-surveys in 2012 revealed that 27 % of the grassland patches had experienced a land-use change. Nearly 90 % of these patches had been converted to maize fields, which were mainly used for bioenergy production. Patch losses solely occurred outside nature reserves. By contrast, nearly all verges still existed in 2012. The increase in mean species richness and expansion of thermophilic habitat generalists with high mobility due to climate change, which had already been observed for the uplands, had also occurred in the northern Münsterland. However, due to the high connectivity of grassland patches within the large-scale nature reserves here also species with low mobility increased (Fig. 43). However, in grassland patches outside nature reserves and in verges this was not the case: Here the Community Farmland Index and Community Temperature Index decreased (Fig. 44). Thus, within the communities now more generalist and thermophilic species dominated. By contrast, in grassland patches inside nature reserves the figures of both indices had not changed. Thus, the grassland patches within nature reserves were more resistant and resilient, respectively, than unprotected patches. Both the continuous low-intensity management and less drainage of the patches and therefore,

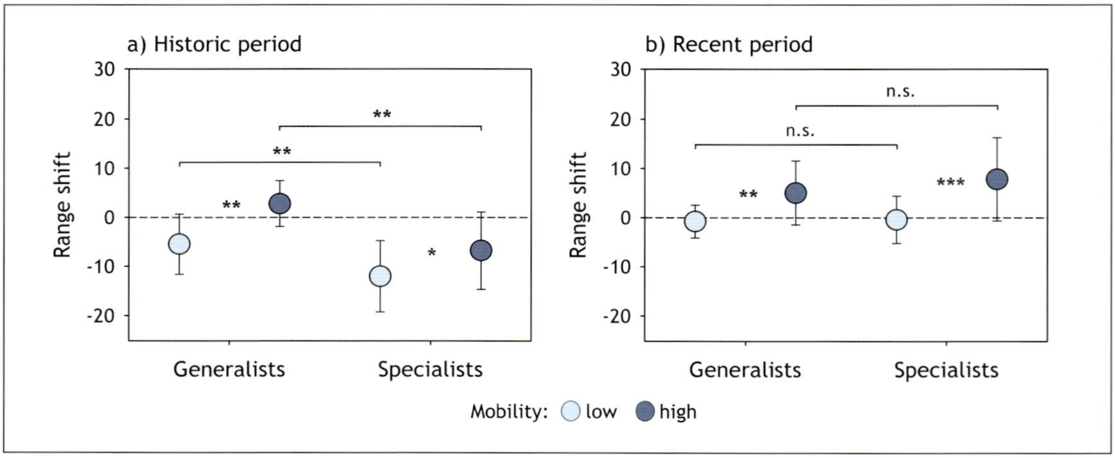

Fig. 35: Range shifts of grasshoppers (N = 58 species) in Germany depending on their habitat specificity and mobility in the historic (pre-1999 vs. 1990–1999) (a) and recent period (1990–1999 vs. 2000–2017) (b). Significance: n.s. = not significant, * $P < 0.05$, ** $P < 0.01$, *** $P < 0.001$. (Source: Poniatowski et al. 2020)

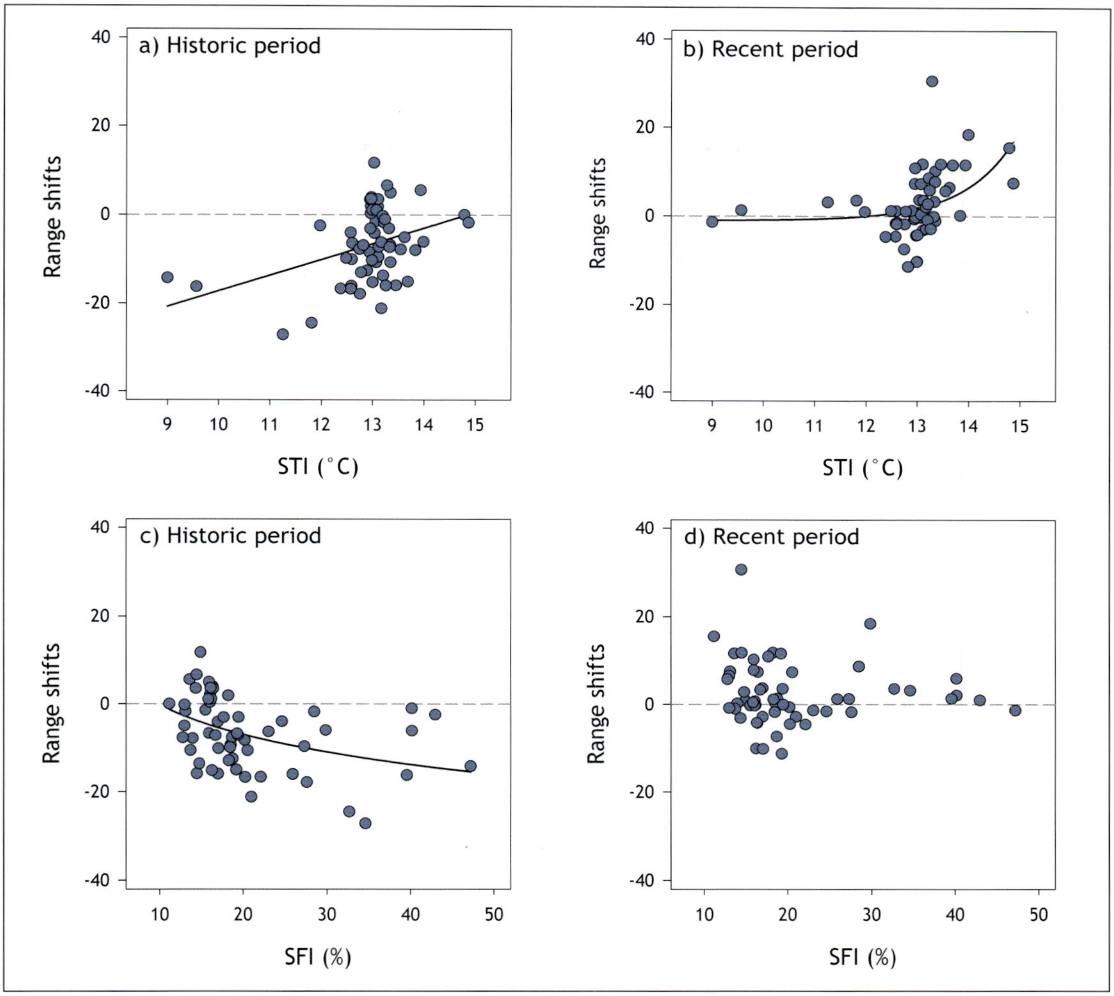

Fig. 36: Relationship between historic (pre-1999 vs. 1990–1999) (a and c) and recent (1990–1999 vs. 2000–2017) (b and d) grasshopper range shifts (N = 58 species) in Germany depending on the Species Temperature Index (STI) (a and b) and Species Farmland Index (SFI) (c and d). The higher the index, the more the species prefers areas characterised by warm summers and large-scale HNV farmland, respectively. A regression curve/line illustrates a significant relationship ($P < 0.01$). (Source: Poniatowski et al. 2020)

despite climate change, a more stable water level seemed to be responsible for this. The species with the strongest losses in the study area was the Common Green Grasshopper (*Omocestus viridulus*). In grassland patches outside nature reserves the frequency decreased by 53% and in verges by 65%. By contrast, in grasslands within nature reserves no changes were observed. The Common Green Grasshopper is equally sensitive to intensification and abandonment of grassland management as well as to increasing drought (Gardiner 2010, Poniatowski et al. 2018b, 2024a, Detzel et al. 2022). A recent study from the same study area still reveals a higher species richness and biomass in grasslands inside nature reserves compared to those outside (Poniatowski et al. 2024b; Fig. 45). This is also true for threatened species.

As the recent drought summers have shown, not only hygrophilic grasshopper species suffer from climate change; those that occur in mesic or dry habitats do as well. In the warmest parts of the Diemel Valley (eastern Westphalia, north-

Fig. 37: The Woodland Grasshopper (*Omocestus rufipes*) is a habitat specialist that strongly declined in the historic period. (Photo credit: Thomas Fartmann)

Fig. 38: The Blue-winged Grasshopper (*Oedipoda caerulescens*) is a mobile habitat generalist whose populations have strongly declined in the past due to land-use change; currently, however, the species is expanding its range rapidly in response to climate change. (Photo credit: Thomas Fartmann)

Fig. 39: The Marsh Grasshopper (*Pseudochorthippus montanus*) suffers strongly from recent land-use and climate change. (Photo credit: Thomas Fartmann)

Fig. 40: The Mottled Grasshopper (*Myrmeleotettix maculatus*) is currently declining due to the loss of early-successional stages. (Photo credit: Thomas Fartmann)

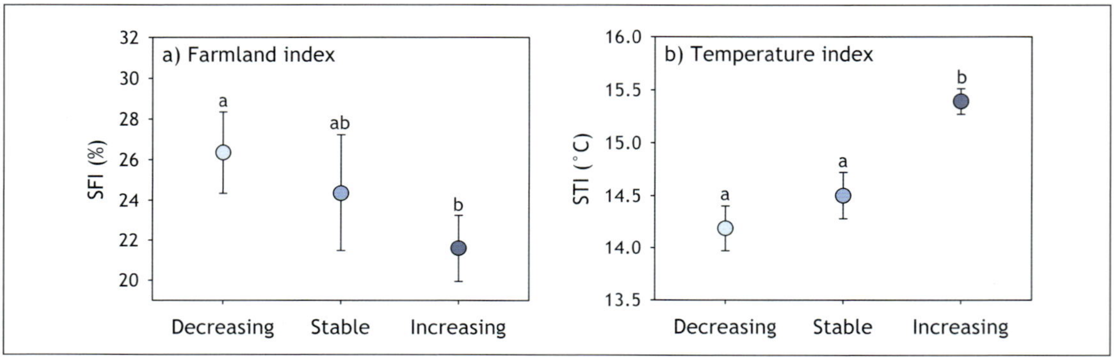

Fig. 41: Current population trend of grasshoppers depending on the mean (± SE) Species Farmland Index (SFI; a) and Species Temperature Index (STI; b). The population trend was derived from the recent red list (Poniatowski et al. 2024a). Species with data deficits or without an SFI/STI were not considered. Different letters indicate significant differences between the parameters ($P < 0.05$). (Source: author's own diagram)

ern Hesse), above-ground vegetation of mesic and semi-dry calcareous grasslands had dried out, which caused food shortage and negatively affected grasshopper abundance (Fartmann et al. 2022a; Fig. 46 and 47).

Conclusion

Until the 1990s, many habitats of grasshoppers had been destroyed due to large-scale land-use change. In particular, this strongly negatively affected habitat specialists with low mobility. In the remaining habitats, it is primarily a gradual deterioration in habitat quality – for example due to the progressive intensification of land use – that is now having a negative impact on grasshopper communities. In addition, climate change has become increasingly important for the distribution of grasshoppers over the last 20 to 30 years. However, due to the strong habitat fragmentation in our landscapes, no grasshopper species is able to keep up with pace of climate change (Fartmann et al. 2021a). Among the species that benefit from the temperature increase are especially thermophilic and simultaneously mobile habitat generalists. Currently, more than 30 % of the native grasshopper species are known to expand their range due to climate change (Poniatowski et al. 2018a). In landscapes that are still characterised by a high habitat connectivity, there are even examples of species with low mobility that expand their range. Among the losers of land-use and climate change are particularly montane/hygrophilic species and many habitat specialists with low mobility (Poniatowski et al. 2018a, 2024a). However, the last drought years have shown that an ever-increasing number of widespread species of mesic and dry habitats suffer from climate change as well (Fartmann et al. 2022a). Overall, the developments described above (decrease of habitat specialists, increase of widespread habitat generalists) lead to an additive biotic homogenisation of our grasshopper assemblages (Thorn et al. 2022).

The negative effects of global change on species richness do not only affect habitats outside protected areas but also inside (Rada et al. 2019, Cooke et al. 2023). One example of this are the large-scale effects of airborne nitrogen deposition (WallisDeVries & Bobbink 2017, Vogels et al. 2023). Nevertheless, nature reserves provide effective area protection and suffer less from fast land-use changes (Schuch et al. 2011, Fartmann 2021a, 2022a, Poniatowski et al. 2024b). Additionally, they are more resistant and resilient against the effects of climate change to a certain degree. In the future, particularly habitat availability and climate change will determine the distribution of grasshoppers in our landscapes (Fartmann et al. 2021b). Habitat availability includes the availability of suitable habitats as well as of dispersal corridors or stepping stones.

Ecological indices

DOMINIK PONIATOWSKI, FRANZ LÖFFLER, FELIX HELBING, FELIX MAXIMILIAN FREIENSTEIN, GREGOR STUHLDREHER & THOMAS FARTMANN

Introduction

Habitat selection of grasshoppers is highly dependent on the vegetation structure and the associated microclimate. Both factors are decisively influenced by land use; overall, a management with a low intensity is generally the key to species- and individual-rich grasshopper communities. Grasshoppers are considered very good indicators of environmental change in open landscapes (Bazelet & Samways 2011, Fartmann 2020). This applies to the effects of both land use and climate change. However, habitat requirements can vary greatly from species to species (Reinhardt et al. 2005, Poniatowski et al. 2020). In order to understand environmental changes using grasshoppers, it is therefore helpful to define species-specific ecological indices or traits (Klaiber et al. 2017, Dvořák et al. 2022). These include, for example, body weight, habitat specificity, microclimatic preference or the mobility of a species. At the landscape level, dependence on a specific macroclimate or preference for landscapes with a high proportion of grassland of high conservation value can function as ecological parameters. In a broader sense, the population trend and the red list status are also understood here as ecological indices as they are also very well suited to answering nature conservation questions. At the same time, the ecological indices can be used as meaningful indicators in the context of monitoring programs (see also chapter Monitoring).

Basic ecological indices for the 83 native grasshopper and praying mantis species in Germany are presented below (Table 1). These are categori-

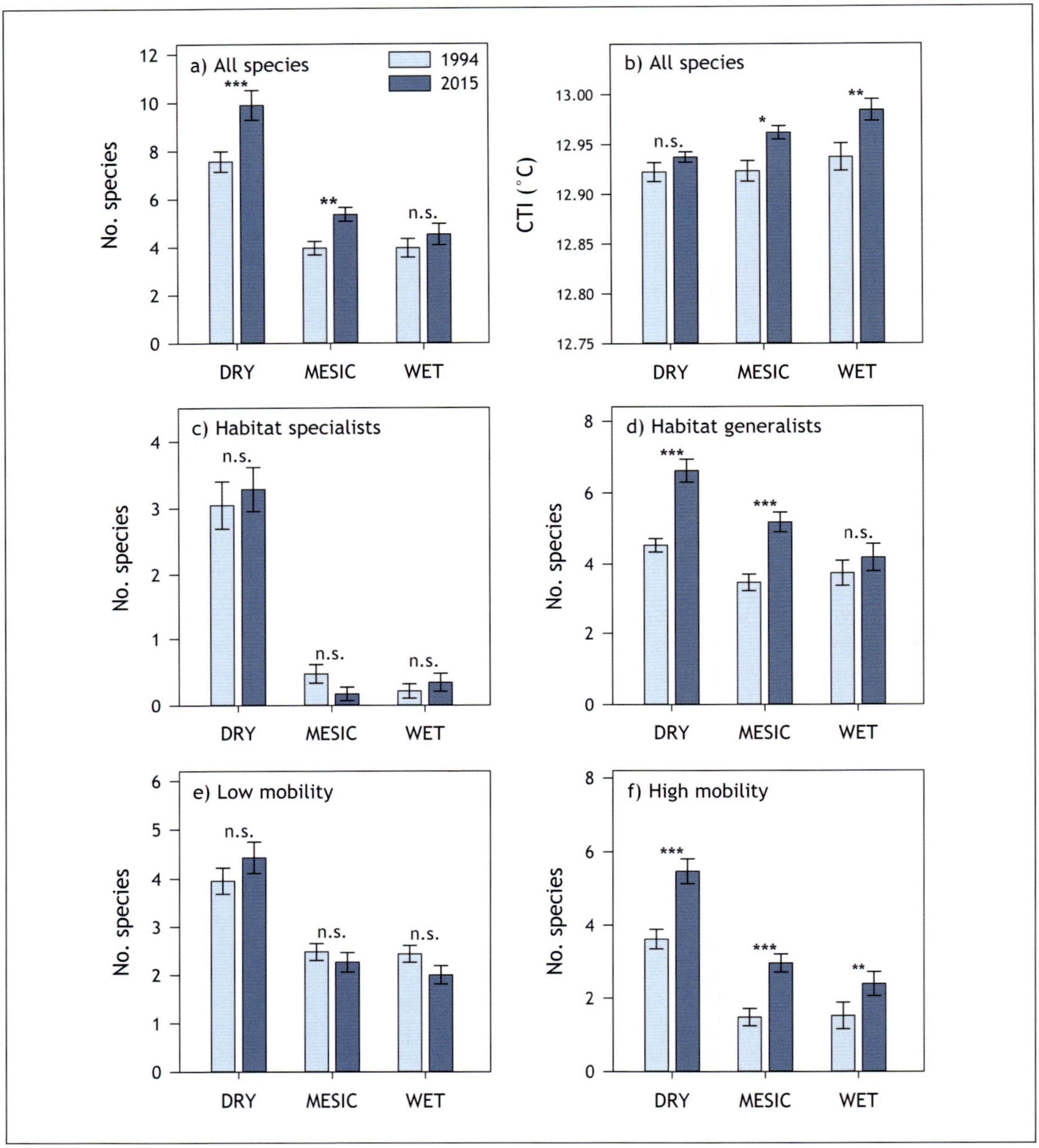

Fig. 42: Mean species richness and Community Temperature Index (CTI) (always ± SE), respectively, of all grasshopper species (a and b), habitat specialists (c), habitat generalists (d), species with low mobility (e) and species with high mobility (f) in semi-dry calcareous (DRY), mesic (MESIC) and wet grasslands (WET) (North Rhine-Westphalia: Eifel Mountains). Significance: n.s. = not significant, * $P < 0.05$, ** $P < 0.01$, *** $P < 0.001$. (Source: Löffler et al. 2019)

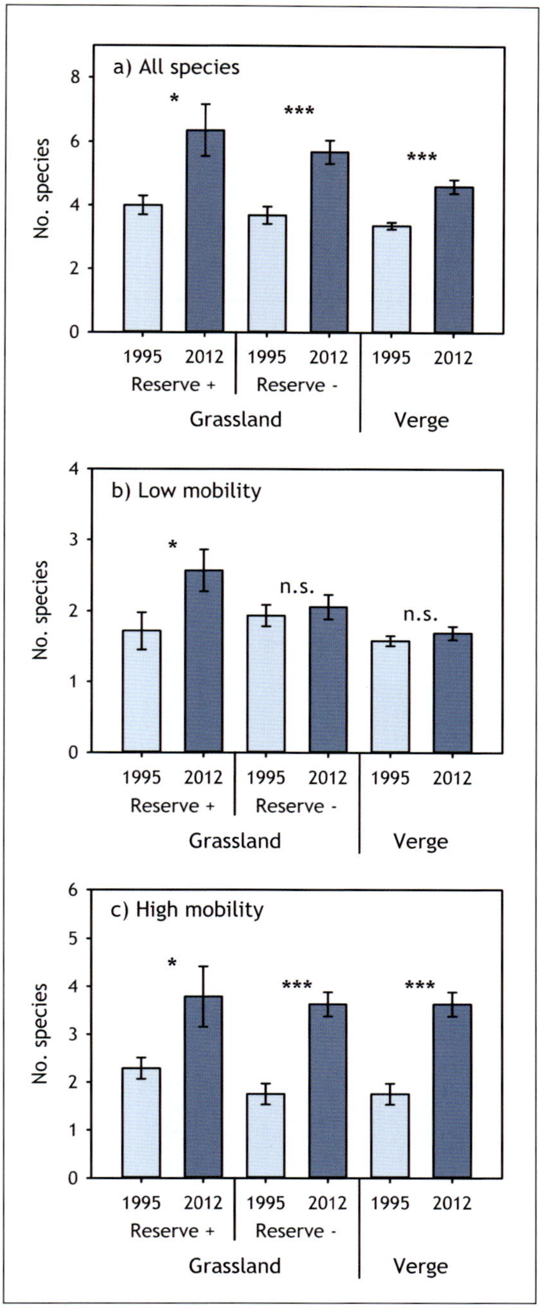

Fig. 43: Mean species richness (± SE) of all grasshopper species (a), those with low mobility (b) and those with high mobility (c) in grassland patches inside (Reserve +) and outside (Reserve -) nature reserves as well as in verges in 1995 and 2012 (North Rhine-Westphalia: Münsterland).
Significance: n.s. = not significant, * $P < 0.05$, *** $P < 0.001$.
(Source: Fartmann et al. 2021b, 2022b)

cal or metric parameters on the ecological preferences and functional characteristics of the species. The values presented are based on literature data, expert assessments or empirical calculations. Indicator values for temperature and moisture are presented here for the first time. They are based on the indicator values known from plants according to Ellenberg et al. (2001). However, all other ecological indices have already been published, at least in a modified form, and successfully used in ecological studies.

Habitat specificity

Habitat specificity describes the degree of specialisation of a species for the colonisation of habitats within their German range. A distinction is made between the two categories 'low' and 'high' (Table 1). Species with low habitat specificity are usually able to colonise numerous and sometimes very different habitats. They are also known as habitat generalists. By contrast, species with high habitat specificity can thrive only in a narrow range of environmental conditions. Those species usually occur in a few habitats with similar ecological characteristics. They are considered habitat specialists. Although there is no fixed border between generalists and specialists and the degree of habitat specialisation can slightly vary from one region to another, a rough classification has successfully been applied in various studies (e. g. Reinhardt et al. 2005, Fartmann et al. 2012, Löffler et al. 2019). The classification presented here largely corresponds to Poniatowski et al. (2020). However, novel insights in the species-specific habitat specificity required a slight adjustment of the classification in some cases.

Mobility

Mobility is a measure of the dispersal ability of grasshopper species under the current environmental conditions. Based on Poniatowski et al. (2020), it is here divided into the two categories 'low' and 'high' (Table 1). Species with low mobility are generally unable to cover larger distances between habitat islands in our agriculture landscapes. Their dispersal potential is correspondingly low. By contrast, highly mobile species can colonise new habitats far away from existing populations. This is either due to their very good

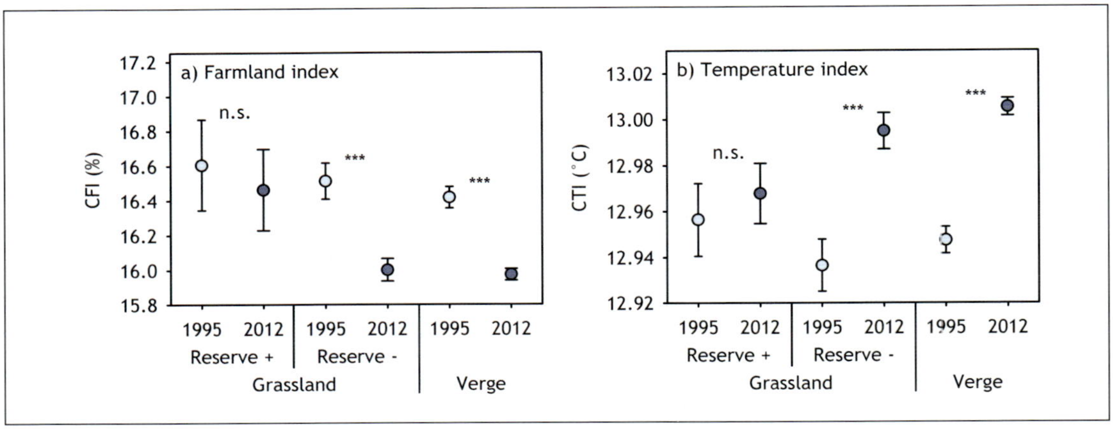

Fig. 44: Mean Community Farmland Index (CFI) and Community Temperature Index (CTI) (always ± SE) in grassland patches inside (Reserve +) and outside (Reserve -) nature reserves as well as in verges in 1995 and 2012 (North Rhine-Westphalia: Münsterland). Significance: n.s. = not significant, *** P < 0,001. (Source: Fartmann et al. 2021b, 2022b)

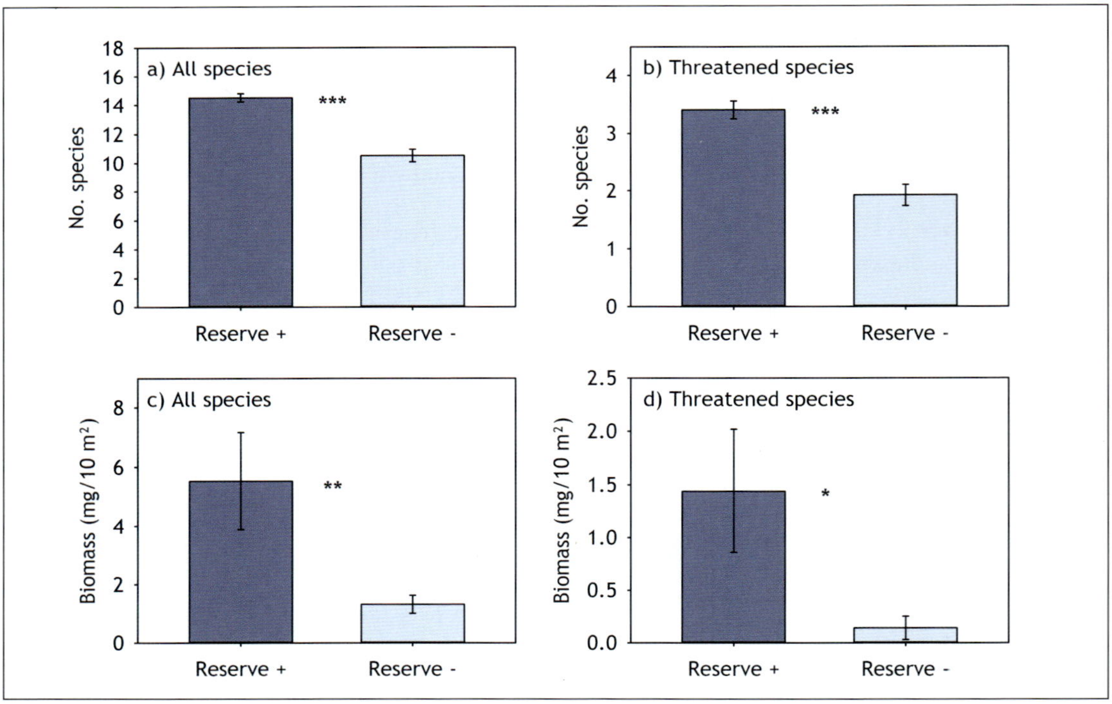

Fig. 45: Number of all and threatened grasshopper species at the landscape level (5 ha) (a and b) as well as biomass in a grassland patch (c and d) (always mean ± SE) inside (Reserve +) and outside (Reserve -) nature reserves (North Rhine-Westphalia: Münsterland). Significance: n.s. = not significant, * P < 0.05, ** P < 0.01, *** P < 0.001. (Source: Poniatowski et al. 2024b)

Fig. 46: Mesic grassland in August 2017 (left) and during a summer drought in the same month in 2018 (right) (Hesse: Diemel Valley). (Photo credits: Dominik Poniatowski)

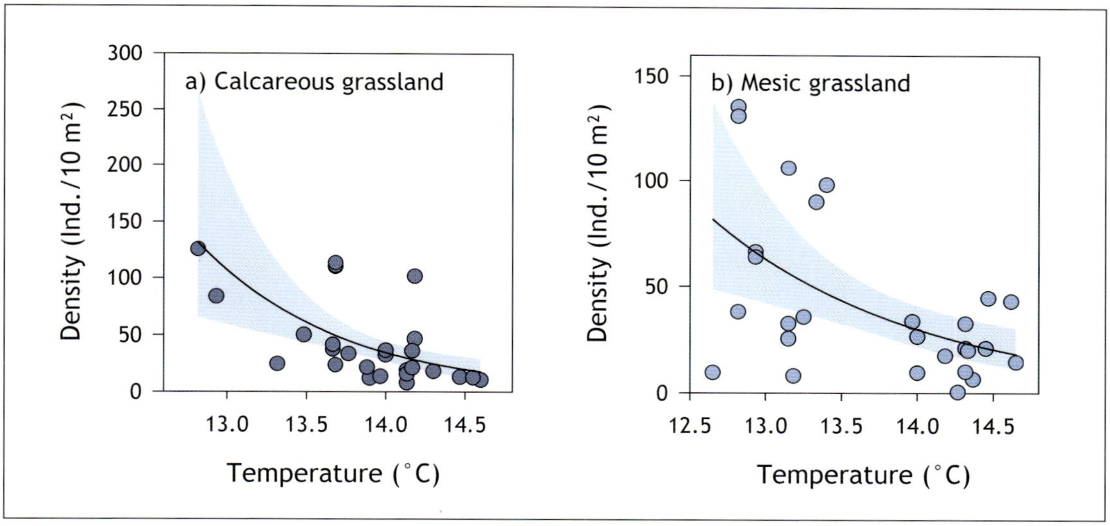

Fig. 47: Grasshopper abundances in semi-dry calcareous grasslands (a) and mesic grasslands (b) during the summer drought 2020 in relation to the long-term mean summer temperature of the investigated plots. (North Rhine-Westphalia/Hesse: Diemel Valley). Regression curve ± 95 % confidence interval. Significance: a) $P < 0.01$; b) $P < 0.001$. (Source: Fartmann et al. 2022a)

flight ability or because they regularly can bridge large distances via passive dispersal vectors. All species that are predominantly short-winged, but in which long-winged (macropterous) individuals regularly occur, were classified as highly mobile. This group includes the Large Gold Grasshopper (*Chrysochraon dispar*) and Roesel's Bush-cricket (*Roeseliana roeselii*). Both species have greatly expanded their range in recent decades (Poniatowski et al. 2012, Löffler at al. 2019, Fartmann et al. 2021b). In addition, species with mobile males but dispersal-poor females were assigned to the category 'low mobility' as males alone cannot establish new populations. These include, for example, the Speckled Buzzing Grasshopper (*Bryodemella tuberculata*), the Rattle Grasshopper (*Pso-*

phus stridulus), the Lesser Mottled Grasshopper (*Stenobothrus stigmaticus*) and the Black-spotted Toothed Grasshopper (*Stenobothrus nigromaculatus*). As in the case of habitat specificity, it was not always easy to assign them to one of the two categories as there is hardly any empirical data on the dispersal distances of grasshoppers. Nevertheless, the rough classification used here has proven to be practicable in previous studies (e.g. Reinhardt et al. 2005, Löffler et al. 2019, Fumy et al. 2020, Fartmann et al. 2021b).

Species Farmland Index (SFI) and Species Temperature Index (STI)

In order to make the effects of land use and climate change on grasshoppers more visible, the Species Farmland Index (SFI) and the Species Temperature Index (STI) can be used (Table 1; e.g. Löffler et al. 2019, Fumy et al. 2020, Riggi & Berggren 2020, Thorn et al. 2022). The calculation of these species-specific indicator values is based on the current distribution of grasshoppers in Germany (Poniatowski et al. 2020). The SFI reflects the proportion of High Nature Value farmland (HNV farmland) in the open landscape over the distribution range of the respective species in Germany. HNV farmland mainly covers nutrient-poor habitats such as wet meadows, heathlands and nutrient-poor grasslands (Paracchini et al. 2008, Veen et al. 2009) (Fig. 7 and 48). It therefore has a high nature conservation value.

Species with high SFI values, such as the Heath Bush-cricket (*Gampsocleis glabra*) (SFI: 38.8), the Large Saw-tailed Bush-cricket (*Polysarcus denticauda*) (SFI: 45.3) and the Rattle Grasshopper (*Psophus stridulus*) (SFI: 35.3), occur mainly in landscapes with large areas of these habitats. However, the SFI is not necessarily a measure of habitat specificity. There are also some grasshopper species with a low SFI, such as the Orange-tipped Grasshopper (*Omocestus haemorrhoidalis*) (SFI: 19.5) and the Lesser Mottled Grasshopper (*Stenobothrus stigmaticus*) (SFI: 20.7), whose occurrences are almost exclusively restricted to nutrient-poor grasslands. Compared to the species mentioned above, however, they can survive in very small relict sites, even if the surrounding landscape is structurally poor and unsuitable as a habitat on a large scale. The SFI is therefore primarily an indicator of landscape quality.

To calculate the STI, the mean temperatures in the summer term (April–September, reference period 1981–2010, DWD 2024) were averaged over the range of the respective species. The STI therefore reflects the macroclimatic requirements of the species. Species with a high STI such as the Slender Green-winged Grasshopper (*Aiolopus thalassinus*) (STI: 16.5) or the Brown-spotted Bush-cricket (*Tessellana tessellata*) (STI: 16.4) are particularly common in the warm regions of Germany. By contrast, species living in higher elevations of the mountains such as the Green Mountain Grasshopper (*Miramella alpina*) (STI: 12.4) and the Alpine Dark Bush-cricket (*Pholidoptera aptera*) (STI: 12.2) usually have a low STI.

The species-specific indicator values can also be averaged across all species within an area, reflecting the preferences of the entire species assemblage. These indices are referred to as the Community Farmland Index (CFI) and the Community Temperature Index (CTI). Recent studies have shown that these indices are suitable proxies of land use (Farmland Index) or climate change (Temperature Index) in order to relate them to range changes of grasshoppers or changes in species composition (Löffler et al. 2019, Fumy et al. 2020, Riggi & Berggren 2020, Thorn et al. 2022).

Indicator values: temperature and moisture

The system of indicator values was introduced for plants as early as 1950 (Ellenberg et al. 2001). It can be used to precisely describe the ecological requirements of a species at its place of growth with regard to various environmental factors such as light, moisture, nutrients or temperature. The indicator values of plants have therefore long been an established tool for characterising plant communities or tracking changes in species composition (e.g. Harásek et al. 2023, Klinkovská et al. 2023, Schüle et al. 2023). There are also various classifications for grasshoppers with regard to temperature and moisture preferences (Ingrisch & Köhler 1998). Often, however, no exact values have been determined, but a range of values is given (e.g. Oschmann 1991, Fröhlich 1994, Klaiber et al. 2017). Although these value ranges describe the ecological requirements of the species very precisely, it is unfortunately not possible to use them in scientific studies – for example to determine an average indicator value for a specific species community. However, Dvořák et al. (2022) recently presented an assessment for the grasshoppers of the Czech Republic that is comparable to Ellenberg's system. However, due to the significantly more continental climate, the situation in the Czech Republic is not quite comparable to the climatic conditions in Germany (more sub-Atlantic or subcontinental climate). In addition, the species spectrum studied by Dvořák et al. (2022) covers only 44 % of our native species. We have therefore made our own attempt to define precise indicator values for the microclimate (temperature and moisture) for as many species as possible that are considered established in Germany (Table 1). All species that live predominantly close to the surface in the soil, on the soil surface or in grassy-herbaceous vegetation were taken into account. Only the predominantly arbusticolous and arboricolous species (*Barbitistes* spp., *Leptophyes punctatissima*, *Meconema* spp. and *Phaneroptera nana*) and the three species that mainly occur in ant nests, buildings or caves (*Acheta domesticus*, *Myrmecophilus acervorum*, *Troglophilus neglectus*) were excluded from the classification. Due to the different stratum (shrub or tree layer) or system isolated from the open landscape (e. g. buildings and caves), the microclimatic conditions in the habitats of the excluded species differ significantly from those of the considered open-land species.

The focus was particularly on the very specific microclimatic preferences of the eggs but also of the nymphs of the respective species. Eggs lack the ability to actively relocate in unfavourable weather conditions, and nymphs are usually only able to do so to a limited extent (see also chapter Habitats). In addition, the majority of native species spend a much longer time in the egg stage than in the nymphal or adult stage. Overall, it should be noted that some species colonise a relatively broad temperature and/or moisture spectrum or have different requirements depending on the region in which they occur. The indicator values we proposed for Germany therefore represent the ecological optimum of the respective species. They should be valid over a wide range and reflect the requirements in nature under competitive conditions (Fig. 49). However, depending on the breadth of the ecological valence, the evaluated species can sometimes also be found on sites with deviating temperature and moisture characteristics. If in individual cases the amplitude was too large or no preference for a certain microclimate was recognisable, we assumed an indifferent behaviour towards the respective environmental parameter. The classifications were based on various sources that dealt with the temperature and moisture preferences of the species: Oschmann (1973, 1991), Brocksieper (1978), Ingrisch (1978a, b, 1983a, 1988), Ingrisch & Boekholt (1983), Bruckhaus (1992), Froehlich (1994), Ingrisch & Köhler (1998), Wünsch et al. (2012), Klaiber et al. (2017) and Dvořák et al. (2022). Following Ellenberg et al. (2001), a nine-level scale was chosen for both the temperature and the moisture indicator value. In the future, it would be desirable to further validate the classifications with empirical data and adjust them if necessary. Similarly, a more precise definition of Ellenberg's indicator values was achieved, which are now generally recognised (Ellenberg et al. 2001).

Fig. 48: The Medebach Bay on the eastern edge of the Rothaar Mountains (Hesse and North Rhine-Westphalia) is an old, heterogeneous cultural landscape characterised by different types of grassland with high nature conservation value (HNV farmland). (Photo credit: Dominik Poniatowski)

Temperature value (= T)

Besides moisture (see below), temperature has a strong influence on the distribution of grasshoppers in their habitat as it is of great importance for egg and nymphal development (Ingrisch 1978a, b). In contrast to the STI, which is a measure of the macroclimatic preferences (see above), the species-specific temperature value is a microclimatic parameter. The Club-legged Grasshopper (*Gomphocerus sibiricus*), for example, has an STI of 11.2 °C, which is very low compared to many other

Wet grassland | Mountain grassland | Dry grassland

Wet grassland

Chrysochraon dispar	5	6
Conocephalus dorsalis	5	7
Pseudochorthippus parallelus	5	5
Pseudochorthippus montanus	5	8
Roeseliana roeselii	5	6
Stethophyma grossum	6	8
Tetrix subulata	6	7
Tetrix undulata	6	6
Tettigonia viridissima	6	x
Mean	**5.4**	**6.6**

Mountain grassland

Chorthippus biguttulus	6	4
Decticus verrucivorus	7	x
Omocestus viridulus	5	6
Pseudochorthippus parallelus	5	5
Roeseliana roeselii	5	6
Tetrix undulata	6	6
Tettigonia cantans	4	6
Mean	**5.4**	**5.5**

Dry grassland

Bicolorana bicolor	7	3
Calliptamus italicus	7	2
Chorthippus apricarius	7	4
Chorthippus brunneus	7	3
Chorthippus mollis	7	3
Gryllus campestris	7	4
Myrmeleotettix maculatus	7	3
Oedipoda caerulescens	8	3
Omocestus haemorrhoidalis	7	3
Phaneroptera falcata	6	x
Platycleis albopunctata	7	3
Stenobothrus lineatus	7	3
Mean	**7.0**	**3.1**

Fig. 49: Examples of the use of indicator values along a moisture gradient: wet grassland – moist to wet site (Hesse: Rothaar Mountains); mountain grassland – mesic site (North Rhine-Westphalia: Rothaar Mountains); sandy dry grassland – dry site (Brandenburg: Uckermark).The translation of the temperature (T) and moisture (M) values is possible using the scales presented below in the book (see also Table 1). (Photo credits: Dominik Poniatowski, Thomas Fartmann)

native grasshopper species, as the occurrence of this species is limited to the Alps – a region with a cool macroclimate. Within the Alps, however, the Club-legged Grasshopper colonises habitats with favourable microclimates such as sparsely vegetated lawns on south-facing slopes (p. 192). The species' temperature value of 7 is correspondingly high. The scale of the temperature value covers the entire spectrum of possible reaction types – from species that colonise cold to moderately warm to hot habitats. The individual temperature levels are:
1 = cold
2 = between 1 and 3
3 = cool
4 = between 3 and 5
5 = moderate warm
6 = between 5 and 7
7 = warm
8 = between 7 and 9
9 = hot.

Moisture value (= M)

The eggs of grasshoppers require different amounts of water for successful development depending on the species (Ingrisch 1983a, 1988). In addition, drought resistance of the eggs varies from species to species. According to Oschmann (1991), the species in level M1 to M3 can be described as xerophilic, those in level M4 to M6 as mesophilic and those in level M7 to M9 as hygrophilic. The individual moisture levels are:
1 = very dry
2 = between 1 and 3
3 = dry
4 = between 3 and 5
5 = mesic
6 = between 5 and 7
7 = moist
8 = between 7 and 9
9 = wet.

Body weight

Grasshoppers can be very abundant in grasslands and accordingly build up a high biomass (Fartmann et al. 2024). They are therefore considered an important food source for many insectivorous taxa (Samways 2005, Zahn et al. 2010). Examples include reptiles, mammals and birds but also predatory invertebrates such as spiders. Changes in grasshopper biomass can therefore have serious consequences for higher trophic levels. Accordingly, Streitberger et al. (2024) suggest that biomass should also be considered in scientific surveys and monitoring programs. The sex-specific weights of grasshopper species reported here were determined by weighing adult individuals in the field using a portable precision scale (Fig. 50, Table 1). To determine the total biomass of grasshopper communities, the abundance of grasshoppers can be recorded using a standardised box quadrat. The abundance of a species can then be multiplied by its specific weight (Fartmann et al. 2024).

Range limit

The occurrence of grasshoppers is determined by various factors such as landscape structure, macroclimate and topography (for details see chapter Distribution). Looking at the species-specific distribution patterns, it is striking that almost two-thirds (63 %) of the established species reach their range limits in Germany. According to Ingrisch & Köhler (1998), this can be explained not only by the ecological factors mentioned above, but also as the result of post-glacial immigration processes from various refugia in southern and southeastern Europe. Currently, 24 species live in Germany at their northern, 22 species at their northwestern, five species at their northeastern and one species at its western range limit (Table 1).

Population trend

The categories for the short-term population trend according to the most recent red list of grasshoppers and praying mantises in Germany were used as a parameter for the population trend of the species (Table 1, Poniatowski et al. 2024a). The population trends represent changes in species distribution over the last 25 to 30 years. A distinction is made between the following categories:
↓↓↓ very strong decline
↓↓ strong decline
↓ moderate decline
(↓) decline of unknown extent
= Stable
↑ significant increase
? data deficient.

For extinct or lost species, the last year of detection is given according to Poniatowski et al. (2024a).

Fig. 50: The sex-specific weights of grasshopper species were determined by weighing adult individuals in the field using a portable precision scale (Kern TGD 50-3C, accuracy = 0.001 g). The scale was protected from wind by a plastic cover. (Photo credit: Dominik Poniatowski)

Red list status (RL status)

The RL status reflects an objective framework for classifying the risk of extinction of the German grasshopper species. The threat categories of the species correspond to the most recent red list of grasshoppers and praying mantises in Germany (Table 1; Poniatowski et al. 2024a). A distinction is made between the following categories, the exact definition of which can be found in Ludwig et al. (2009):
EX = Extinct
CR = Critically endangered
EN = Endangered
VU = Vulnerable
R = Extremely rare
NT = Near threatened
DD = Data deficient
LC = Least concern.

Table 1: Overview of the ecological indices of the grasshopper and mantis species native to Germany. Red list status (RL) and population trend (Trend) according to Poniatowski et al. (2024a); habitat specificity and mobility as well as Species Temperature Index (STI) and Species Farmland Index (SFI) according to Poniatowski et al. (2020); indicator values according to own expert assessment: T = Temperature, M = moisture, x = indifferent behaviour, for more information see text; weight (= living body weight); range limit according to Schlumprecht & Waeber (2003), Maas et al. (2011), Pfeifer et al. (2011): N = north, NE = northeast, NW = northwest and W = west; – = not evaluated/no value.

Species	RL	Trend	Habitat specificity	Mobility	STI (°C)	SFI (%)	Indicator value		Ø Weight (g)		Range limit
							T	M	♀♀	♂♂	
Acheta domesticus	DD	?	low	high	–	–	–	–	–	–	–
Aiolopus thalassinus	NT	↑	high	high	16.54	18.74	8	7	–	–	N
Arcyptera fusca	CR	↓	high	low	–	–	7	3	–	–	N
Arcyptera microptera	EX	† 1957	high	low	–	–	–	–	–	–	NW
Barbitistes constrictus	LC	?	low	low	14.03	–	–	–	–	–	NW
Barbitistes serricauda	LC	?	low	low	14.34	–	–	–	–	–	N
Bicolorana bicolor	NT	↓	high	high	14.97	23.18	7	3	0.399	0.342	NW
Bryodemella tuberculata	CR	↓↓	high	low	10.80	–	8	2	–	–	W
Calliptamus italicus	LC	↑	high	high	15.55	20.54	7	2	0.749	0.189	N
Chorthippus albomarginatus	LC	=	low	high	14.79	17.32	5	6	0.207	0.070	–
Chorthippus apricarius	NT	↓	low	low	14.89	16.65	7	4	0.230	0.086	–
Chorthippus biguttulus	LC	=	low	high	14.74	18.60	6	4	0.183	0.086	–
Chorthippus brunneus	LC	=	low	high	14.76	18.99	7	3	0.302	0.104	–
Chorthippus dorsatus	LC	↑	low	high	14.84	19.70	6	4	0.202	0.095	–
Chorthippus eisentrauti	R	=	high	high	–	–	7	2	–	–	N
Chorthippus mollis	LC	=	high	high	15.12	16.43	7	3	0.120	0.062	–
Chorthippus pullus	CR	↓	high	low	12.95	35.36	7	2	–	–	NW
Chorthippus vagans	VU	↓↓	high	low	15.14	23.07	9	2	0.238	0.093	–
Chrysochraon dispar	LC	↑	low	high	14.88	20.51	5	6	0.379	0.101	NW
Conocephalus dorsalis	LC	=	high	low	14.91	16.41	5	7	0.192	0.124	–
Conocephalus fuscus	LC	↑	low	high	15.11	18.67	6	×	0.176	0.131	N
Decticus verrucivorus	VU	↓	high	low	14.44	28.08	7	×	3.058	1.937	–
Epacromius tergestinus	EX	† 1941	high	high	–	–	–	–	–	–	N
Ephippiger diurnus	EN	=	high	low	15.45	15.12	7	3	–	–	NO

Species	RL	Trend	Habitat specificity	Mobility	STI (°C)	SFI (%)	Indicator value		Ø Weight (g)		Range limit
							T	M	♀♀	♂♂	
Eumodicogryllus bordigalensis	LC	↑	low	high	16.27	24.11	7	3	–	–	N
Euthystira brachyptera	LC	=	low	low	14.45	26.30	6	×	0.302	0.093	NW
Gampsocleis glabra	CR	=	high	low	14.91	38.76	7	3	0.857	0.606	N
Gomphocerippus rufus	LC	=	low	low	14.65	24.10	6	4	0.202	0.095	–
Gomphocerus sibiricus	LC	=	high	low	11.19	51.42	7	3	–	–	N
Gryllotalpa gryllotalpa	VU	↓	high	high	15.01	23.52	7	6	–	–	–
Gryllus campestris	LC	↑	low	low	14.89	23.26	7	4	–	–	N
Isophya kraussii	VU	↓	low	low	14.50	25.11	5	5	–	–	NW
Leptophyes albovittata	LC	=	high	low	15.19	16.14	6	4	–	–	NW
Leptophyes punctatissima	LC	↑	low	low	15.06	15.79	–	–	–	–	–
Locusta migratoria	EX	† 1932	low	high	–	–	–	–	–	–	–
Mantis religiosa	LC	↑	low	high	15.65	19.43	8	3	–	–	N
Meconema meridionale	LC	↑	low	high	15.27	–	–	–	–	–	N
Meconema thalassinum	LC	↓	low	low	14.89	–	–	–	–	–	–
Mecostethus parapleurus	LC	↑	low	high	14.88	34.17	6	6	0.547	0.139	N
Metrioptera brachyptera	NT	↓	high	low	14.37	23.29	6	5	0.365	0.254	–
Miramella alpina	NT	(↓)	low	low	12.36	46.52	3	6	–	–	N
Modicogryllus frontalis	CR	↓↓↓	high	low	–	–	8	3	–	–	NW
Montana montana	EX	† 2000	high	low	–	–	–	–	–	–	NW
Myrmecophilus acervorum	LC	=	low	low	15.12	10.65	–	–	–	–	NW
Myrmeleotettix maculatus	NT	↓↓	high	low	14.77	19.83	7	3	0.137	0.059	–
Nemobius sylvestris	LC	=	low	low	14.93	–	7	3			NO
Oecanthus pellucens	LC	↑	low	high	15.58	14.99	7	4			N
Oedipoda caerulescens	LC	↑	high	high	15.18	17.73	8	3	0.449	0.159	–

Table 1 (Continuation)

Species	RL	Trend	Habitat specificity	Mobility	STI (°C)	SFI (%)	Indicator value		Ø Weight (g)		Range limit
							T	M	♀♀	♂♂	
Oedipoda germanica	EN	↓	high	low	15.08	17.48	9	1			N
Omocestus haemorrhoidalis	VU	↓↓	high	low	15.05	19.45	7	3	0.245	0.089	NW
Omocestus rufipes	EN	↓↓	high	low	14.54	27.68	7	3	0.273	0.091	–
Omocestus viridulus	NT	↓↓	low	low	14.54	21.21	5	6	0.295	0.106	–
Phaneroptera falcata	LC	↑	low	high	15.03	17.84	6	×	0.342	0.225	N
Phaneroptera nana	LC	↑	low	high	16.26	18.27	–	–	–	–	N
Pholidoptera aptera	LC	=	low	low	12.21	47.50	×	×	–	–	N
Pholidoptera griseoaptera	LC	=	low	low	14.77	18.85	5	6	0.455	0.397	–
Platycleis albopunctata	LC	↑	high	high	15.16	19.03	7	3	0.556	0.426	–
Podisma pedestris	EN	(↓)	high	low	12.74	54.80	8	2	–	–	NW
Polysarcus denticauda	EN	↓	high	low	13.65	45.29	4	5	–	–	N
Pseudochorthippus montanus	VU	↓↓	high	low	14.64	21.82	5	8	0.229	0.104	–
Pseudochorthippus parallelus	LC	=	low	low	14.74	18.56	5	5	0.204	0.095	–
Psophus stridulus	EN	↓	high	low	13.67	35.32	7	3	–	–	NW
Pteronemobius heydenii	NT	↑	high	low	14.97	46.95	6	9	–	–	N
Roeseliana roeselii	LC	=	low	high	14.74	18.70	5	6	0.485	0.295	–
Ruspolia nitidula	LC	↑	low	high	15.56	32.84	6	5	0.514	0.408	N
Sphingonotus caerulans	LC	↑	high	high	15.38	16.23	9	1	0.536	0.172	–
Stauroderus scalaris	VU	=	high	high	13.47	54.50	7	4	–	–	NW
Stenobothrus crassipes	EN	=	high	low	–	–	7	3	–	–	NW
Stenobothrus lineatus	LC	=	high	high	14.72	21.86	7	3	0.380	0.133	–
Stenobothrus nigromaculatus	EN	↓↓	high	low	14.34	30.33	7	3	0.344	0.113	NW
Stenobothrus stigmaticus	VU	↓	high	low	14.58	20.66	7	3	0.143	0.062	NO
Stethophyma grossum	LC	↑	high	high	14.75	20.32	6	8	0.724	0.194	–

Species	RL	Trend	Habitat specificity	Mobility	STI (°C)	SFI (%)	Indicator value		Ø Weight (g)		Range limit
							T	M	♀♀	♂♂	
Tessellana tessellata	EN	↑	high	low	16.43	29.29	7	2	0.254	0.192	NO
Tetrix bipunctata	EN	↓↓	high	low	14.25	25.50	7	3	0.081	0.055	NW
Tetrix ceperoi	VU	?	high	high	15.59	11.52	7	7	–	–	NO
Tetrix subulata	LC	=	high	high	14.85	18.56	6	7	0.043	0.023	–
Tetrix tenuicornis	NT	↓	high	low	14.81	19.58	7	3	–	–	NW
Tetrix tuerki	CR	↓	high	low	11.43	–	7	7	–	–	NW
Tetrix undulata	LC	↓	low	low	14.76	18.67	6	6	0.050	0.028	–
Tettigonia cantans	LC	↓	low	low	14.41	21.01	4	6	–	–	NW
Tettigonia caudata	VU	(↓)	low	low	15.65	15.25	5	5	–	–	NW
Tettigonia viridissima	LC	=	low	high	14.82	18.21	6	×	2.174	1.667	–
Troglophilus neglectus	R	=	high	low	–	–	–	–	–	–	N

Monitoring

MERLE STREITBERGER, GREGOR STUHLDREHER, FRANZ LÖFFLER, DOMINIK PONIATOWSKI & THOMAS FARTMANN

There are different methodological approaches to detect long-term trends of the grasshopper fauna (Fartmann et al. 2021a). Re-surveys provide insight into shifts in grasshopper assemblages by the comparison of historical with recent data. The most valuable baselines for re-surveys are old data sets for which the sampling methods and study sites are well documented. This enables the repetition of the exact study design (i. e. the use of the same sampling methods and study sites) and, thus, the collection of comparative data which can be used for the detection of shifts in grasshopper assemblages. At best, environmental data is available for the respective time spans, which allows the analysis of the drivers of changes in grasshopper assemblages. Data on habitat structure and land usage of the analysed sites is especially valuable for this purpose. Regional climate data can be obtained from the German Meteorological Service for different time spans. As old, well-documented data sets on grasshoppers are mostly available for specific geographical areas, most re-surveys have a regional focus. The first re-surveys with respect to grasshoppers were carried out on a small number of sampling sites (Schuch et al. 2011, 2012). By now, several re-surveys based on a high number of samples are also available. These studies were carried out in the Westphalian Bay (Fartmann et al. 2021b, 2022b), the Eifel region (Löffler et al. 2019), Rhineland-Palatinate (Ogan et al. 2022), the Black Forest (Fumy et al. 2020) and the Hassberge region in northern Bavaria (Thorn et al. 2022). By means of these surveys, knowledge on the effects of land-use and climate change on grasshoppers has been substantially improved (see also chapter Land-use and climate change). A general pattern discovered by these studies is the additive biotic homogenization of grasshopper assemblages since the 1980s. While the diversity of habitat specialists has decreased during the past decades, the numbers of thermophilic and mobile species has increased. As a result, overall species richness has increased in most habitats and, nowadays, species assemblages are much more similar to each other. By means of an ongoing research project ('Retrospective Data Analysis'; 2022–2025) supported by the Federal Agency for Nature Conservation with funds from the Federal Ministry for the Environment, changes in grasshopper assemblages are being analysed with the help of study sites distributed across all of Germany. This project includes the compilation of available old data sets on grasshoppers and their comparison with recent data recorded in the scope of the project.

A further approach to reveal changes in the grasshopper fauna is the modelling of range shifts of grasshopper species (Buse & Griebeler 2011, Poniatowski et al. 2020, Engelhardt et al. 2022). For this purpose, detailed data on the distribution of grasshoppers is necessary. In Germany, grasshoppers represent a well-studied species group (Poniatowski et al. 2024a). Therefore, extensive data on the historical and recent distribution of grasshoppers is available for this country (see this book). However, the data stems from different sources (e. g. data provided by nature conservation agencies; Maas et al. 2002). So far, systematically recorded data on the historic and recent distribution of grasshoppers, which would be an ideal prerequisite for modelling range shifts, is not available in Germany. When range shifts of species are modelled by using unsystematically recorded data, there is a risk of a systematic error as the different regions and the respective periods were usually not sampled with equal intensity. However, innovative modelling approaches enable the analysis of range shifts with unstandardized data by considering the sampling intensity of the spatial units (e. g. grid cells) on which the modelling of ranges is based in the analysis. By means of such an approach, Poniatowski et al. (2020) were able to analyse range shifts of 58 grasshopper species in Germany during two periods (pre-1990 vs. 1990–1999 [historical]; 1990–1999 vs. 2000–2017 [recent]) based on unstandardized distribution data of grasshopper species in Germany (see also chapter Land-use and climate change). The study revealed that ranges of habitat specialists retracted during the historical time span due to land-use change. By contrast, mobile and thermophilic species expanded their ranges in the recent period because of global warming. The latter effect was also shown by a recent modelling study from Bavaria (Engelhardt et al. 2022).

However, the best way to detect changes in the grasshopper fauna at an early stage is standardized monitoring, i. e. the regular sampling of grasshoppers, ideally in short time intervals. By detecting changes at an early stage, counter measures can be implemented timely, if required, and are therefore particularly promising. An important prerequisite for monitoring grasshoppers is the application of standardized sampling methods to generate comparable data sets in the time series. Moreover, monitoring should be based on a sufficiently large number of sampling sites distributed representatively across the study area. In Germany, grasshopper monitoring has been carried out for a few years in some federal states in the framework of the German Insect Monitoring Scheme.

The German Insect Monitoring Scheme has a modular structure and consists of numerous modules that highlight specific aspects of insect diversity. One module comprises the monitoring of grasshoppers in grasslands. The aim of this module is to document the status and derive long-term trends of the grasshopper fauna in grasslands of the wider countryside. For the implementation of this module, the nationwide representative sampling sites are used (Mitschke et al. 2005). These sites represent an established network of sampling sites ($N = 1,000$ in the basic program), which have been used for other nationwide monitoring programs with respect to birds or High Nature Value (HNV) farmland for several years (Sudfeldt et al. 2012, Benzler et al. 2015). The joint use of sampling sites is advantageous for the German Insect Monitoring Scheme because it enables the inclusion of data from the other monitoring programs in the analyses of the insect data. For example, the HNV Farmland Monitoring Scheme provides valuable information on the nature conservation value of the sampled grassland patches (see below) as this scheme monitors the conservation value of all agriculturally used patches within the nationwide representative sampling sites.

The nationwide representative sampling sites have a size of 1 km^2 and are assigned to different land-use types (e. g. forest, grassland, settlement) according to the land use in the centre of the sites. For monitoring grasshoppers in grasslands, the 200 sampling sites that are assigned to the land-use type 'grassland' are used (Fig. 51). In each site, grasshopper sampling is carried out in the central grassland patch by using a box quadrat. This quadratic tool has sidewalls composed of fabric sheets and is open on the bottom and top (Fig. 52; see also Fartmann et al. 2024). For sampling grasshoppers, the box quadrat is placed on the ground and all grasshopper individuals caught in the quadrat are counted and their species identity is determined. In the German Insect Monitoring Scheme, a box quadrat with an area of 2 m^2 is used. It is placed at 15 randomly selected locations in each grassland patch so that, overall, an area of 30 m^2 is sampled per patch. Most methods for grasshopper or insect sampling in general are strongly influenced by weather or vegetation structure and, therefore, density values derived by these methods vary depending on these factors (Gardiner et al. 2005, Samways et al. 2010). This also holds true for sweep net sampling or the acoustic and visual sampling of grasshoppers along transects (Gardiner et al. 2005). By contrast, sampling with a box quadrat is less strongly influenced by these factors and generates fully quantitative data, i. e. abundance data with an accurate areal reference (Ingrisch & Köhler 1998, Gardiner et al. 2005, Gardiner & Hill 2006, Badenhausser et al. 2009, Fartmann et al. 2024). To detect species that occur locally within a grassland patch, an additional search for further species is carried out in the structures not sampled with the box quadrat. In addition to grasshopper sampling, habitat-related environmental variables are recorded within each grassland patch. Amongst these are variables with respect to land use and vegetation structure. Due to the low effort of the sampling methods, monitoring of grasshoppers in grasslands should be carried out annually on all sampling sites by the federal sates (for more information on the national monitoring standard see BfN 2023). In the scope of further modules of the German Insect Monitoring Scheme, it is also planned to monitor grasshoppers in rare habitats on monitoring sites of the EU Habitats Directive by using comparable methods.

The federal states are responsible for the implementation of the German Insect Monitoring Scheme and, thus, also for the monitoring of grasshoppers. So far, eight federal states have started with the monitoring of grasshoppers in grasslands (Fig. 51). Beyond that, there are state-specific monitoring activities with respect to single species such as the monitoring of the Heath Bush-cricket (*Gampsocleis glabra*) in Lower Saxony. Supplementary to the monitoring activities of the federal states, monitoring of grasshoppers in grasslands was

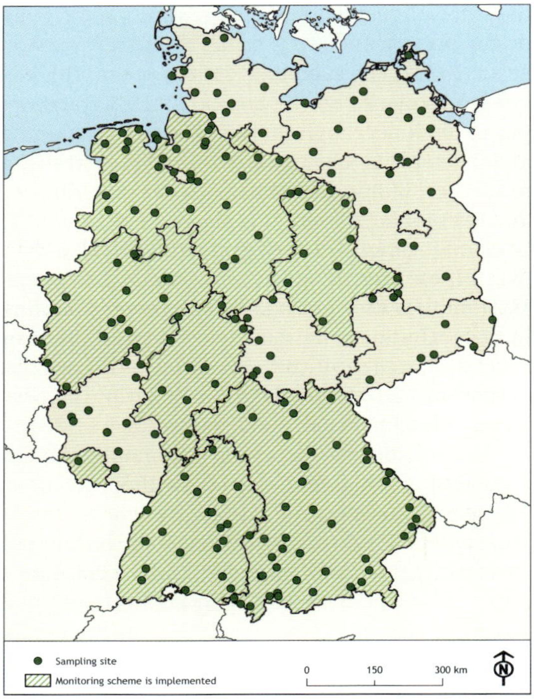

Fig. 51: Distribution of the sampling sites of the module 'Grasshoppers in grasslands' of the German Insect Monitoring Scheme and status-quo of the implementation of this module in the federal states. (Source: authors's own diagramm)

Fig. 52: The box quadrat is used for sampling grasshoppers in the German Insect Monitoring Scheme. (Photo credit: Lisa Holtmann)

carried out nationwide in two consecutive years (2020-2021) during the pilot phase of the German Insect Monitoring Scheme (research and development project 'Bundesweites Insektenmonitoring: Pilotphase') by the Department of Biodiversity and Landscape Ecology of Osnabrück University. The aim of the pilot phase was to test the practicability of the sampling methods and the study design of the grasshopper monitoring. In 2020, 35 grasshopper species were detected in total within the analysed grassland patches (N = 172; Fig. 53). Amongst the detected species, seven were endangered. The most common species, with a frequency of over 60 %, were *Chorthippus albomarginatus*, *Pseudochorthippus parallelus*, *Roeseliana roeselii* and *Chorthippus biguttulus*. Endangered species rarely occurred (frequency <10 %). On average, five species were detected per grassland patch. The average density was 17 adult individuals per 10 m^2 (see also Streitberger et al. 2024). The analysis of the grasshopper data in combination with recent data provided by the HNV Farmland Monitoring Scheme revealed that the number of grasshopper species was dependent on the nature conservation value of the sampled grassland patch and highest in patches of very high HNV farmland quality (Fig. 54).

Data basis

FELIX HELBING & DOMINIK PONIATOWSKI

An extensive dataset from a variety of sources was used to create the distribution maps. The atlas data from Maas et al. (2002) and the data compiled for the now outdated red list of 2007 served as a basis (Maas et al. 2011). In addition, numerous literature sources were consulted, including diploma and master theses as well as scientific papers. Much of the current distribution data (2007 to 2023) was provided by federal state authorities and regional grasshopper societies in Germany. Other important data sources were various online platforms. These included the web portals ArtenFinder Rheinland-Pfalz, Faunistisch-Floristisches Informationsportal Saarland (FFIpS), Insekten Sachsen, Observation, and OrthopteraWeb. Data from the open-access platforms iNaturalist (limited to 'research quality' classified data), Naturgucker (targeted search for records of rare species with verifiable photographic evidence) and the Global Biodiversity Information Facility (GBIF; records of *Mantis religiosa* only) were also partially used. In total, the combination of data from these sources amounted to almost 950,000 individual observations.

While the species records in Maas et al. (2002, 2011) and from most federal state offices were provided as 10 × 6 geographic minute grid data, the distribution data in the web portals were often pinpoint accurate. Data from literature sources contained both levels of spatial accuracy. Distribution maps were created with the aim of providing a standardised and clear representation of species distribution. Therefore, all data with different spatial resolutions were converted to grid data, resulting in a homogeneous dataset. Grid cells were classified as occupied when at least one observa-

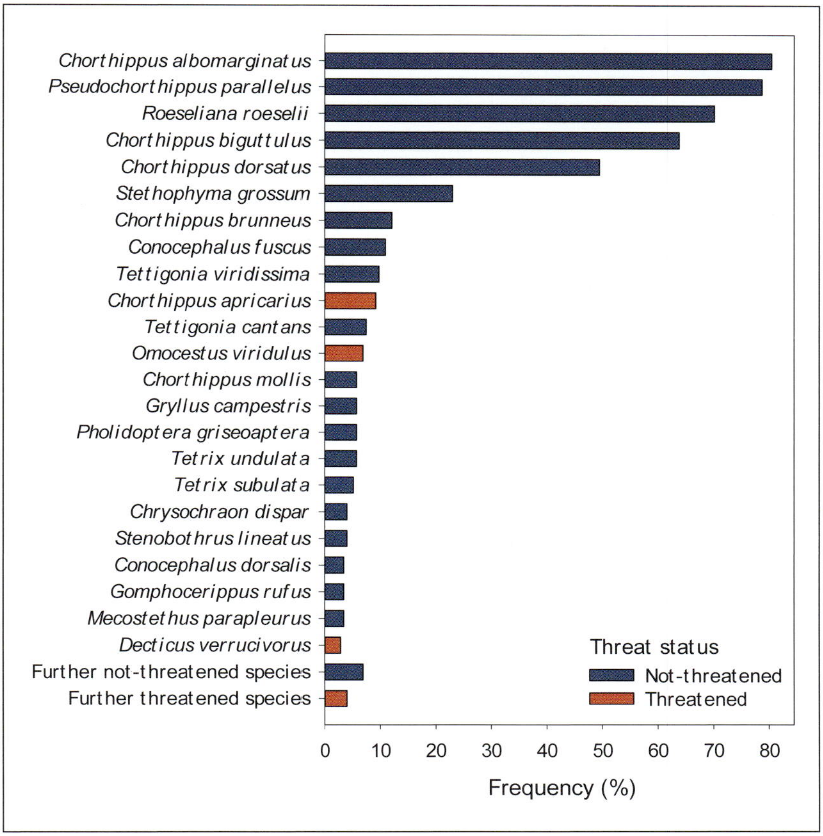

Fig. 53: Frequencies of grasshopper species in the sampled plots (N = 172) in 2020. Threatened species: species of the categories 1, 2, 3, V and G according to Poniatowski et al. (2024a). Species with a frequency < 2 % were merged to the categories 'further not-threatened species' and 'further threatened species', respectively. Further not-threatened species: *Calliptamus italicus, Euthystira brachyptera, Leptophyes albovittata, L. punctatissima, Mantis religiosa, Oedipoda caerulescens, Phaneroptera falcata* and *Platycleis albopunctata*; further threatened species: *Bicolorana bicolor, Pseudochorthippus montanus, Tetrix bipunctata* and *Stauroderus scalaris*. (Source: Streitberger et al. 2024)

tion in the considered time interval (see chapter Structure of species profiles) was available (cf. Termaat et al. 2019, van Strien et al. 2019, Poniatowski et al. 2020). Redundant information, i.e. multiple observations of a species within a single grid cell during a given time period, was removed from the dataset, and any synonyms of species names were summarised.

The inclusion of data from online platforms required either confirmed quality or the availability of verifiable photographic evidence. Previously published data could generally be considered validated. However, despite the initial verification, a multi-stage quality control was conducted on the entire dataset. This was necessary because there were still individual errors in the data. The first step was to remove any obvious errors. A preliminary distribution map was then created for each species. Once these working maps had been reviewed for further errors, partly in consultation with regional stakeholders and co-authors, the final maps were generated.

Overall, more than 390,000 aggregated grid data were available for the distribution maps. The data set covers the majority of Germany very well. Only in the federal states of Mecklenburg-Western Pomerania and Brandenburg are there some gaps in the data. In addition, the Common Ant-cricket (*Myrmecophilus acervorum*), the European Mole Cricket (*Gryllotalpa gryllotalpa*) and the arbusticolous or arboricolous species (*Barbitistes* spp., *Meconema* spp. and *Leptophyes punctatissima*) are under-recorded due to their hidden way of life. Both should be considered when using the distribution maps.

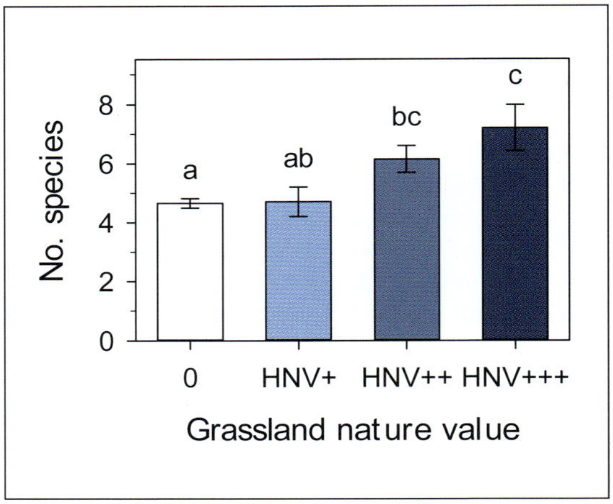

Fig. 54: Mean (± SE) grasshopper species richness in grasslands of different categories of nature value in 2020. The nature value of the grasslands in which the plots were located was derived from the German High Nature Value (HNV) Farmland Monitoring Scheme: 0 (N = 125) = no HNV farmland; HNV+ (N = 17) = moderate HNV farmland quality; HNV++ (N = 20) = high HNV farmland quality; HNV+++ (N = 10) = very high HNV farmland quality. Different letters indicate significant differences between categories ($P < 0.05$). (Source: Streitberger et al. 2024)

Range of species, nomenclature and taxonomy

DOMINIK PONIATOWSKI & THOMAS FARTMANN

The atlas covers all 82 native grasshopper species according to Poniatowski et al. (2024a). Of these, 78 species can currently be found in the wild. The remaining four species had historical occurrences in Germany. They are now considered lost or have been extinct for over 50 years. The species spectrum of grasshoppers is completed by the European Praying Mantis (*Mantis religiosa*). It is the only Mantodea with autochthonous populations in Germany. All 83 mantis and grasshopper species meet the establishment criteria according to Ludwig et al. (2009). They can therefore be viewed as native.

In recent years, several other species of both taxonomic groups have been observed in Germany. In many cases, these are species that have been introduced or drifted with the wind from southern or southeastern Europe, such as *Anacridium aegyptium*, *Eupholidoptera megastyla*, *Gryllomorpha dalmatina*, *Pteronemobius lineolatus* and *Rhacocleis annulata* (e.g. Maas et al. 2002, Pfeifer 2020, 2024, Sändig & Fritze 2022). It is unknown whether these species will form native populations in the next few years. At the moment, they do not meet the establishment criteria according to Ludwig et al. (2009) and must therefore be viewed as inconsistent (see also Poniatowski et al. 2024a). Four other species that are often used by terrarium owners as reptile food were not included in this book. These are *Gryllodes sigillatus*, *Gryllus assimilis*, *Gryllus bimaculatus* and *Schistocerca gregaria*. Occasionally, escaped animals from these species can be observed in the wild (Bellmann et al. 2019, Fischer et al. 2020, Detzel et al. 2022). However, they are unable to reproduce or build stable populations over a longer period of time.

Tachycines asynamorus is also one of the non-established species. It presumably comes from subtropical areas of East Asia (Detzel 1998, Bellmann et al. 2019). In Germany, it has so far been detected in numerous garden centres and tropical houses (Fischer et al. 2020). *Tachycines asynamorus* is very thermophilic and requires high humidity. Therefore, the species cannot survive outside heated buildings. This also applies in a similar way to *Phlugiola dahlemica*. The species occurred in the palm house in Berlin-Dahlem and the greenhouses in Babelsberg likely until 1945 (Harz 1957). Since then, no further records have been made in Germany (Poniatowski et al. 2024a).

The status of *Antaxius pedestris* is currently unclear. So far, in addition to an anthropogenic population outside its native range (Röller & Sturm 2017), there has only been a single observation by Guggemoos (2023) in a potentially natural habitat. It is a scree slope in the Bavarian Limestone Alps near Mittenwald at 975 m a.s.l. The nearest known populations of the species are in Austria, about 12 km away. If the assumption of a native population in the Bavarian Limestone Alps is confirmed in the next few years, a further species would have to be added to the list of the German grasshopper fauna.

In the second half of the 19th and early 20th century, some grasshopper species were reported that may have been part of the historical grasshopper fauna of Germany. However, the observations are usually not documented with specimens and must therefore be classified as uncertain (see also Poniatowski et al. 2024a). This applies, for example, to the observations of *Acrotylus longipes*, *Celes variabilis* and *Oedaleus decorus* (Harz 1960, Köhler 2001, Fischer et al. 2020). In addition, there are some historical species reports that can almost certainly be attributed to location confusion or misidentification (Detzel 2001, Köhler 2001, Maas et al. 2002, 2011).

The scientific nomenclature follows Poniatowski et al. (2024a). For the English names, Cigliano et al. (2023), Rutschmann & Roesti (2024) and Sutton et al. (2024) were used as a basis. The systematics and therefore the taxonomic ranks of the species presented in this book are based on the Orthoptera Species File (Cigliano et al. 2023).

Structure of species profiles

DOMINIK PONIATOWSKI & THOMAS FARTMANN

The species profiles follow a uniform scheme and always contain a double page for each species. At the top of the left-hand side, there is a photo of the species (male or female) as well as a typical habitat picture. This is followed by a running text that is divided into three sections: 'distribution', 'habitat' and 'population trend and threats.'

Species text

In the section 'distribution', the German range of the given species is briefly presented – for example by mentioning the main strongholds and partly also the elevational distribution. The seven major biogeographical regions and subordinate units serve as a reference (see book cover). The focus of the description is on the situation in Germany and can be traced on the distribution map. Moreover, those environmental drivers are explained that determine the observed distribution pattern. Particularly, the macroclimate often plays a crucial role for many species (see chapter Distribution; cf. Fig. 6) and is described accordingly.

The next section presents the most important habitats of the respective species. An essential part of this section is also the characterisation of the habitat requirements. The focus here is on the species-specific demands concerning microclimate and vegetation structure (see also chapter Habitats).

The section 'population trend and threats' provides insights into the short-term and partly also the long-term population development of the species. If changes have occurred, the causes are described. Here, the focus is on land-use and climate change due to their prime importance for changes in the mantis and grasshopper fauna (see also chapter Land-use and climate change).

Each profile also contains a box with various ecological indices. Information on the current threat status and the short-term population trend was taken from the new red list of grasshoppers and praying mantises in Germany (Poniatowski et al. 2024a). As a reference for the revised parameters 'habitat specificity' and 'mobility' (two categories each: low and high), the classification in Poniatowski et al. (2020) was used. Further information on threat status, habitat specificity, mobility and range shift can be found in the chapter 'Ecological indices' (p. 33).

Overview of the categories of the current threat status and short-term population trend shown in the boxes according to Poniatowski et al. (2024a).

Threat status		Trend	
EX	Extinct	↓↓↓	Very strong decline
CR	Critically endangered	↓↓	Strong decline
EN	Endangered	↓	Moderate decline
VU	Vulnerable	(↓)	Decline of unknown extent
R	Extremely rare	=	Stable
NT	Near threatened	↑	Significant increase
DD	Data deficient	?	Data deficient
LC	Least concern	†	No trend – species extinct

Two other ecological indices that are increasingly used in entomological studies are the Species Temperature Index (STI) and the Species Farmland Index (SFI) (e. g. Löffler et al. 2019, Fumy et al. 2020, Riggi & Berggren 2020, Thorn et al. 2022). While the STI shows whether the respective species prefers a warm or a cool macroclimate, the SFI illustrates the extent to which the respective species depends on landscapes with a high proportion of High Nature Value farmland. With the indices, for example, relationships between range shifts and climate or land-use change can be shown (Poniatowski et al. 2020). Further information – including how to calculate and use the indices – can be found in the chapter 'Ecological indices' (p. 33). The SFI is not shown for the tree or forest dwellers (genera *Barbitistes* and *Meconema* as well as *Nemobius sylvestris*) and the inhabitants of dynamic river floodplains (*Bryodemella tuberculata* and *Tetrix tuerki*) since the presence of grassland is not crucial for the occurrence of these species. In the case of extinct, synanthropic and very rare species, the box only illustrates the threat status, habitat specificity, mobility and population trend.

For the sake of space, the most important sources are listed separately for each species in a service section at the end of the book (p. 222).

Distribution map

The distribution map can be found on the right-hand side of the species profiles. It is based on a topographical map with eight different elevational levels (Fig. 55).

For better orientation, the largest bodies of water in Germany and the borders of the federal states are also shown. Superimposed, there is a grid based on 10 × 6 geographic minute grid cells.

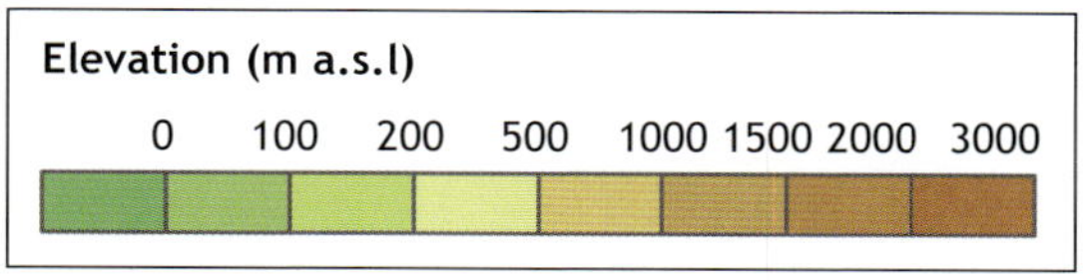

Fig. 55: Elevational levels of the distribution maps. (Source: author's own diagram)

The number of the grid cell can be determined by combining the row and column labels. For example, grid cell 3446 covers parts of Berlin. The distribution map contains information on historical and current records of the respective species. A distinction is made between four time periods:

- since 2010
- 1990–2009
- 1950–1989
- until 1949.

If a species has been recorded in a grid cell for several periods, the most recent record is shown at the top. The QR code in the top left-hand corner of the map can be used to call up the latest distribution data for the respective species from the homepage https://www.heuschrecken-monitoring.de

You can find an input form for the web code at https://www.ulmer/webcode
or by scanning this QR code:

Species Profiles of Grasshoppers and Praying Mantises in Germany

Female Praying Mantis (*Mantis religiosa*).
(Photo credit: Dominik Poniatowski)

A mosaic of open and scrubby dry grasslands ranks among the typical habitats of Praying Mantis (*Mantis religiosa*) at the southern Upper Rhine Valley (Baden-Württemberg: Kaiserstuhl).
(Photo credit: Dominik Poniatowski)

Praying Mantis – *Mantis religiosa* (Linnaeus, 1758)

Distribution In Germany, the Praying Mantis was originally almost completely restricted to the southern Upper Rhine Valley, which is characterised by extraordinarily warm summers. Additionally, there were historic records from adjacent regions: the northern Upper Rhine Valley, the Middle Rhine and Moselle Valley, Main-Franconia as well as the Danube Valley near Passau. Originating from the southern Upper Rhine Valley, the species has, particularly since the 2003 heat wave, expanded its range northwards and to warm zones of the Southwestern and Western Uplands. Moreover, a second stronghold has been established, stretching from the Thuringian Basin to the southern part of the Northeast German Lowland, which is also characterised by warm summers. Besides active dispersal, passive dispersal and release of individuals have contributed to the range expansion. The strong dependence of the Praying Mantis on warm summers is reflected by an STI of 15.7 °C.

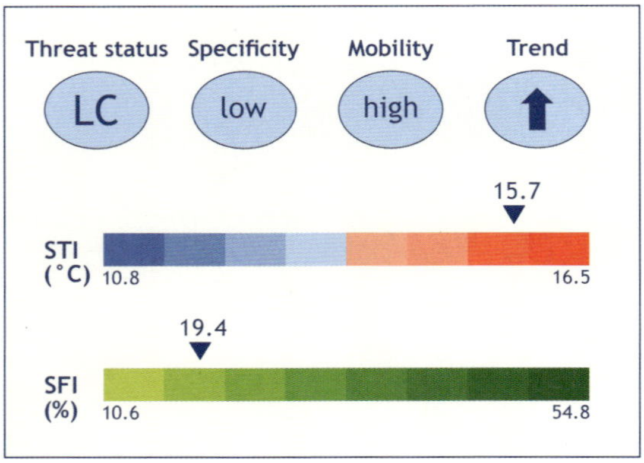

Habitat In former times, at the southern Upper Rhine Valley, the Praying Mantis inhabited calcareous and sandy grasslands, dry ruderal habitats as well as vineyard fallows and slopes. However, due to climate change, the species has broadened the range of colonised habitats to now include gardens, orchards and former mining areas or quarries. In the course of the development from egg to adult, the species undergoes a microhabitat shift: the oothecae are deposited in short, sparsely vegetated sunny swards; young nymphs also live there. By contrast, older, larger nymphs and particularly adults mainly colonise tall, dense structures.

Population trend and threats Climate change has resulted in a positive population trend and a strong range expansion of the Praying Mantis in Germany. Accordingly, the species is considered of least concern.

STEFAN STÜBING

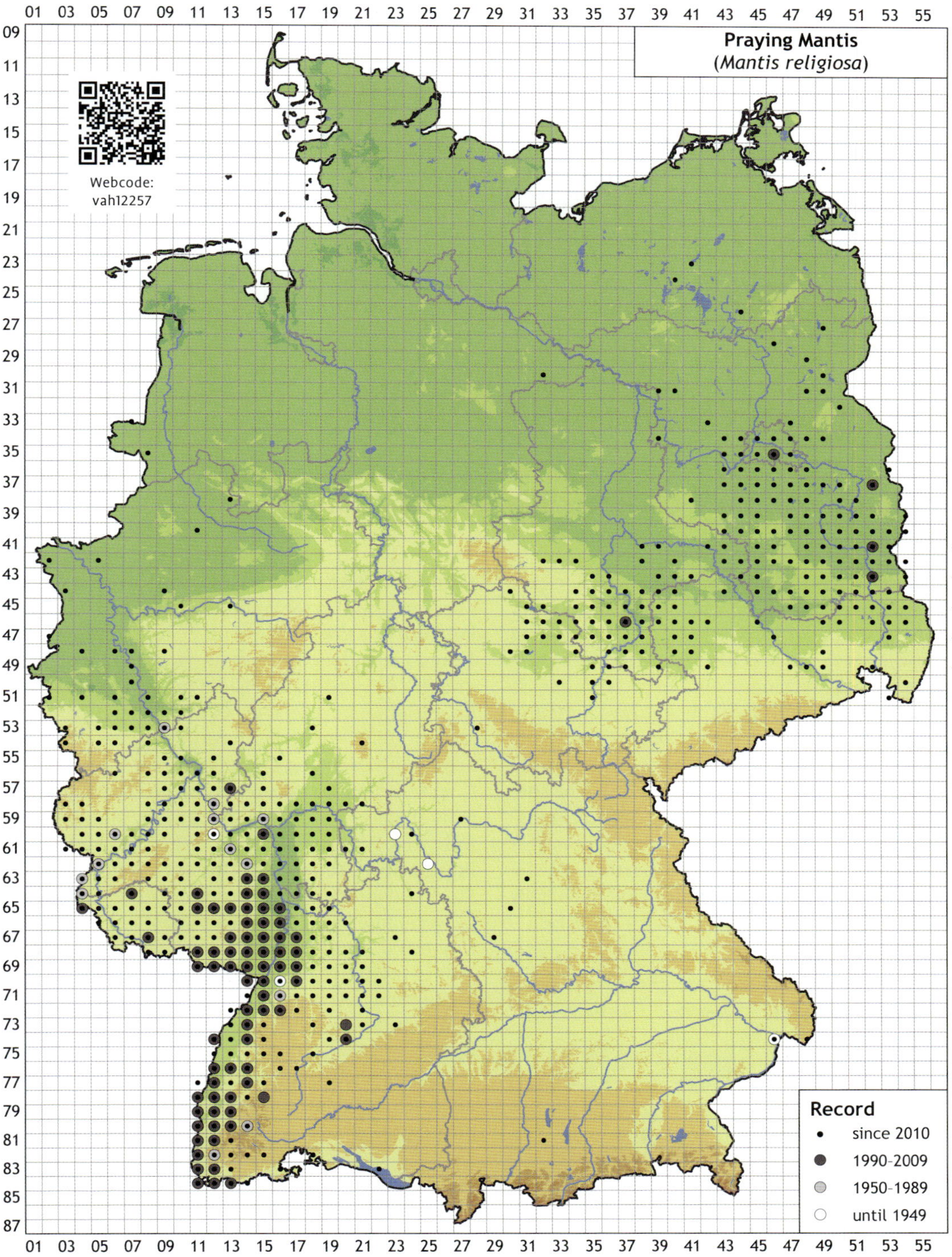
Praying Mantis
(*Mantis religiosa*)
Webcode:
vah12257
Record
since 2010
1990-2009
1950-1989
until 1949

Abandoned semi-dry grasslands are typical habitats of the Common Sickle Bush-cricket (*Phaneroptera falcata*) (Baden-Württemberg: Upper Danube Valley).
(Photo credit: Dominik Poniatowski)

Female Common Sickle Bush-cricket (*Phaneroptera falcata*).
(Photo credit: Thomas Fartmann)

Common Sickle Bush-cricket – *Phaneroptera falcata* (Poda, 1761)

Distribution The Common Sickle Bush-cricket ranks among the more thermophilic grasshopper species in Germany; its STI is 15.0 °C. For a long time, it was restricted to those regions that are characterised by the warmest summers: strongholds were the lower elevations of the Southwestern and Western Uplands as well as the Cologne Bay, Lake Constance region, Thuringian Basin and parts of the Leipzig Bay. However, since the 1990s, the species has strongly expanded its range due to climate change. The Common Sickle Bush-cricket is still absent solely in the coolest regions, particularly the most oceanic northwest of Germany, and the higher elevations of the low mountain ranges and Alps.

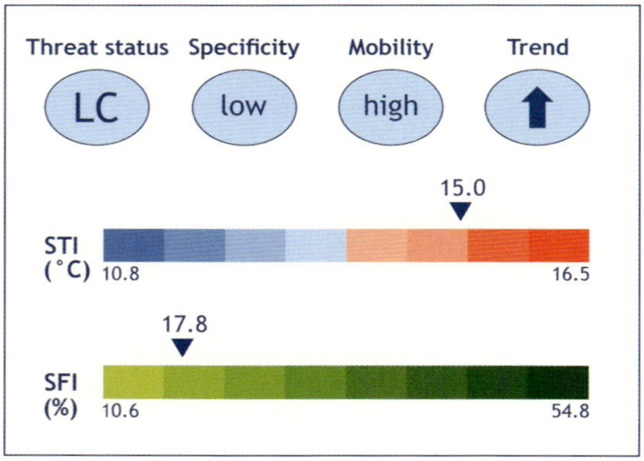

Habitat The Common Sickle Bush-cricket prefers open and semi-open habitats characterised by no land use or a low land-use intensity. Consequently, fringes, abandoned grasslands or shrubby semi-dry grasslands rank among the typical habitats. Within settlements, the species also occurs in urban brownfields. Moreover, in warm regions, wetlands are colonised by the Common Sickle Bush-cricket. Characteristic components of the strongly vertically structured habitats are usually stands of tall forbs and grasses with some admixed shrubs.

Population trend and threats The Common Sickle Bush-cricket is considered of least concern in Germany. Due to its long wings, it is capable of flight and highly mobile. Consequently, the species is regularly be found far away from known populations. As a profiteer of climate change, the Common Sickle Bush-cricket will further expand its range. By contrast, local population losses as a result of reintroduction of management or habitat destruction have so far not affected the overall positive population trend.

ANDREAS KRONSHAGE

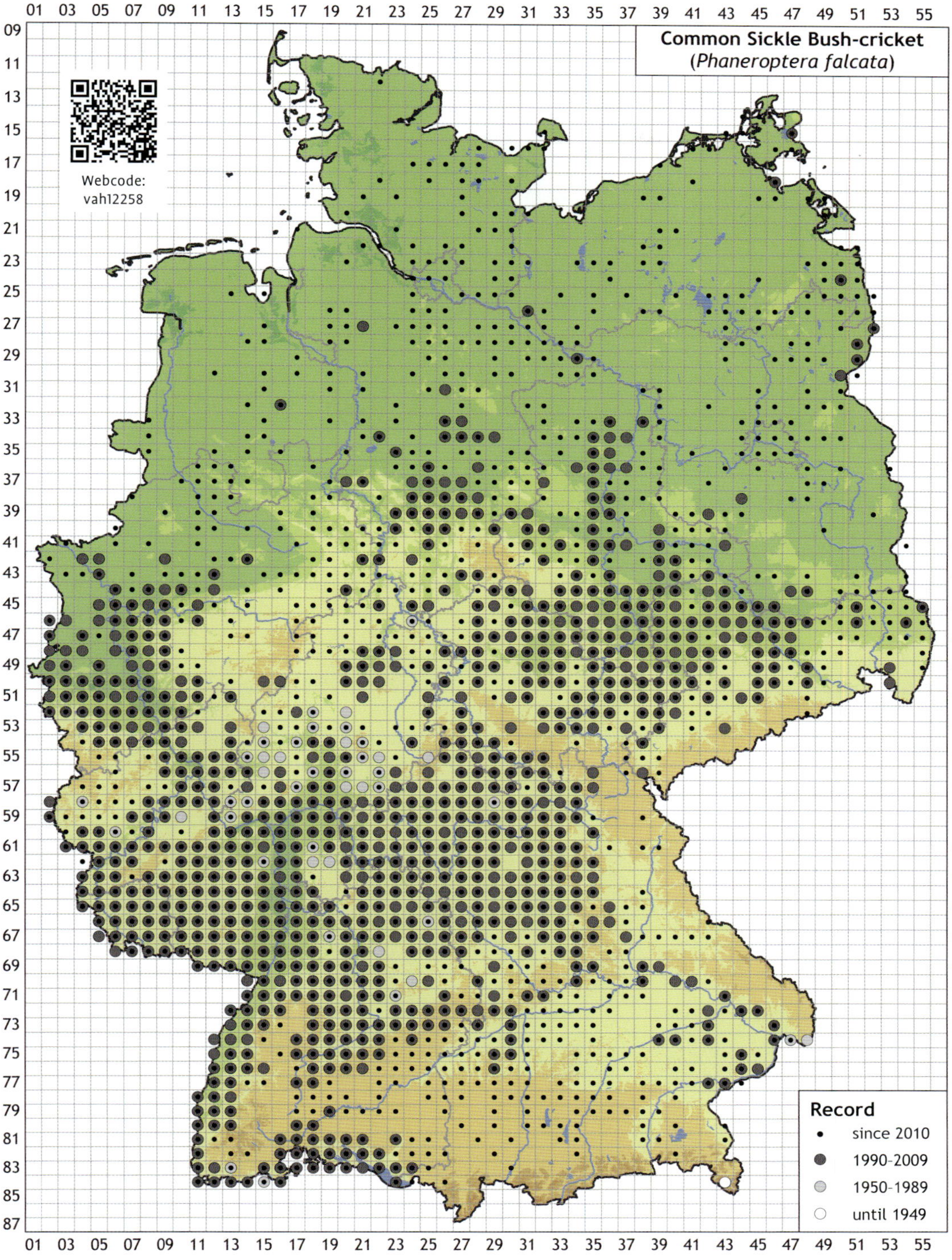
Common Sickle Bush-cricket
(*Phaneroptera falcata*)
Webcode:
vah12258
Record
since 2010
1990-2009
1950-1989
until 1949
01 03 05 07 09 11 13 15 17 19 21 23 25 27 29 31 33 35 37 39 41 43 45 47 49 51 53 55
09 11 13 15 17 19 21 23 25 27 29 31 33 35 37 39 41 43 45 47 49 51 53 55 57 59 61 63 65 67 69 71 73 75 77 79 81 83 85 87

Woody plants in settlements are currently the main habitat of the Southern Sickle Bush-cricket (Rhineland-Palatinate: northern Upper Rhine Valley).
(Photo credit: Manfred Alban Pfeifer)

Female Southern Sickle Bush-cricket (*Phaneroptera nana*).
(Photo credit: Jürgen Fischer)

Southern Sickle Bush-cricket – *Phaneroptera nana* Fieber, 1853

Distribution The Pontic-Mediterranean and very thermophilic (STI: 16.3 °C) Southern Sickle Bush-cricket was first recorded in Germany in 2003 in the Upper Rhine Valley north of Basel. Since then, the species has expanded its range northwards. This has especially been the case along the very warm Rhine Valley up to the Cologne Bay. The Southern Sickle Bush-cricket recently arrived in Bavaria from the southeast and is now spreading to the north. Additionally, the species has been observed far away from these two dispersal corridors, partly also further north.

Habitat The Southern Sickle Bush-cricket exhibits a strong dependence on woody plants, whether evergreen or deciduous. Since urban areas are heat islands and woody plants are common there, such areas have been the strongholds of the species within the already very warm Rhine Valley. However, an expansion is currently taking place from the settlements to the wider countryside, for example, to the southern Palatinate Forest. Characteristic habitats of the Southern Sickle Bush-cricket in urban areas and their suburbs are gardens, hedges, industrial brownfields, market gardens, parks and tree nurseries. In the surrounding landscape, the species occurs in orchards as well as in shrubby dry grasslands and vineyards.

Population trend and threats The Southern Sickle Bush-cricket is highly capable of flight. Hence, due to climate-change driven temperature increases, it is actively expanding its range in Germany northwards and also to higher elevations. Moreover, passive dispersal through plant material from market gardens and tree nurseries is very likely also contributing to the colonisation of new areas.

MANFRED ALBAN PFEIFER

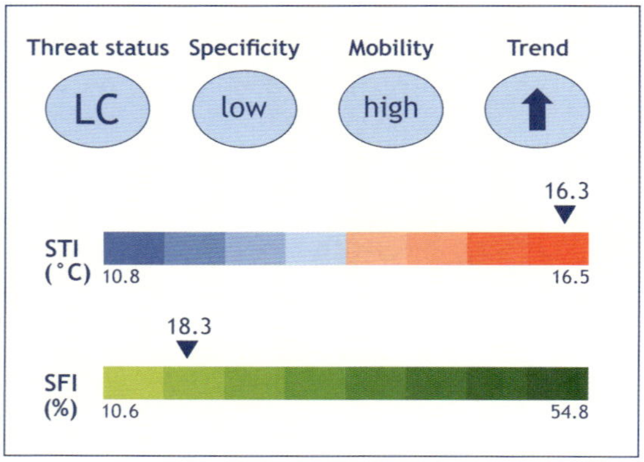

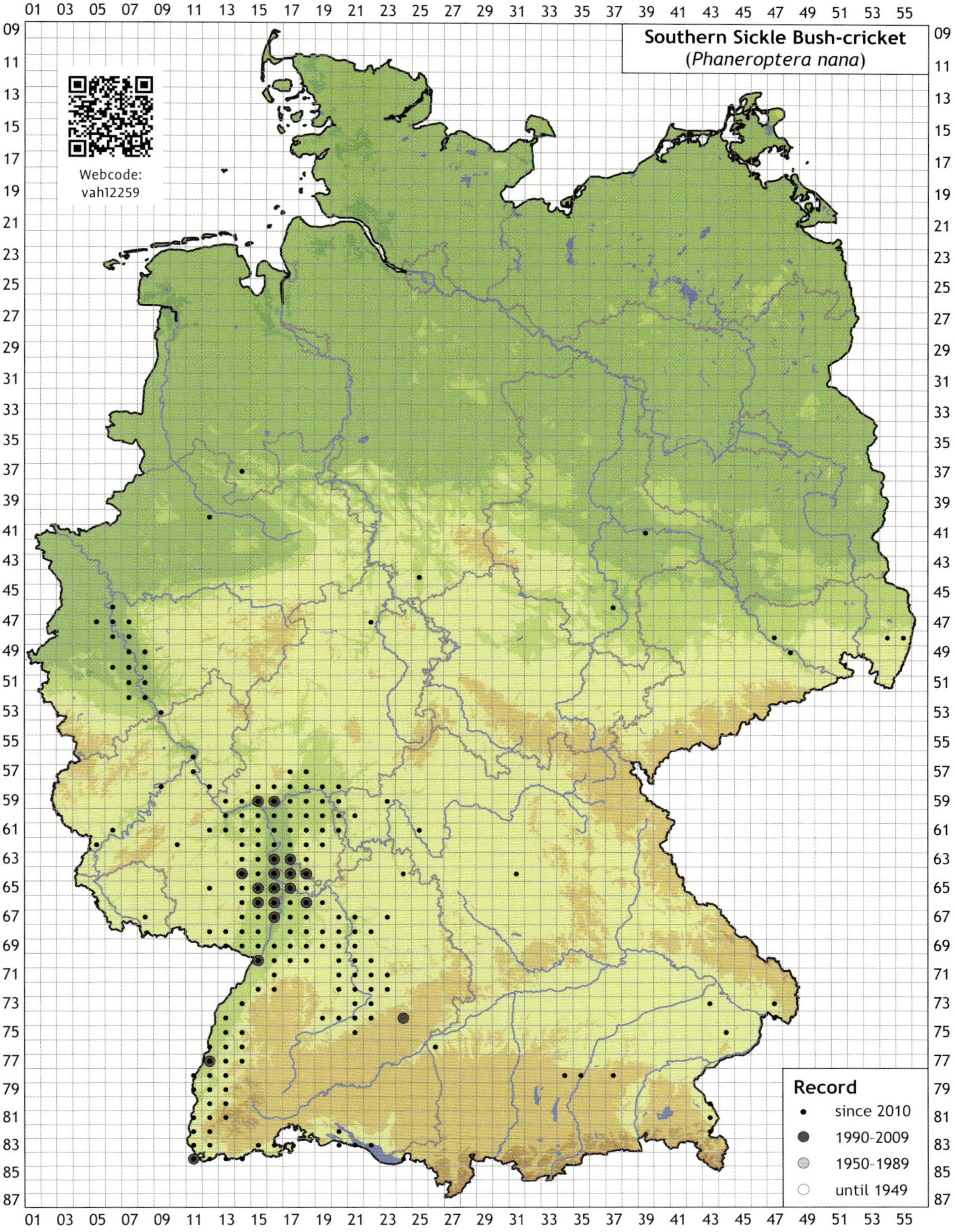
Southern Sickle Bush-cricket
(*Phaneroptera nana*)
Webcode:
vah12259
Record
since 2010
1990-2009
1950-1989
until 1949

Male Speckled Bush-cricket (*Leptophyes punctatissima*).
(Photo credit: Jürgen Fischer)

Sunny hedges and adjacent fringes in semi-dry calcareous grasslands are typical habitats of the Speckled Bush-cricket (Baden-Württemberg: Kaiserstuhl).
(Photo credit: Marcel Kettermann)

Speckled Bush-cricket – *Leptophyes punctatissima* (Bosc, 1792)

Distribution The strongholds of the Speckled Bush-cricket in Germany have always been the warm lowlands and urban agglomerations, which exhibit a warmer climate than the surrounding landscapes. The species' dependence on a warm climate is reflected by an above average STI of 15.1 °C. Since the 1990s, the Speckled Bush-cricket has strongly expanded its range and now also increasingly colonises the coastal regions, higher elevations of the low mountain ranges and even the Alps. The expansion of the flightless species has mainly been the result of passive dispersal of eggs, nymphs or adults (e. g. by woody plants from garden centres) by humans. Due to climate change, these individuals have been able to build viable populations at the new, formerly climatically unsuitable locations.

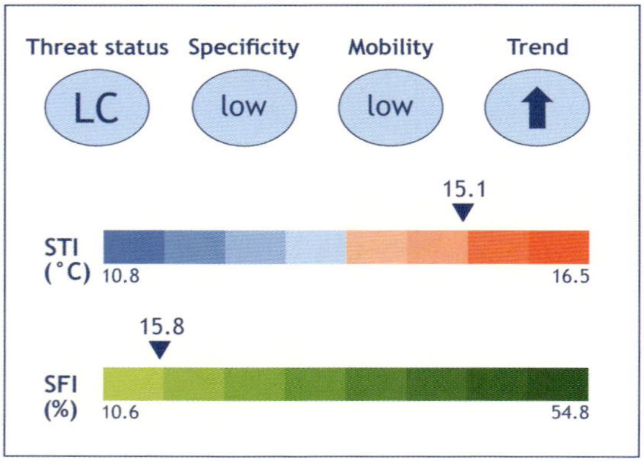

Habitat Due to the pronounced small-scale habitat shift, the Speckled Bush-cricket is characteristic of ecotones. For oviposition, woody plants with soft or cracked bark are used. By contrast, nymphs develop in the adjacent herbaceous vegetation close to the ground. The adults can later again be found in the woody plants. Accordingly, typical habitats of the Speckled Bush-cricket are sunny clear-cuts, dry grasslands, ruderal habitats and stands of tall forbs with some woody plants or forest edges and orchards. Urban areas, cemeteries, gardens and parks are also among the typical habitats.

Population trend and threats Both the long- and short-term population trends of the Speckled Bush-cricket in Germany are positive. Currently, the species is unthreatened. Since 15 % of the global populations and 10 % of the core range are located in Germany, the country has a high responsibility for the long-term survival of the species.

THORSTEN MÜNSCH

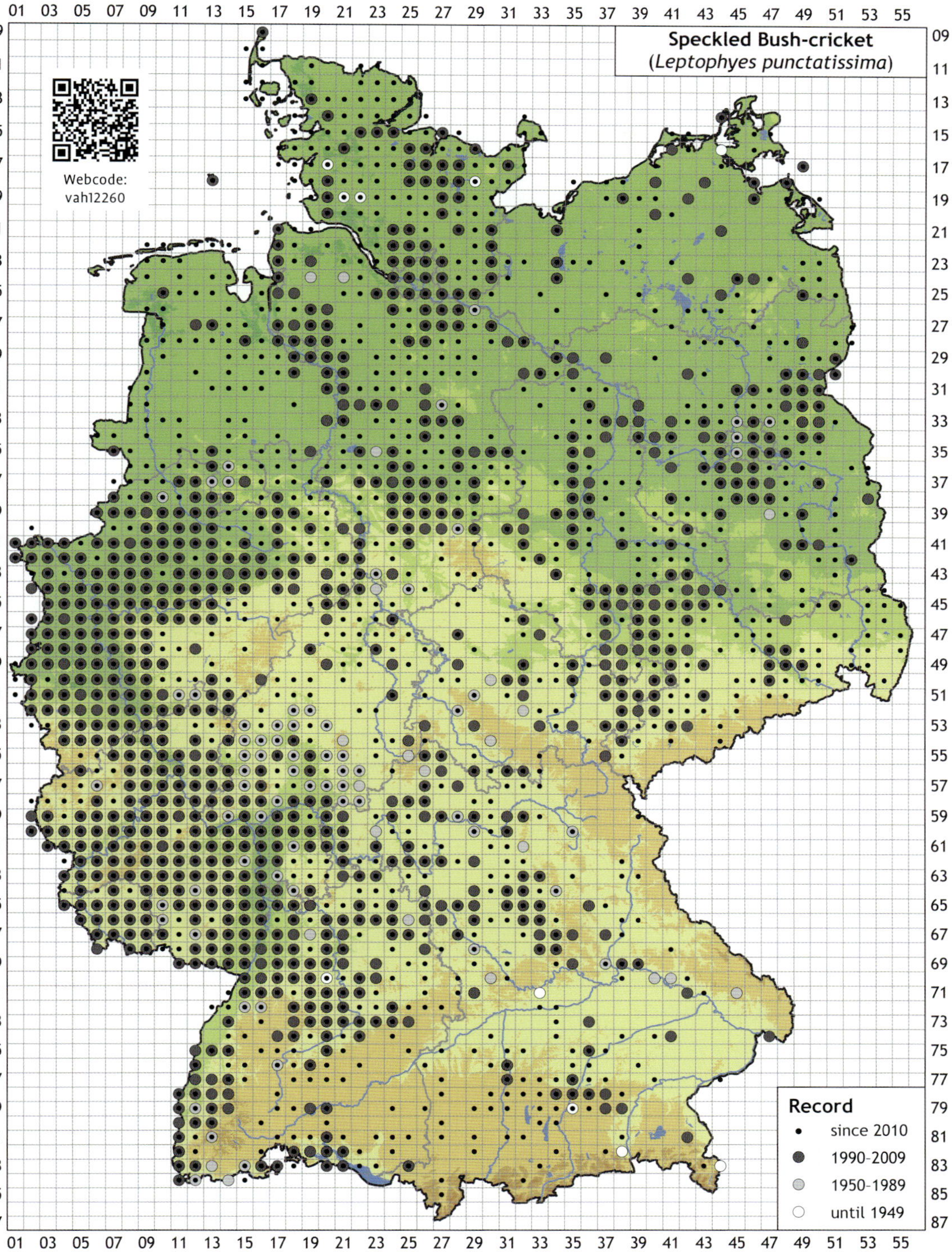
Speckled Bush-cricket
(Leptophyes punctatissima)
Webcode:
vah12260
Record
since 2010
1990-2009
1950-1989
until 1949
01 03 05 07 09 11 13 15 17 19 21 23 25 27 29 31 33 35 37 39 41 43 45 47 49 51 53 55
09 11 13 15 17 19 21 23 25 27 29 31 33 35 37 39 41 43 45 47 49 51 53 55 57 59 61 63 65 67 69 71 73 75 77 79 81 83 85 87

Female Striped Bush-cricket (*Leptophyes albovittata*).
(Photo credit: Thomas Fartmann)

Abandoned mesic grasslands rich in herbs are typical habitats of the Striped Bush-cricket (Saxony: Floodplain of the river Mulde).
(Photo credit: Marcel Kettermann)

Striped Bush-cricket – *Leptophyes albovittata* (Kollar, 1833)

Distribution The Striped Bush-cricket reaches its northwestern range limit in Germany. Its strongholds are the southern and central Franconian Jura, Main-Franconia and the entire Valley of the river Elbe. It also occurs frequently along the Bavarian part of the river Danube. Since all these regions are characterised by warm summers, the STI is very high at 15.2 °C. In addition, there are several isolated records across the country. It is striking that the majority of the populations is located along the floodplains of large rivers.

Habitat The Striped Bush-cricket depends on sunny, tall-growing vegetation. It inhabits abandoned mesic grasslands, shrubby calcareous and sandy grasslands and stands of tall forbs. Thermophilic fringes and forest edges also represent typical habitats. The bush-crickets often stay on the flowers of herbs such as Common Yarrow (*Achillea millefolium*), Oxeye Daisy (*Leucanthemum vulgare*) or Tansy (*Tanacetum vulgare*). However, individuals are also regularly found on shrubs.

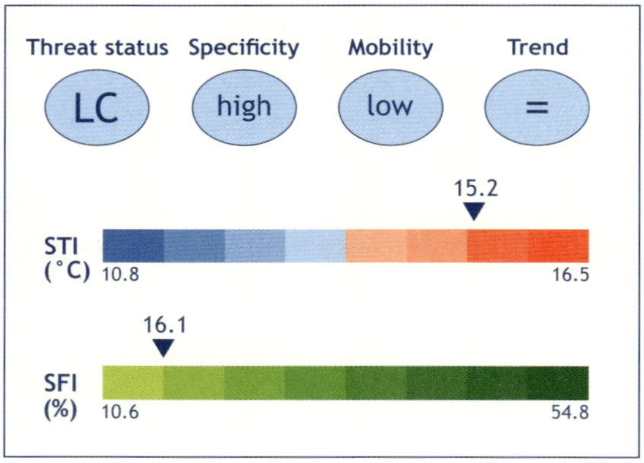

Population trend and threats In Germany, the Striped Bush-cricket is rare with stable populations. The species is flightless due to its short wings and, therefore, has a limited natural dispersal ability. Nevertheless, it has spread slightly in the last few years. New records have often been found isolated from previously known populations. This is likely a result of passive dispersal along rivers by ships or via train and car. In addition, eggs attached to plant parts can be dispersed through flooding. Local threats, on the other hand, can be both the intensification of land use and the long-term abandonment of habitats leading to microclimatically unfavourable forest stages. Moreover, dike elevation measures have a negative impact on populations of the Striped Bush-cricket along the river Elbe.

FELIX HELBING

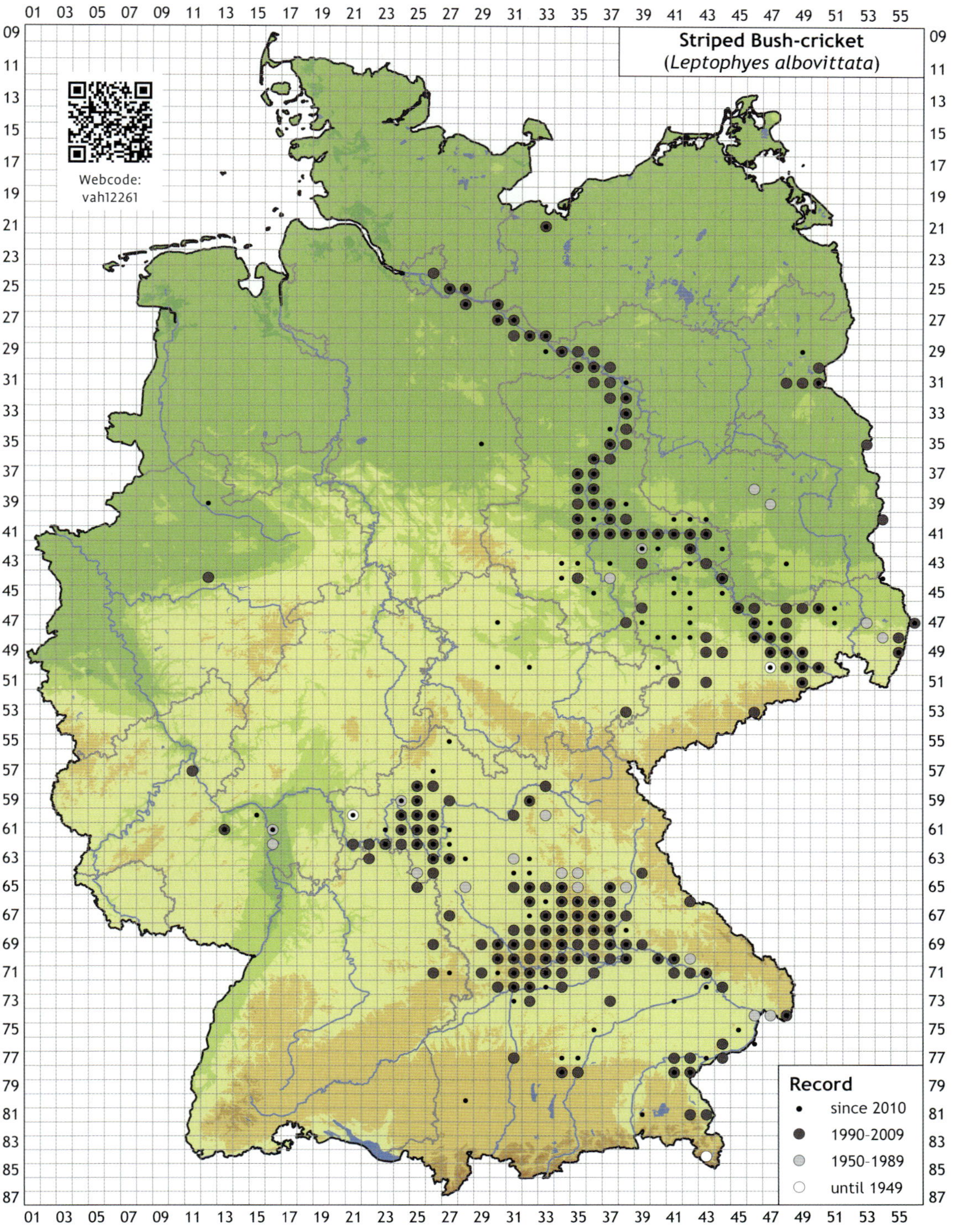
Striped Bush-cricket
(*Leptophyes albovittata*)
Webcode:
vah12261
Record
since 2010
1990-2009
1950-1989
until 1949
01 03 05 07 09 11 13 15 17 19 21 23 25 27 29 31 33 35 37 39 41 43 45 47 49 51 53 55
09 11 13 15 17 19 21 23 25 27 29 31 33 35 37 39 41 43 45 47 49 51 53 55 57 59 61 63 65 67 69 71 73 75 77 79 81 83 85 87

In the Alps, light montane mixed woodlands are colonised by the Common Saw Bush-cricket (Bavaria: Berchtesgaden Alps). (Photo credit: Sebastian Seibold)

Male Common Saw Bush-cricket (*Barbitistes serricauda*).
(Photo credit: Jürgen Fischer)

Common Saw Bush-cricket – *Barbitistes serricauda* (Fabricius, 1794)

Distribution In Germany, the Common Saw Bush-cricket has a scattered distribution in the hilly and mountainous regions. The occurrence is almost completely restricted to the Western and Southwestern Uplands, Alpine Foreland and Alps. Here, the strongholds are located at the Moselle and Middle Rhine Valley, Upper Gäue, Swabian and Franconian Jura, southern Bavarian Alpine Foreland and Alps.

Habitat The Common Saw Bush-cricket inhabits a broad spectrum of light forests, particularly mixed deciduous and coniferous forests. However, it is absent from small and isolated woodlands. In the course of the ontogenesis, the Common Saw Bush-cricket undergoes a microhabitat shift: nymphs occur mainly in the herbaceous layer and feed on leaves of herbs. By contrast, adults colonise adjacent taller shrubs and especially trees at woodland edges, paths and glades. Adults use leaves and needles of different species of woody plants as their food source. Oviposition occurs inside cracked barks of shrubs and trees. Pure beech forests are avoided since the leaves are inedible and the smooth bark cannot be used for egg laying.

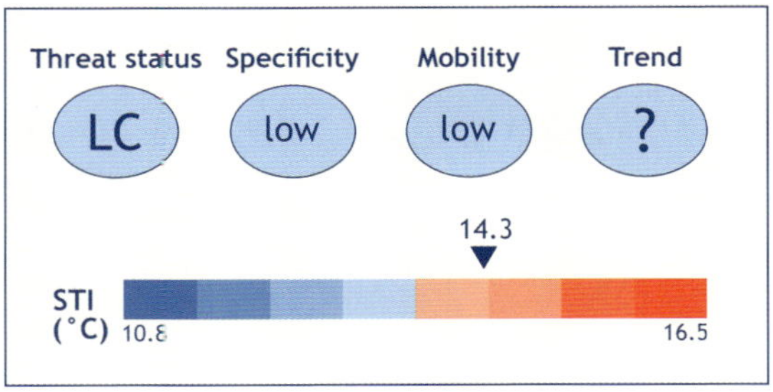

Population trend and threats The long-term trend of the Common Saw Bush-cricket in Germany is considered stable. By contrast, the short-term trend is vague. In some regions, no changes have been observed, but in others, available data indicated declines. However, real losses are hard to detect in this species since the Common Saw Bush-cricket may not be found every year in woodlands even with confirmed occurrences, mostly due to low population densities and fluctuations in population size. Germany harbours slightly less than one third of the global population. Moreover, the country is located at the distribution boundary and partly within the distribution centre of the species. As a result, Germany has a high responsibility for the long-term survivability of the Common Saw Bush-cricket.

SIGFRID INGRISCH

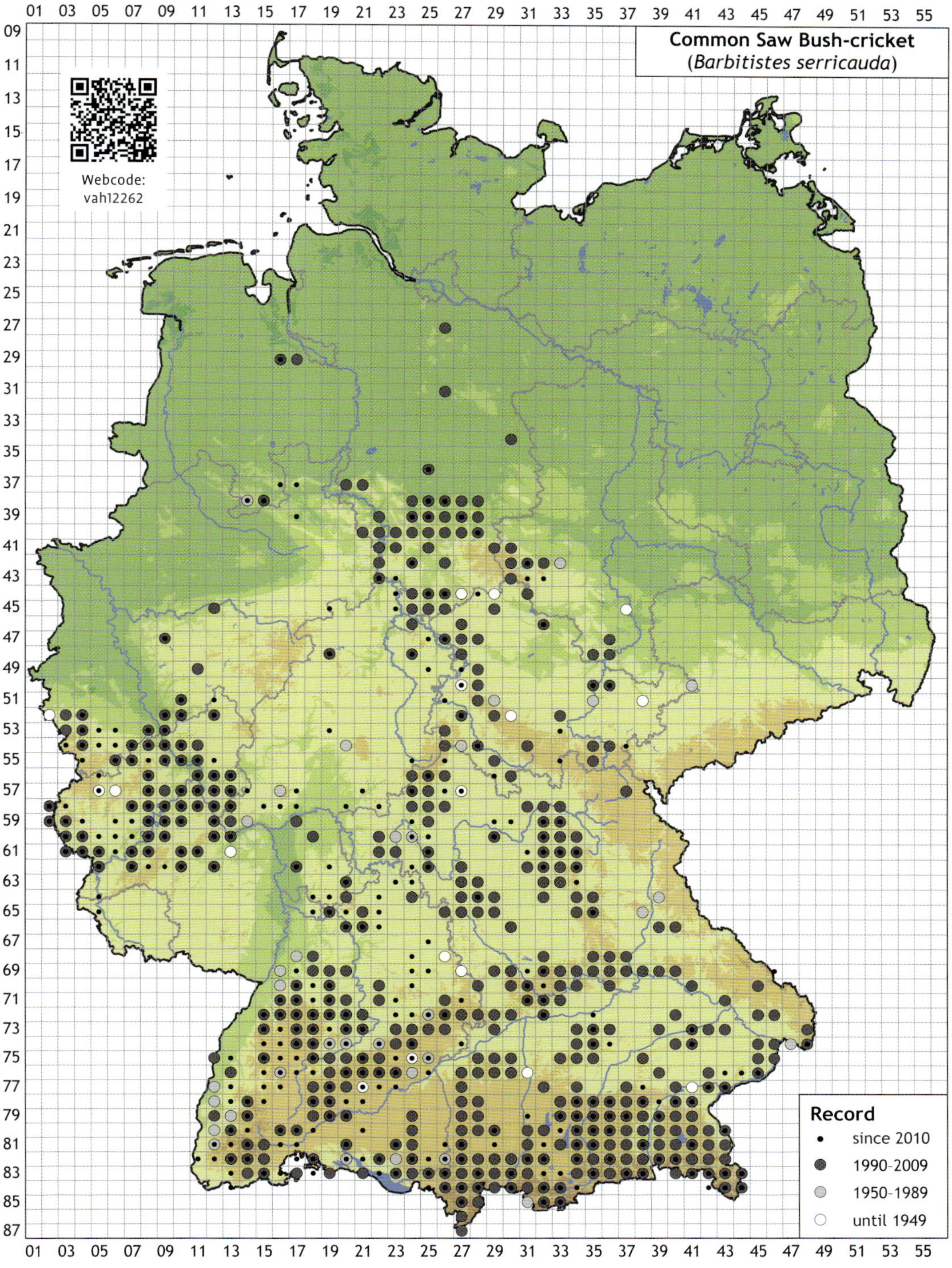
Common Saw Bush-cricket
(*Barbitistes serricauda*)
Webcode:
vah12262
Record
since 2010
1990-2009
1950-1989
until 1949
01 03 05 07 09 11 13 15 17 19 21 23 25 27 29 31 33 35 37 39 41 43 45 47 49 51 53 55
09 11 13 15 17 19 21 23 25 27 29 31 33 35 37 39 41 43 45 47 49 51 53 55 57 59 61 63 65 67 69 71 73 75 77 79 81 83 85 87

Light coniferous forests are characteristic habitats of the Eastern Saw Bush-cricket (Thuringia: foothills of the Thuringian Forest). (Photo credit: Marcel Kettermann)

Male Eastern Saw Bush-cricket (*Barbitistes constrictus*). (Photo credit: Jürgen Fischer)

Eastern Saw Bush-cricket – *Barbitistes constrictus*
Brunner von Wattenwyl, 1878

Distribution The Eastern Saw Bush-cricket mainly occurs in Germany in the Eastern Uplands and in the east of the Southwestern Uplands. Here, it reaches its western range margin. The strongholds of the species are located in a wide belt from the Bavarian Forest through the Franconian Jura and the Thuringian Forest to the Ore Mountains. Less frequently, the species also has been detected in the adjacent lowland regions and in the Bavarian Alps. The strong dependence of the Eastern Saw Bush-cricket on uplands results in a clearly lower than average STI of 14.0 °C.

Habitat The Eastern Saw Bush-cricket inhabits spruce and pine forests and their adjacent fringes and woodland edges. If conifers are present, mixed forests are also frequently colonised. Occasionally, the species also occurs in juniper-rich calcareous grasslands or in settlements. While the adults can mainly be found in the canopy of trees, larvae usually occur in the herb and shrub layer.

Population trend and threats In the long-term, the population trend of the Eastern Saw Bush-cricket in Germany is stable. While some authors have reported a westward expansion, the few recent records of the species indicate a short-term population decline. However, it is unclear whether the Eastern Saw Bush-cricket is undergoing a true decline or is only poorly recorded. In general, the species is common in coniferous forests within its range. For successful nymphal development, herb-rich margins within forests should be maintained. So far, there is no knowledge on how the increasing dieback of spruce monocultures due to extreme weather events and bark beetle outbreaks in recent years affect the populations of the species.

FRANZ LÖFFLER

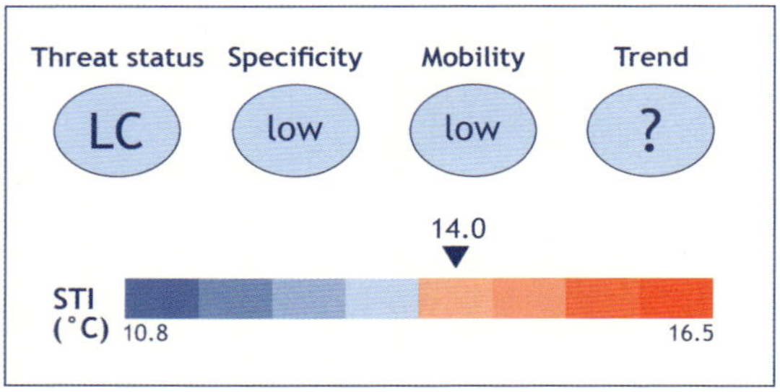

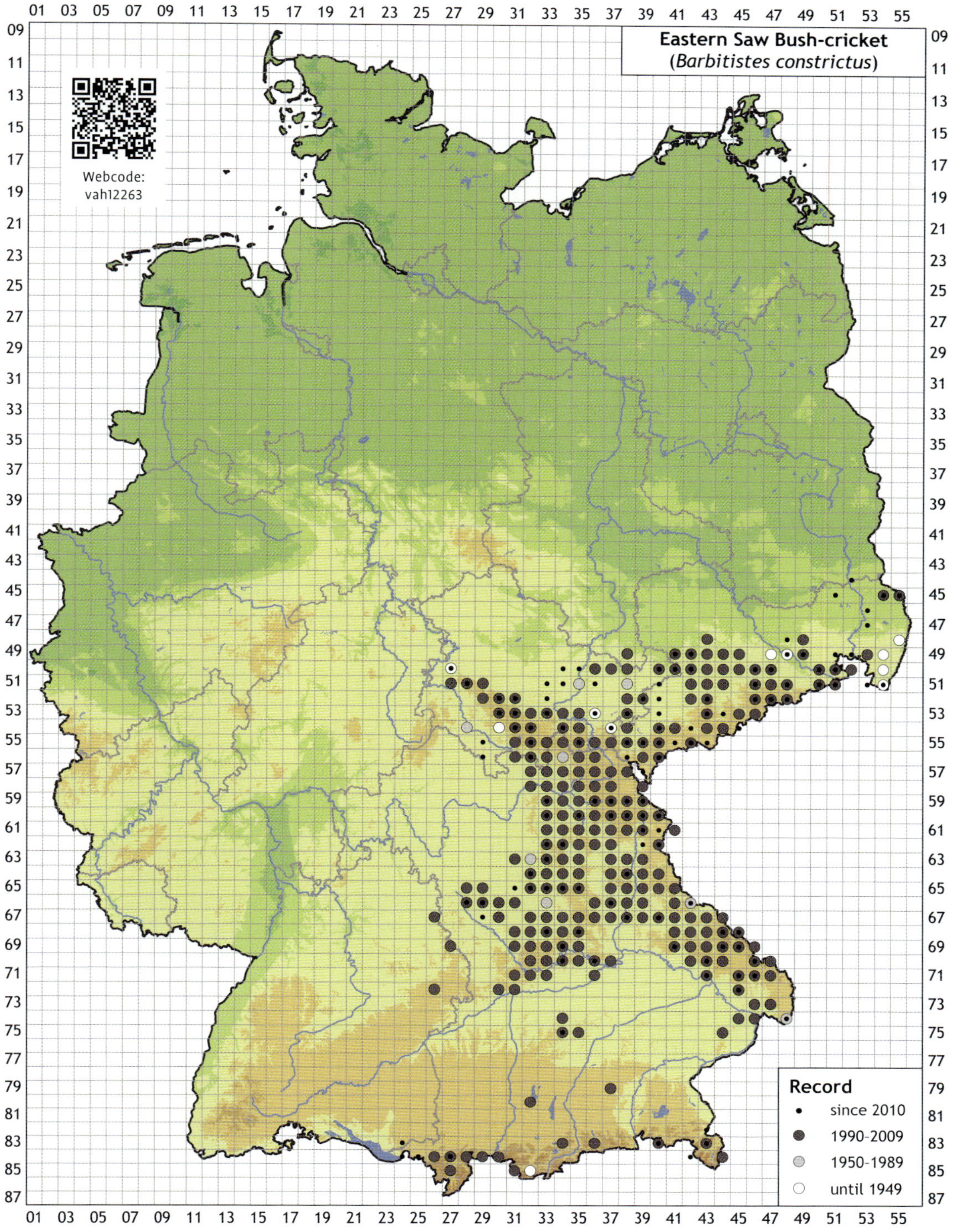
Eastern Saw Bush-cricket
(Barbitistes constrictus)
Webcode:
vah12263
Record
since 2010
1990-2009
1950-1989
until 1949

Female Krauss's Bush-cricket (*Isophya kraussii*).
(Photo credit: Dominik Poniatowski)

Montane, acidic grasslands are among the habitats of the Krauss's Bush-cricket (Saxony: Ore Mountains).
(Photo credit: Günter Köhler)

Krauss's Bush-cricket – *Isophya kraussii*
Brunner von Wattenwyl, 1878

Distribution The occurrence of the Krauss's Bush-cricket in Germany is nearly exclusively restricted to the colline and montane zone of the Southwestern and Eastern Uplands. Most populations are found across a belt of base-rich/calcareous substrate from the Swabian Jura through the Franconian Jura and Rhön Mountains to the Thuringian Forest and its northern foothills.

Habitat The habitats of the Krauss's Bush-cricket are highly diverse; however, they mostly have in common a dense vegetation rich in tall grasses and forbs. Typical habitats are dry and mesic grasslands rich in fringe species, wet tall-forb or needle vegetation and light parts of forests. The nymphs are sensitive to cold, since some already hatch in late April. Therefore, depressions where cold air pools and late frosts occur more regularly are avoided, whereas warm slopes are preferred. Within grasslands, the species can only persist under low stocking densities or late mowing (July/August), after the adult peak. Consequently, the SFI of 25.1 % is relatively high.

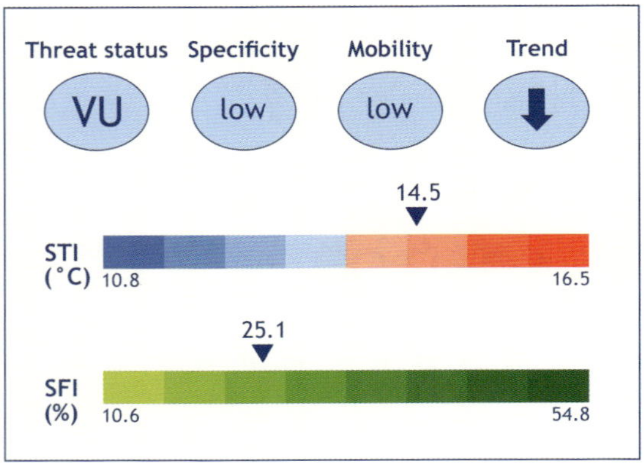

Population trend and threats Due to land-use change, many habitats of the Krauss's Bush-cricket such as nutrient-poor grasslands rich in fringe species, abandoned mesic grasslands and stands of tall forbs have been lost. Hence, the long- and short-term trends are assumed to be moderately negative. Since the Krauss's Bush-cricket is flightless, dispersal strongly depends on a high habitat connectivity. Germany has a high responsibility for the long-term viability of the species since more than a quarter of the global population occurs here.

GÜNTER KÖHLER

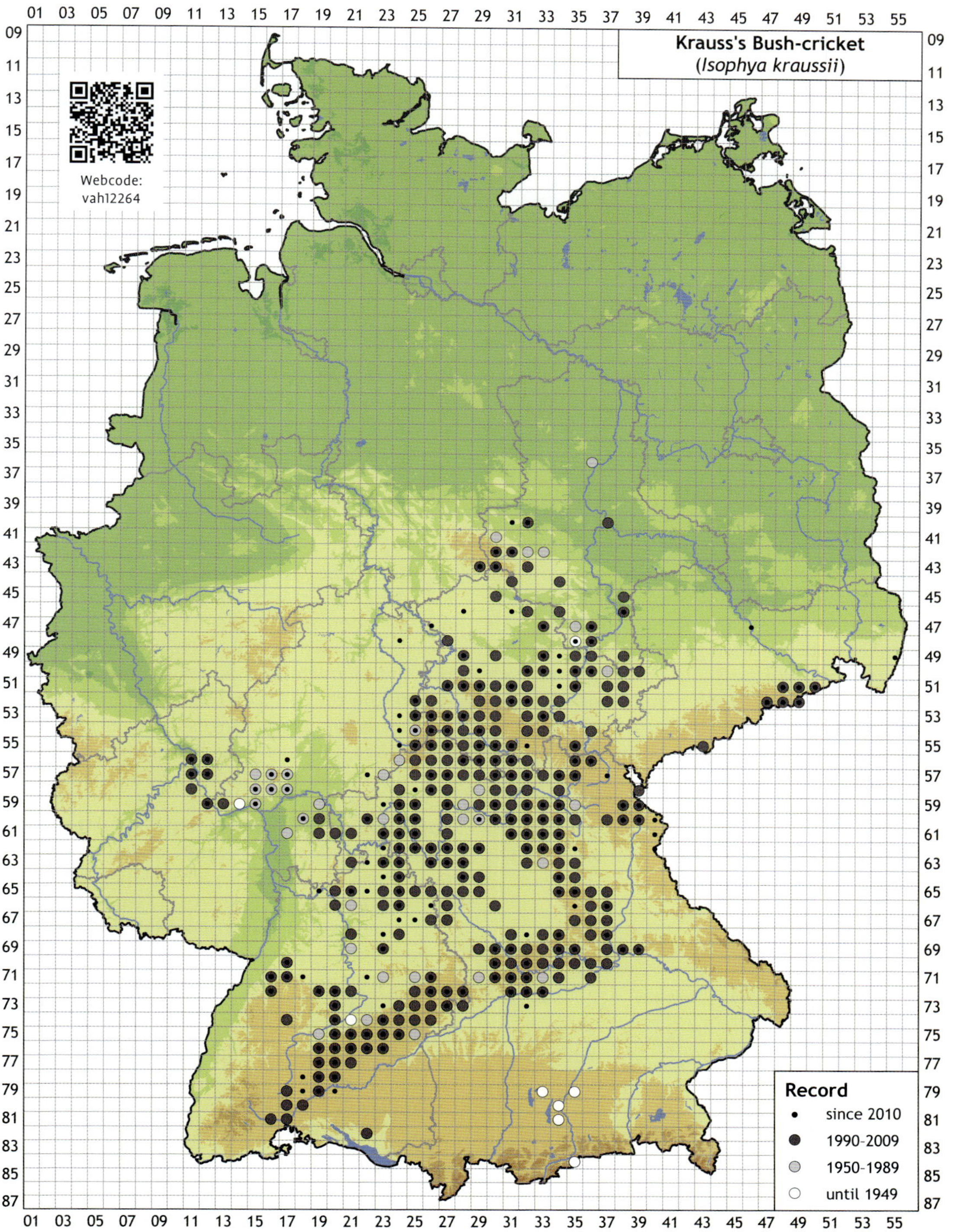
Krauss's Bush-cricket
(Isophya kraussii)
Webcode:
vah12264
Record
since 2010
1990-2009
1950-1989
until 1949

Montane species-rich hay meadow – typical habitat of the Large Saw-tailed Bush-cricket (Baden-Württemberg: Swabian Jura). (Photo credit: Arnold Kleiner)

Male Large Saw-tailed Bush-cricket (*Polysarcus denticauda*).
(Photo credit: Dominik Poniatowski)

Large Saw-tailed Bush-cricket – *Polysarcus denticauda* (Charpentier, 1825)

Distribution Almost all populations of the Large Saw-tailed Bush-cricket in Germany occur in the colline and montane zone of the two strongholds Baar and Swabian Jura together with its foreland on the one hand and the Rhön Mountains and Grabfeld on the other hand. Additionally, isolated populations exist at the southern edge of the Thuringian Basin and the Grünten in the Allgäu, here even at more than 1,700 m a.s.l. The strong dependence on cool biogeographical regions is reflected by a below average STI of 13.7 °C. It remains unclear whether or not the recently discovered population in the Franconian Jura is indigenous.

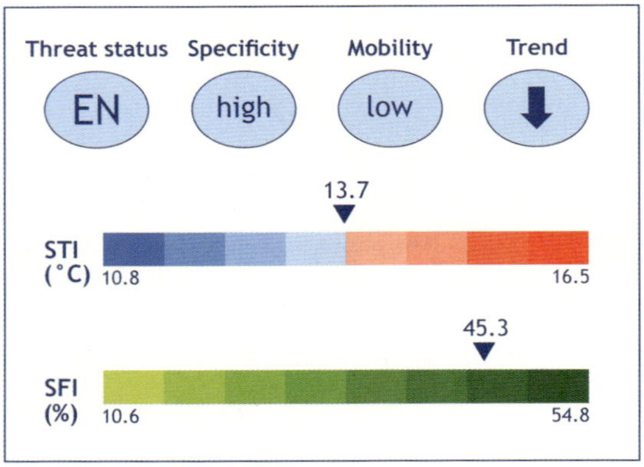

Habitat The Large Saw-tailed Bush-cricket colonises very sunny, preferentially moderately dry to moderately moist sites exhibiting herb-rich vegetation. It is a typical species of mesic meadows that are characterised by low land-use intensity. As a result, the SFI is extremely high with 45.3 %. Further habitats are fringes and wet abandoned grasslands. In general, a successful reproduction depends on tall vegetation of sufficient size until midsummer.

Population trend and threats The long- and short-term population trends of the Large Saw-tailed Bush-cricket in Germany are negative. Many former populations have become extinct, and abundance declines have been observed for still existing populations. Intensification of grassland management, including more cuttings per year, large-scale cutting at the same time and increased fertilizer application are considered the main threats to the species. Moreover, the flightless, sedentary species suffers from landscape homogenization and fragmentation of the habitats.

JÜRGEN TRAUTNER & SEBASTIAN SÄNDIG

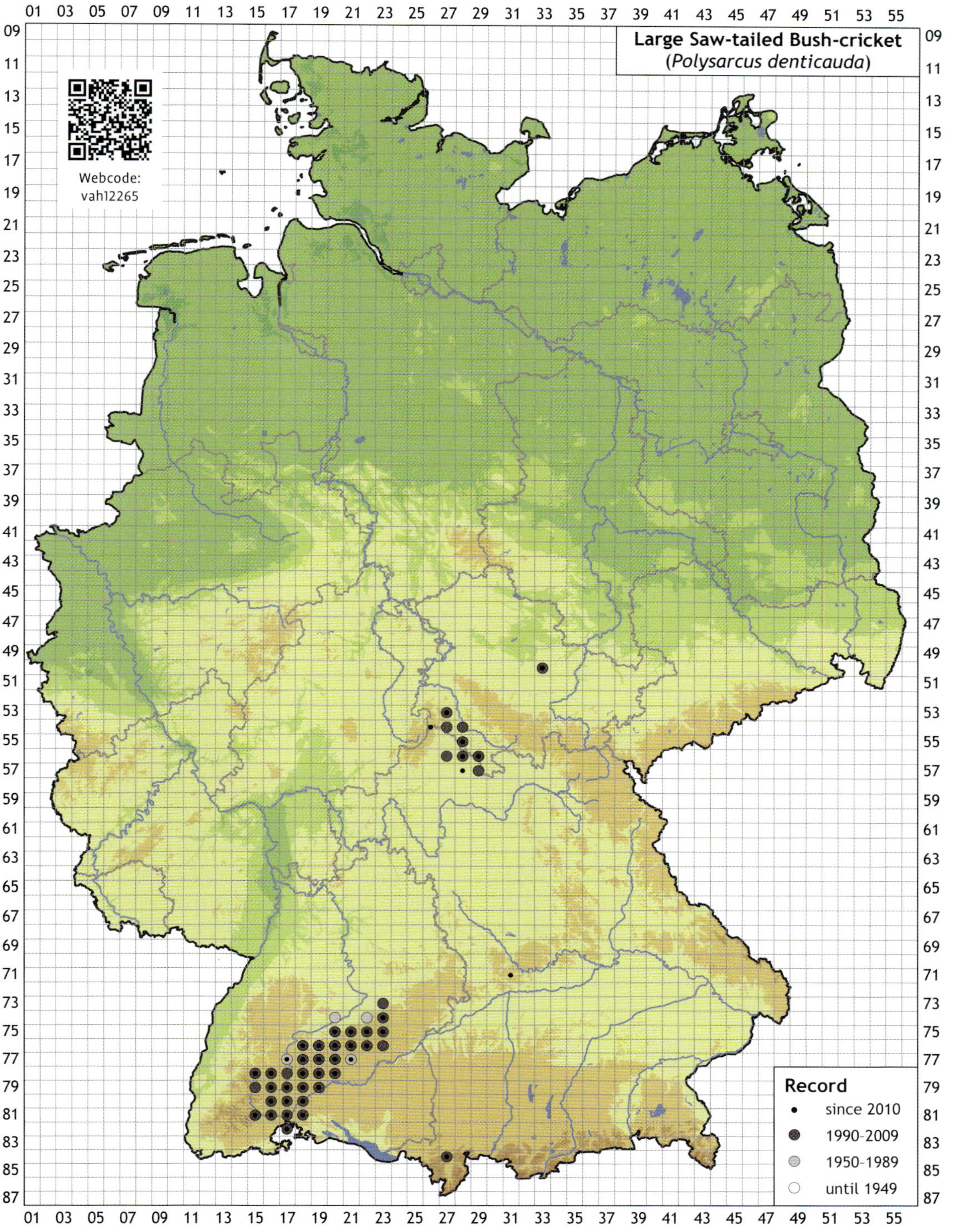
Large Saw-tailed Bush-cricket
(Polysarcus denticauda)
Webcode:
vah12265
Record
since 2010
1990-2009
1950-1989
until 1949

Oak-hornbeam woodlands are regularly colonised by the Oak Bush-cricket (North Rhine-Westphalia: Münsterland).
(Photo credit: Gregor Stuhldreher)

Female of the Oak Bush-cricket (*Meconema thalassinum*).
(Photo credit: Thomas Fartmann)

Oak Bush-cricket – *Meconema thalassinum* (De Geer, 1773)

Distribution The Oak Bush-cricket is widespread in Germany and only absent from cooler, higher elevations of the low mountain ranges and Alps. Besides the unfavourable climate, the lack of deciduous forests may also contribute to the formation of these elevational range limits.

Habitat The Oak Bush-cricket mainly lives on trees in closed deciduous forests and at forest edges. To establish a small population, even a few trees are sufficient. Therefore, the species also occurs in avenues, gardens or parks. The range of colonised deciduous trees is wide: however, all have in common a cracked bark, which is used for oviposition, and large leaves, the bottoms of which serve as shelters during the day. The nymphs and adults are cool stenothermic, which explains the pronounced avoidance of direct sunlight. The Oak Bush-cricket is nocturnal and attracted by light. Consequently, it can regularly be found at and in houses.

Population trend and threats Recent observations indicate that the Oak Bush-cricket has moderately declined in some regions across Germany. So far, it remains unclear if interspecific competition with the strongly expanding Southern Oak Bush-cricket may have contributed to this. By contrast, climate change seems among the drivers of recent population losses of the species. Particularly, in warm and dry biogeographical regions, the Oak Bush-cricket is now often restricted to cool north- and east-facing woodland edges or has even completely disappeared. The further development within the next years should be carefully monitored. This is particularly true as Germany is largely responsible for the global conservation of the species. Currently, the Oak Bush-cricket is considered of least concern.

ANDREAS KRONSHAGE

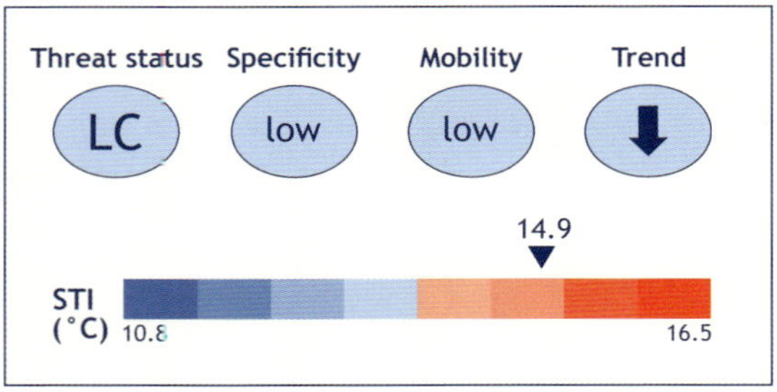

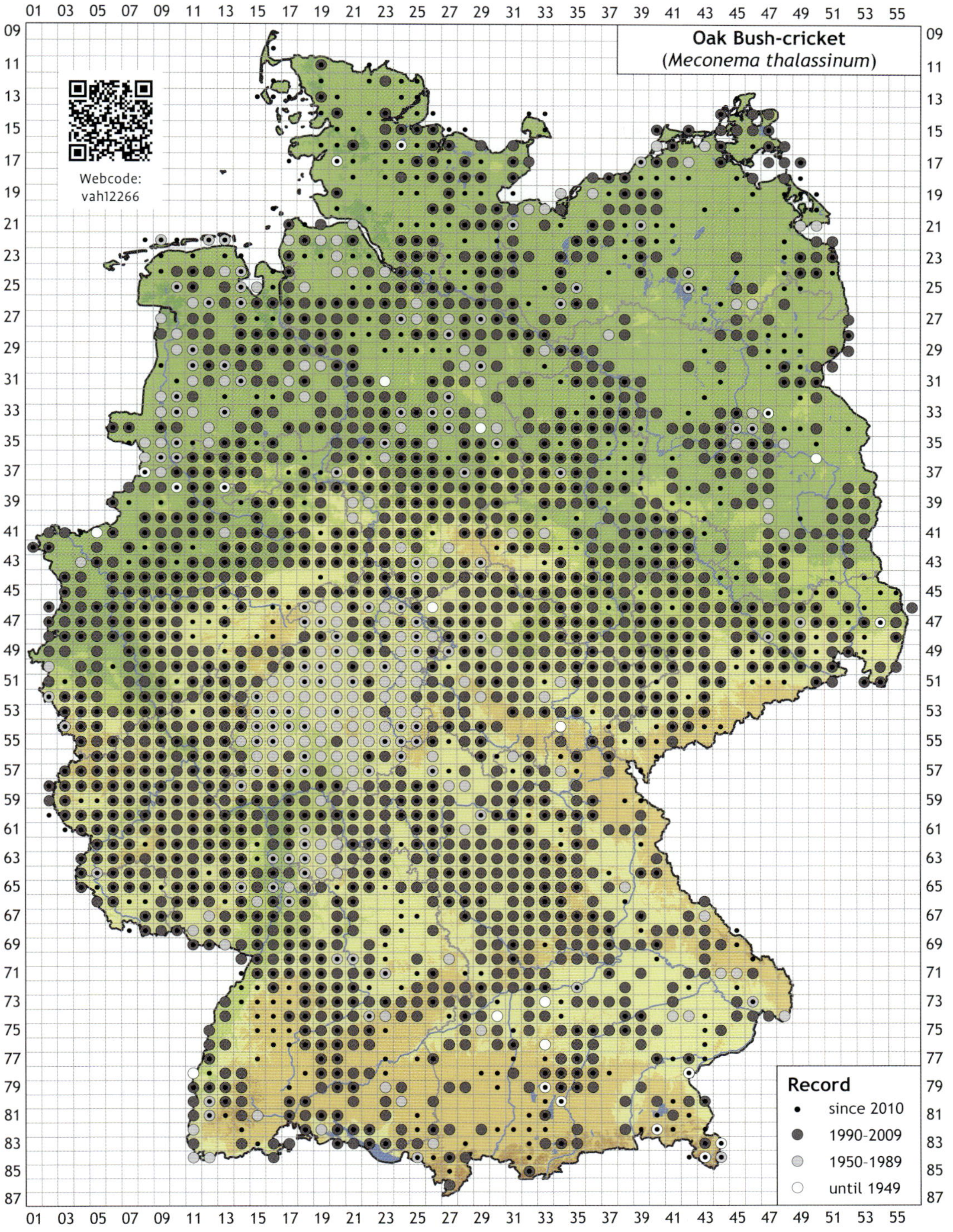
Oak Bush-cricket
(*Meconema thalassinum*)
Webcode:
vah12266
Record
since 2010
1990-2009
1950-1989
until 1949

Cemeteries are also among the many urban habitats of the Southern Oak Bush-cricket (Baden-Württemberg: southern Upper Rhine Valley, Freiburg). (Photo credit: Marcel Kettermann)

Male Southern Oak Bush-cricket (*Meconema meridionale*). (Photo credit: Marcel Kettermann)

Southern Oak Bush-cricket – *Meconema meridionale* Costa, 1860

Distribution The Southern Oak Bush-cricket was first recorded in Germany near Freiburg in 1958. In subsequent years, further records were limited to the warm regions of the Lake Constance Basin and along the Upper Rhine Valley. It was not until the 1980s that the species began expanding northwards along the river Rhine and some of its tributaries. Moreover, numerous records have been reported from urban agglomerations and larger cities, which are known as urban heat islands. In the past decade, this expansion has accelerated. Today, the Southern Oak Bush-cricket can also be regularly found outside of settlements and across all seven major biogeographical regions. Yet, the species still mainly occurs at climatically favoured, anthropogenic habitats, which is well reflected by its high STI value of 15.3 °C.

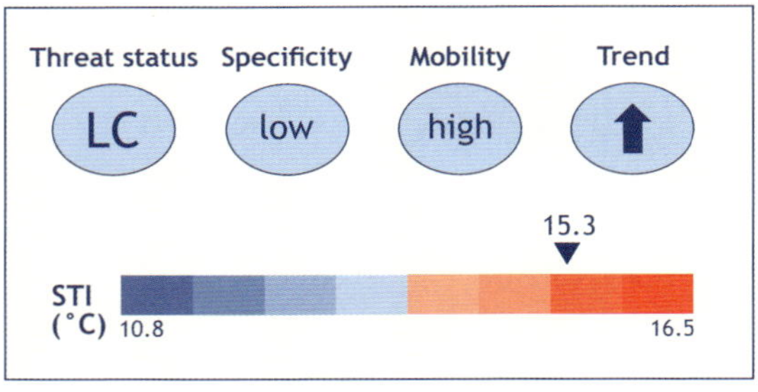

Habitat The Southern Oak Bush-cricket is primarily observed in urban environments such as city parks, gardens and green facades. Outside of settlements, the species occurs in shrublands, sun-exposed forest edges and light forests. It inhabits a wide range of tree and shrub species, with a preference for those with rough or cracked bark for oviposition. Occasionally, it is also found in herbaceous vegetation. At night, individuals are often attracted to heat and artificial lighting, causing them to move towards exterior walls, building interiors and warm vehicles.

Population trend and threats The Southern Oak Bush-cricket is spreading rapidly and is not threatened in Germany. Besides the positive effects of global warming, anthropogenic dispersal by vehicles such as cars or trains is a key driver of its range expansion. This allows the flightless bush-crickets to cover distances of up to several hundred kilometres passively. As a consequence, new occurrences of the species are regularly recorded at sites, particularly in cities, that are far away from the nearest known populations.

FELIX HELBING

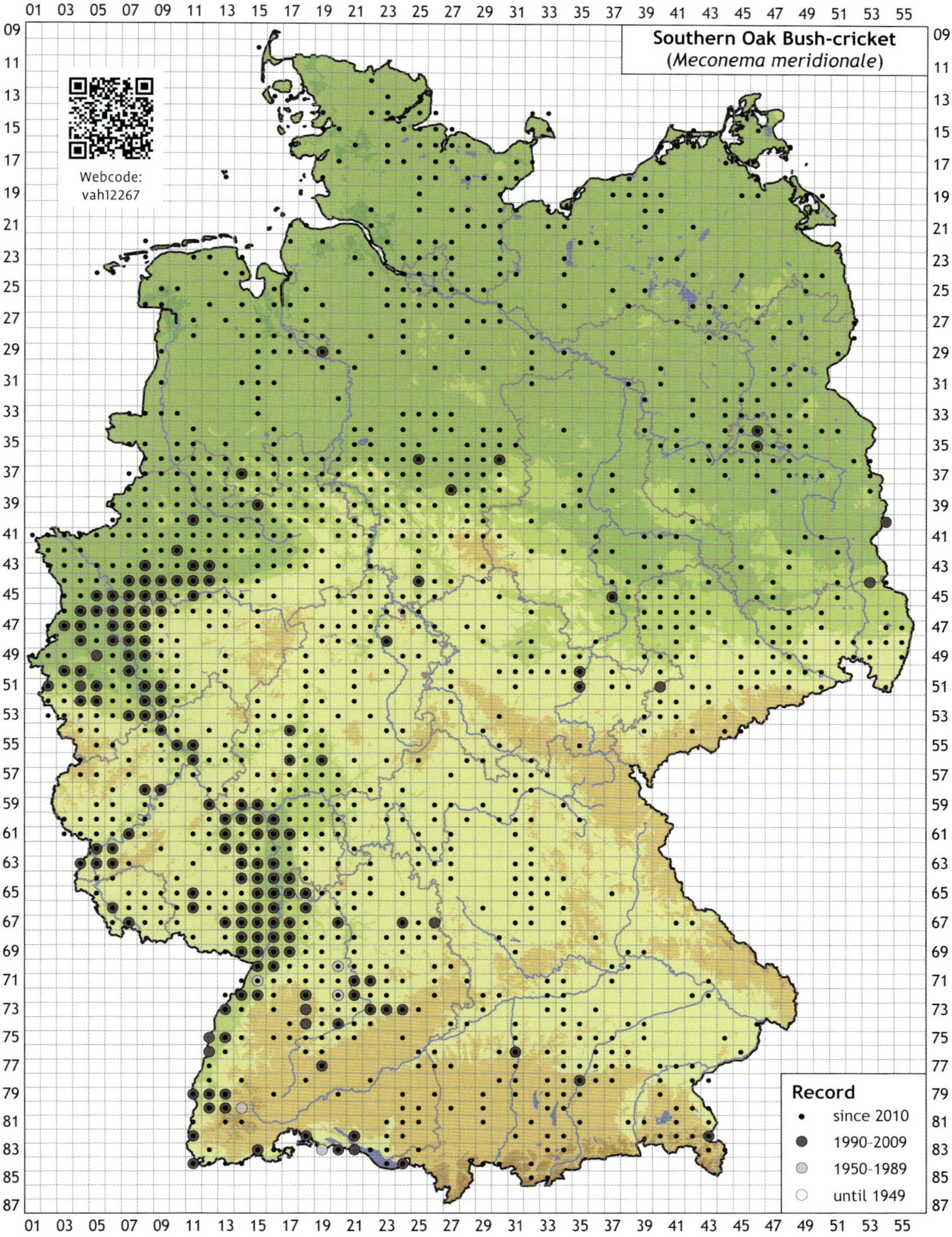
Southern Oak Bush-cricket
(Meconema meridionale)
Webcode:
vah12267
Record
since 2010
1990-2009
1950-1989
until 1949

Wet grasslands exhibiting tall swards are important habitats of the Long-winged Conehead (Baden-Württemberg: Upper Danube Valley).
(Photo credit: Dominik Poniatowski)

Male Long-winged Conehead (*Conocephalus fuscus*).
(Photo credit: Dominik Poniatowski)

Long-winged Conehead – *Conocephalus fuscus* (Fabricius, 1793)

Distribution Historically, the Long-winged Conehead was restricted to the warmest biogeographical regions in Germany. Almost all the lower elevations of southern Germany, the Rhineland and the southern half of the Northeast German Lowland were densely populated. However, due to climate change, the species has strongly expanded its range northwards and to higher elevations over the last 30 years. Thus, the Long-winged Conehead now also regularly occurs further north in the North German Lowland. By contrast, it is still rare at higher elevations of the uplands. The dependence on a warm climate is reflected by an above average STI of 15.1 °C.

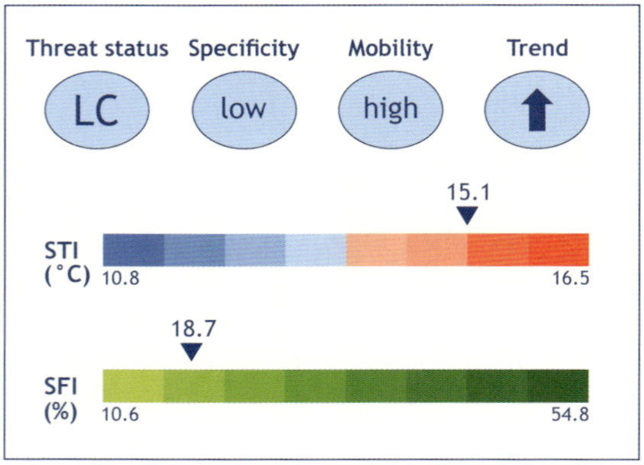

Habitat The Long-winged Conehead inhabits a fairly wide range of tall-growing open habitats of varying moisture. While in southern Germany the species mainly occurs in wet grasslands and moist stands of tall forbs, further north, it can also be found in drier habitats. These include abandoned grasslands, field margins, roadside verges and stands of the Wood Small-reed (*Calamagrostis epigejos*). Since the eggs are laid in plant stems and overwinter there, the Long-winged Conehead is highly sensitive to intensive land use, particularly mowing. The species is capable of flight. Furthermore, its dispersal is favoured by hydrochory.

Population trend and threats The climate-driven range expansion of the Long-winged Conehead during recent decades has resulted in a positive population trend in Germany. Nevertheless, drainage of wet habitats, the loss of grassy ecotones and intensive grassland management may lead to a loss of suitable habitats. Currently, the species is considered of least concern.

FRANZ LÖFFLER

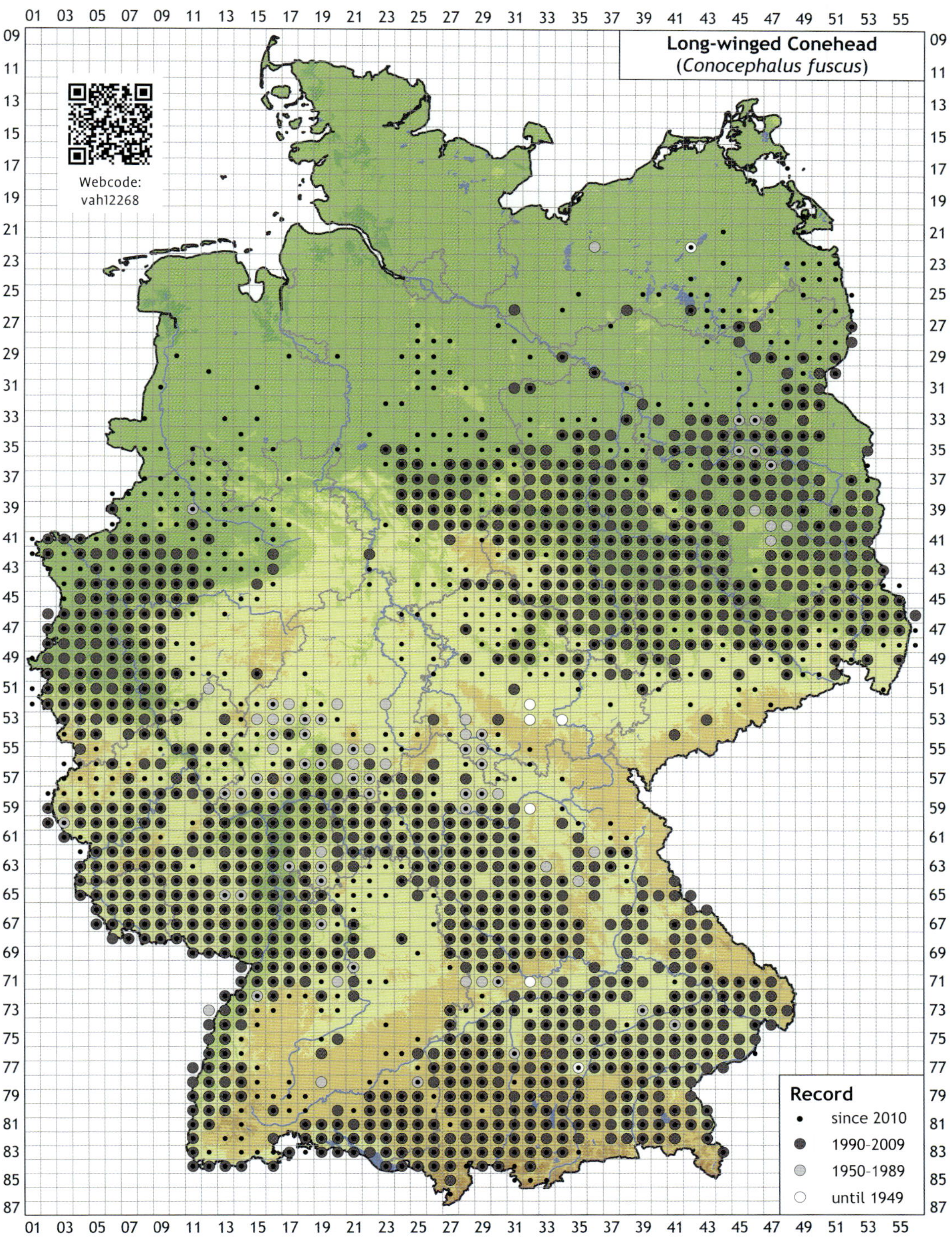
Long-winged Conehead
(*Conocephalus fuscus*)
Webcode:
vah12268
Record
since 2010
1990-2009
1950-1989
until 1949
01 03 05 07 09 11 13 15 17 19 21 23 25 27 29 31 33 35 37 39 41 43 45 47 49 51 53 55
09 11 13 15 17 19 21 23 25 27 29 31 33 35 37 39 41 43 45 47 49 51 53 55 57 59 61 63 65 67 69 71 73 75 77 79 81 83 85 87

Male Short-winged Conehead (*Conocephalus dorsalis*).
(Photo credit: Dominik Poniatowski)

Abandoned wet grasslands may host large populations of the Short-winged Conehead (Schleswig-Holstein: Elbmarschen).
(Photo credit: Christian Winkler)

Short-winged Conehead – *Conocephalus dorsalis* (Latreille, 1804)

Distribution The Short-winged Conehead occurs in all German federal states. However, it is only common in the North German Lowland. At lower elevations of the Western, Southwestern and Eastern Uplands as well as in the Bavarian Alpine Foreland, it occurs at least locally regularly.

Habitat The hygrophilic and slightly thermophilic Short-winged Conehead prefers sunny, damp to wet open habitats with tall, dense vegetation of grasses and herbs. Egg-laying occurs in pithy plant stems, leaf sheaths, bark or dead wood. The species especially colonises stands of rushes, sedges or tall forbs in wet grasslands characterised by low-intensity grazing or abandonment and at the riparian zone of water bodies. The habitats also include coastal areas influenced by salt water, salty inland habitats, fens and bogs. Especially in oceanic regions in northern Germany, the species also occurs in fresh to dry habitats.

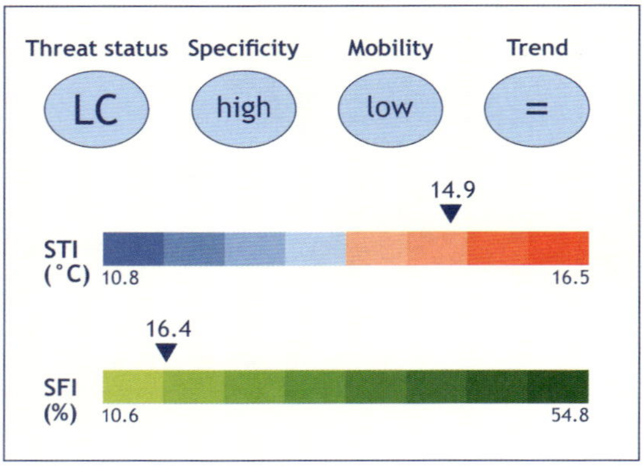

Population trend and threats In the past, the Short-winged Conehead experienced moderate declines in Germany. These losses were attributed to land-use change, in particular drainage of the habitats, intensification of grassland management and conversion of grasslands into arable land. Recently, the populations are considered stable, although the trend may differ among regions: in the southern part of Germany, the species partly continues to decline. By contrast, it is increasing at the local level particularly in the north and northwest of Germany.

CHRISTIAN WINKLER

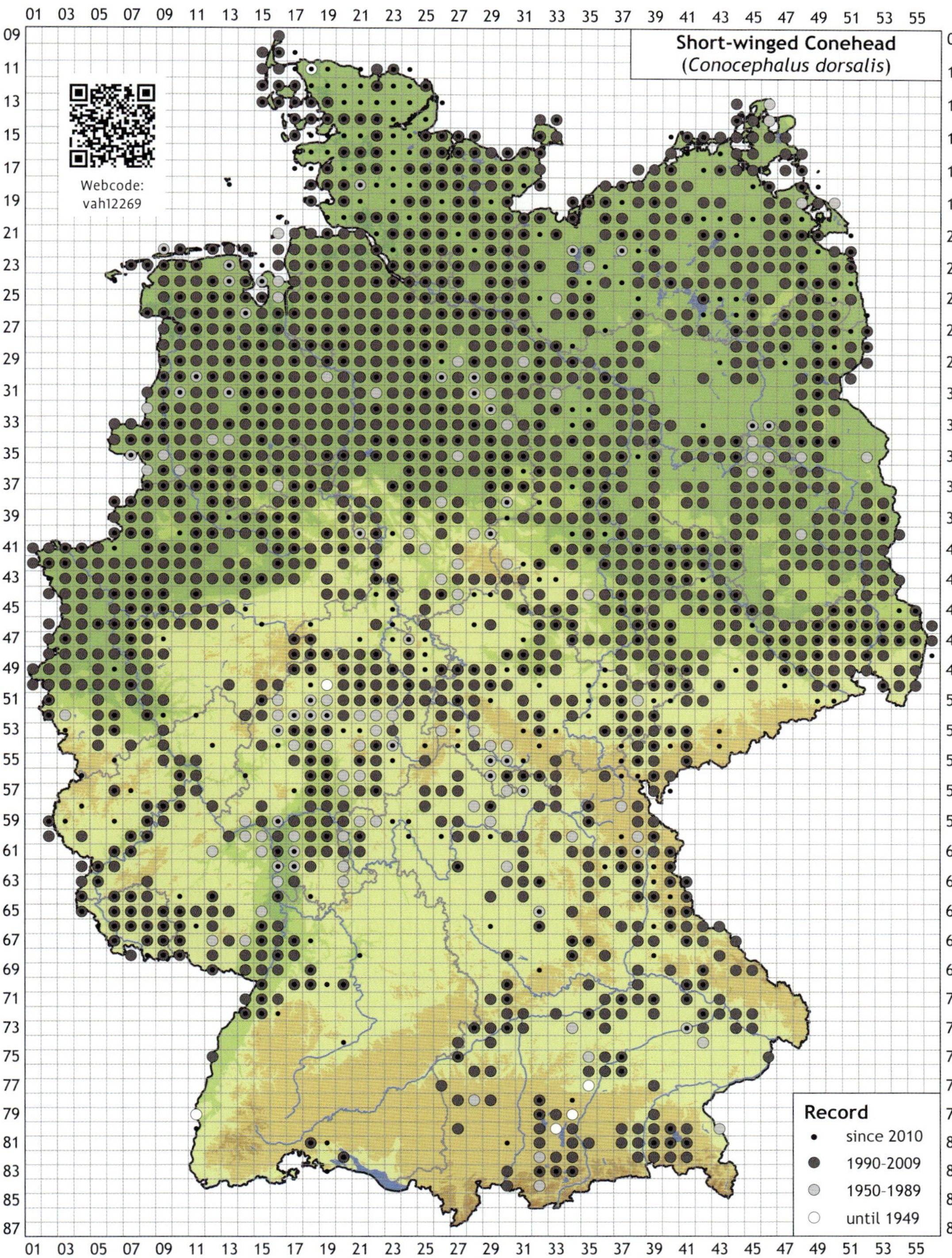
Short-winged Conehead
(Conocephalus dorsalis)
Webcode:
vah12269
Record
since 2010
1990-2009
1950-1989
until 1949
01 03 05 07 09 11 13 15 17 19 21 23 25 27 29 31 33 35 37 39 41 43 45 47 49 51 53 55
09 11 13 15 17 19 21 23 25 27 29 31 33 35 37 39 41 43 45 47 49 51 53 55 57 59 61 63 65 67 69 71 73 75 77 79 81 83 85 87

The species occurs in very different habitats – here, the margins of a ditch are colonised (Baden-Württemberg: Kaiserstuhl region). (Photo credit: Marcel Kettermann)

Female Large Conehead (*Ruspolia nitidula*). (Photo credit: Jürgen Fischer)

Large Conehead – *Ruspolia nitidula* (Scopoli, 1786)

Distribution The Large Conehead is a highly thermophilic species, which, for a long time, only occurred at the Lake Constance Basin within Germany. In the early 1990s, it was even considered extinct. However, in 1995, it was rediscovered at both the Austrian and German side of Lake Constance. Subsequently, due to climate change, a strong and rapid range expansion took place, particularly in southwestern Germany. This is facilitated by the very high STI of the species of 15.6 °C. After the Lake Constance region, the Upper Rhine Valley, which is also characterised by very warm summers, is now the second stronghold. As of today, the northernmost record in Germany originates from the Lower Rhine Lowland in 2021. Due to the high temperature requirements, the Large Conehead mainly occurs at lower elevations. However, by now, single records of the mobile species that is capable of flight have been recorded in the Black Forest at more than 1,000 m a.s.l.

Habitat The Large Conehead has become a habitat generalist within open and semi-open landscapes. It colonises a large variety of habitats as long as they are characterised by a higher cover of grasses or herbs. These include different types of grasslands, gardens or verges along ditches, roads or shrubland.

Population trend and threats Due to climate change, the Large Conehead is strongly expansive in Germany. Moreover, suitable habitats are currently largely available in our landscapes and will most likely continue to be so in the future. Accordingly, the species is considered of least concern.

JÜRGEN TRAUTNER

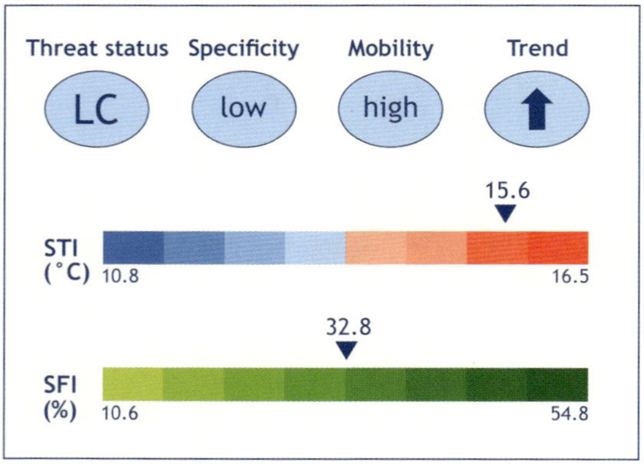

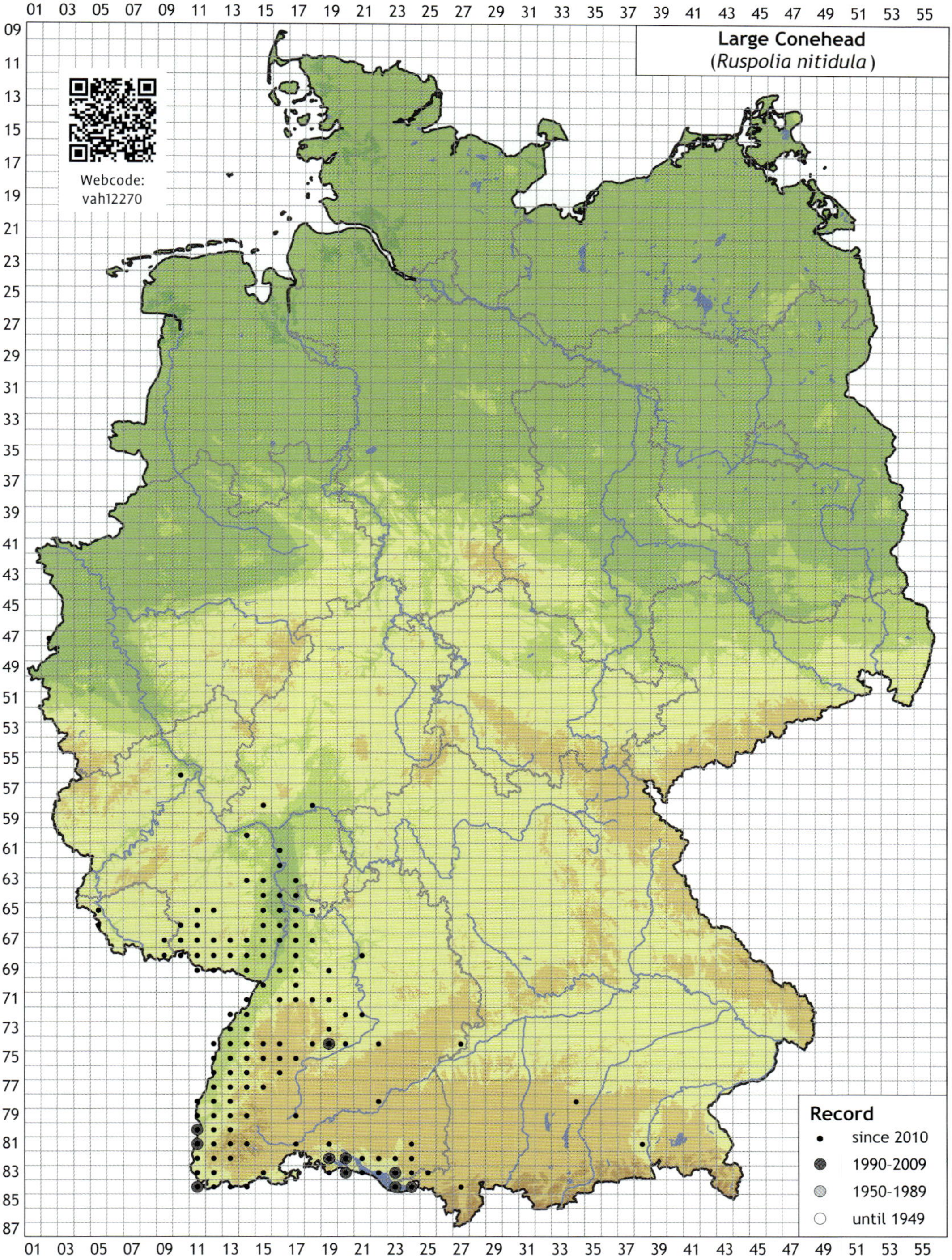
Large Conehead
(*Ruspolia nitidula*)
Webcode:
vah12270
Record
since 2010
1990-2009
1950-1989
until 1949
01 03 05 07 09 11 13 15 17 19 21 23 25 27 29 31 33 35 37 39 41 43 45 47 49 51 53 55
09 11 13 15 17 19 21 23 25 27 29 31 33 35 37 39 41 43 45 47 49 51 53 55 57 59 61 63 65 67 69 71 73 75 77 79 81 83 85 87

High-growing, dense abandoned grasslands – here on wet soils – are highly attractive for adults of the Great Green Bush-cricket (North Rhine-Westphalia: Münsterland).
(Photo credit: Thomas Fartmann)

Male Great Green Bush-cricket (*Tettigonia viridissima*).
(Photo credit: Thomas Fartmann)

Great Green Bush-cricket – *Tettigonia viridissima* (Linnaeus, 1758)

Distribution The Great Green Bush-cricket is among the most widespread grasshopper species in Germany. For a long time, it was largely absent only from the cool higher elevations of the uplands, Alpine Foreland and Alps. However, due to climate change, the species is now expanding its range into these regions too. The weak preference for a warm climate is reflected by a STI of 14.8 °C, which is slightly above average.

Habitat The range of open and semi-open habitats colonised by the Great Green Bush-cricket is very wide and characterises the species as a pronounced habitat generalist. Accordingly, the SFI of 18.2 % is below average. Fallows, forest edges, gardens, grasslands, parks, ruderal habitats, stands of tall forbs and waysides are among the typical habitats. The Great Green Bush-cricket undergoes the strongest microhabitat shift during ontogenesis of all native grasshopper species: adult females often use sparsely vegetated, warm microhabitats for oviposition. Hatchlings and young nymphs also occur there. With increasing size, older nymphs and particularly adults move to areas with denser and taller vegetation, which offers shelter against predators. Additionally, especially adult males are found in shrubs and trees. This guarantees a good audibility of their calling songs over larger distances and that they can stridulate also at night, when temperatures are low near the ground but still warm in the canopy.

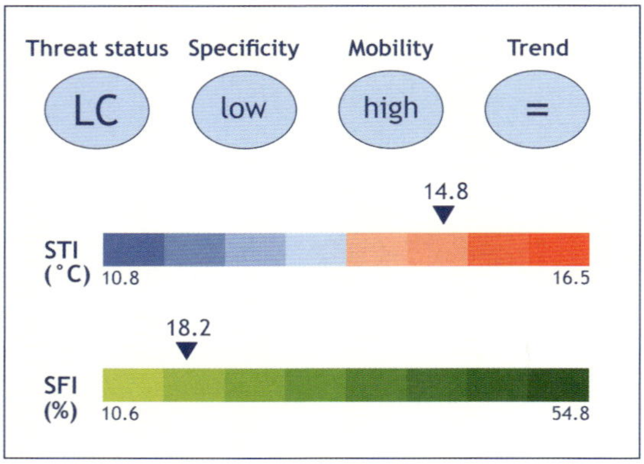

Population trend and threats Both, the long-term and short-term trend of the Great Green Bush-cricket in Germany is considered stable. Since the adults are capable of flight and suitable habitats are largely available in our landscapes, the species can rapidly respond to environmental change. The recent expansion of the species due to climate change confirms this. Consequently, the Great Green Bush-cricket is considered of least concern.

SIGFRID INGRISCH

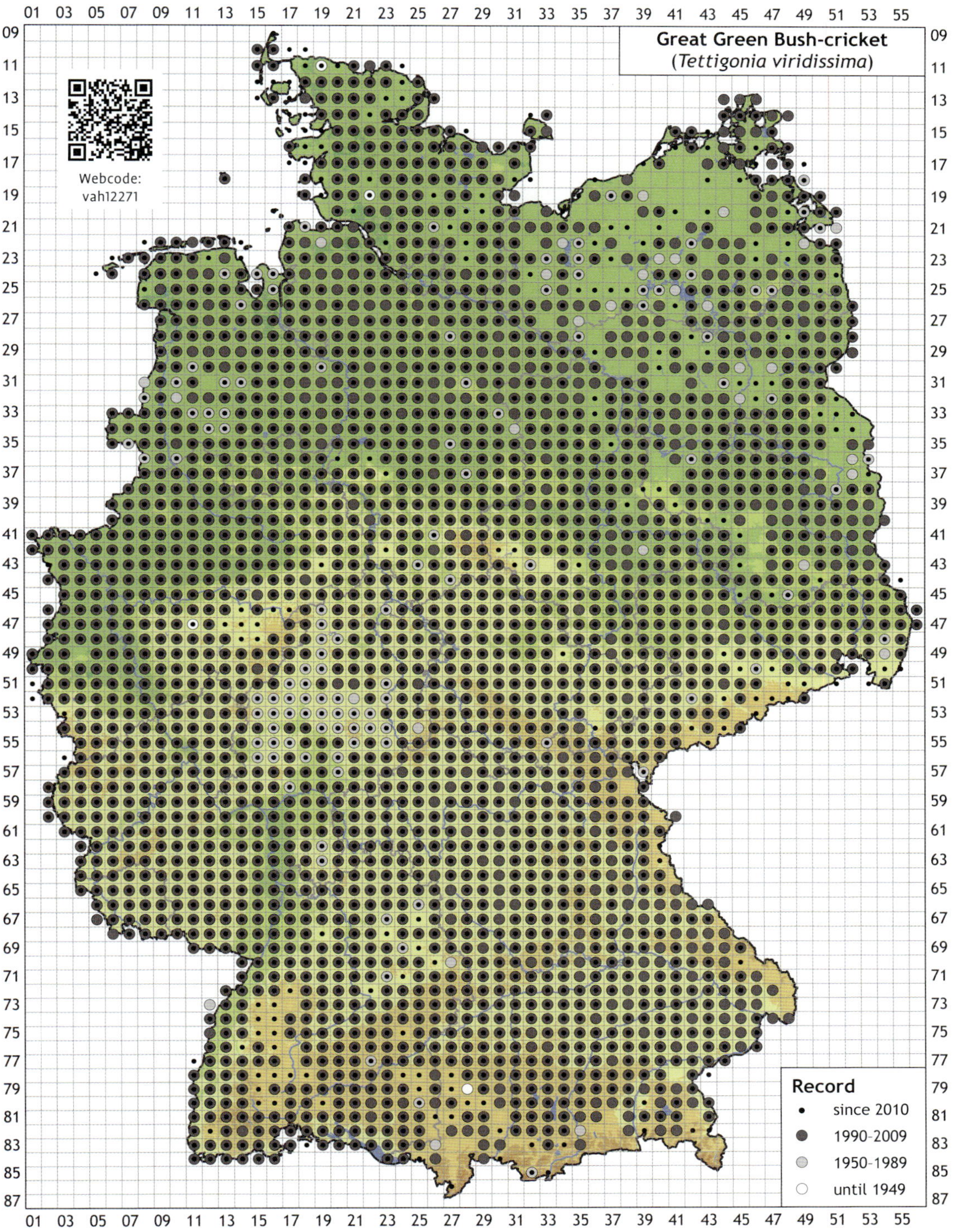
Great Green Bush-cricket
(Tettigonia viridissima)
Webcode:
vah12271
Record
since 2010
1990-2009
1950-1989
until 1949
01 03 05 07 09 11 13 15 17 19 21 23 25 27 29 31 33 35 37 39 41 43 45 47 49 51 53 55
09 11 13 15 17 19 21 23 25 27 29 31 33 35 37 39 41 43 45 47 49 51 53 55 57 59 61 63 65 67 69 71 73 75 77 79 81 83 85 87

Abandoned wet grasslands in the uplands are regularly colonised by the Upland Green Bush-cricket (Lower Saxony: Harz Mountains).
(Photo credit: Dominik Poniatowski)

Female Upland Green Bush-cricket (*Tettigonia cantans*).
(Photo credit: Jürgen Fischer)

Upland Green Bush-cricket – *Tettigonia cantans* (Fuessly, 1775)

Distribution The Upland Green Bush-cricket occurs in all seven major biogeographical regions of Germany. However, there are large areas where the species is missing. Overall, the occurrence of the species depends on soils with sufficient humidity. As a result, core areas of distribution are higher elevations in the uplands, Alpine Foreland and Alps that are rich in precipitation. However, the Upland Green Bush-cricket occurs locally also in the Northwest and Northeast German Lowland: here, depressions with high water levels and regions with extensive occurrence of cohesive soils, exhibiting a high water-holding capacity, are colonised. The latter comprises loess areas or young moraine landscapes with boulder clay and till.

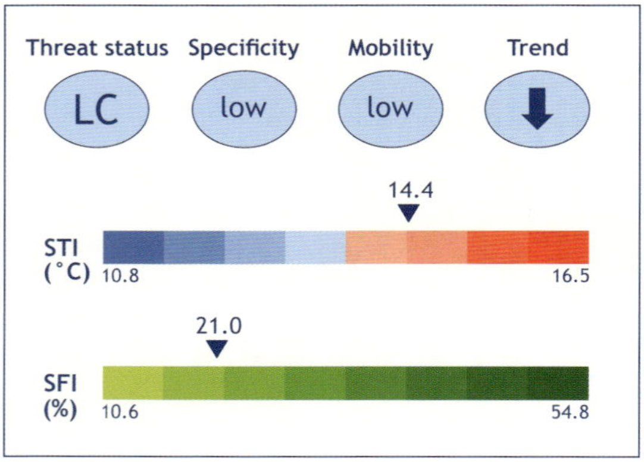

Habitat The Upland Green Bush-cricket inhabits many different open and semi-open habitats. However, in contrast to the Great Green Bush-cricket, it prefers moist and cool habitats. Since the species only exhibits six nymphal instars (as opposed to the Great Green Bush-cricket, which has seven), it also occurs in regions with short growing seasons. The strong moisture dependence, also with regard to the large-scale distribution, may result from the high desiccation sensitivity of the eggs, which are deposited into the soil. In general, typical habitats are characterised by dense and tall vegetation, often admixed with tall forbs or shrubs. Characteristic habitats are forest clearings, fringes, mires, stands of tall forbs and wet grasslands, including their fallows. By contrast, semi-dry grasslands are only colonised in rainy regions.

Population trend and threats Both for the long-term and short-term trend of the Upland Green Bush-cricket in Germany, a moderate decline is assumed. In the past, land-use change was responsible for the decrease due to the destruction of habitats. However, recently, local population losses have been documented in response to climate change. Since the species is still common, it is considered of least concern.

SIGFRID INGRISCH

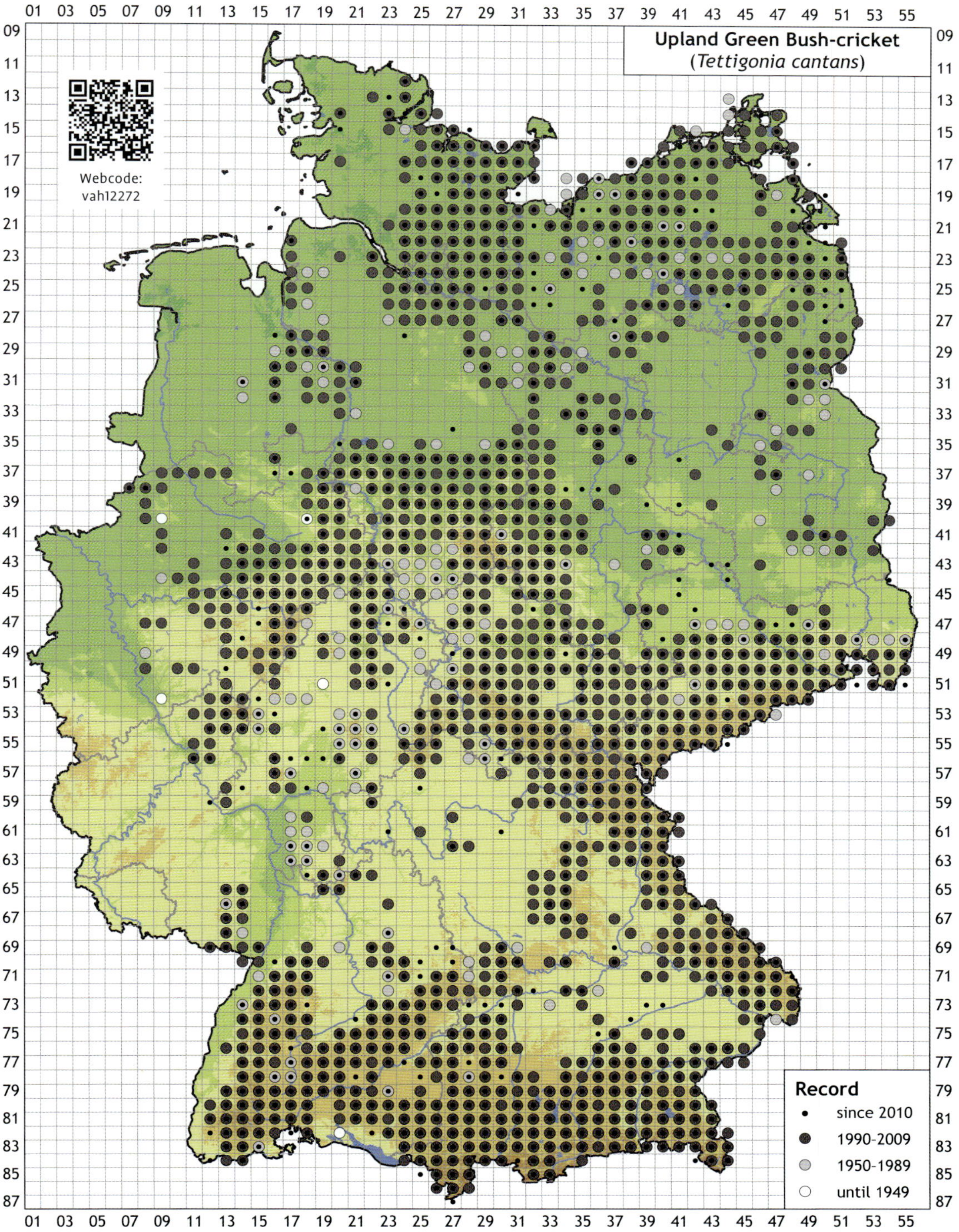
Upland Green Bush-cricket
(*Tettigonia cantans*)
Webcode:
vah12272
Record
since 2010
1990-2009
1950-1989
until 1949

Recently, perennial flower strips adjacent to arable fields have become a valuable habitat for the Eastern Green Bush-cricket (Brandenburg: Oderbruch). (Photo credit: Thomas Fartmann)

Male Eastern Green Bush-cricket (*Tettigonia caudata*). (Photo credit: Thomas Fartmann)

Eastern Green Bush-cricket – *Tettigonia caudata* (Charpentier, 1842)

Distribution In Germany, the Eastern Green Bush-cricket is restricted to the most continental regions of the Northeast German Lowland, which are characterised by a mean annual precipitation of less than 600 mm and mean temperatures in the warmer half of the year of above 15 °C. Accordingly, the STI of 15.7 °C is clearly above average. The species occurs in eastern Brandenburg, eastern Berlin together with its southern surroundings up to the Havelland, the southern Magdeburger Börde and the Valley of the river Elbe north of Dresden.

Habitat Typical habitats of the Eastern Green Bush-cricket are characterised by fresh, permeable soils and a mainly tall, dense and herb-rich vegetation. They are mostly located in agricultural landscapes and comprise especially field and road margins but also fallows, ruderal areas and recently even perennial flower strips. Prior to harvest, adjacent arable fields are also colonised. However, due to regular tillage and since the eggs, which are deposited into the soil, need at least two years for development, they are not suitable for reproduction. The Eastern Green Bush-cricket is well-adapted to a dry and hot macroclimate as well as to fresh soils. The eggs are xerophilic, in contrast to those of most other native bush-crickets. Moreover, since the species has the longest ovipositor of all three Green Bush-cricket species, the eggs can be laid very deep into the soil, which additionally protects them against desiccation during summer. By contrast, moist soils cannot be colonised since high humidity suppresses egg development after the initial diapause.

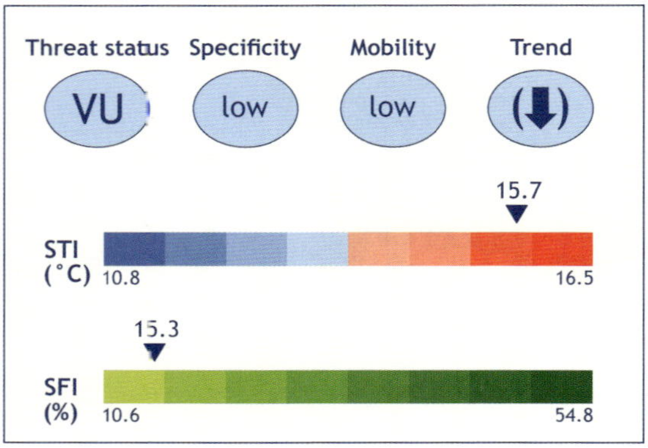

Population trend and threats Many arable fallows, which have become important habitats for the Eastern Green Bush-cricket in eastern Germany since the German reunification, are now once again being used for agriculture. The general loss of other fallows and ruderal areas has contributed to population declines as well. However, the extent of the decline is currently unclear.

THOMAS FARTMANN

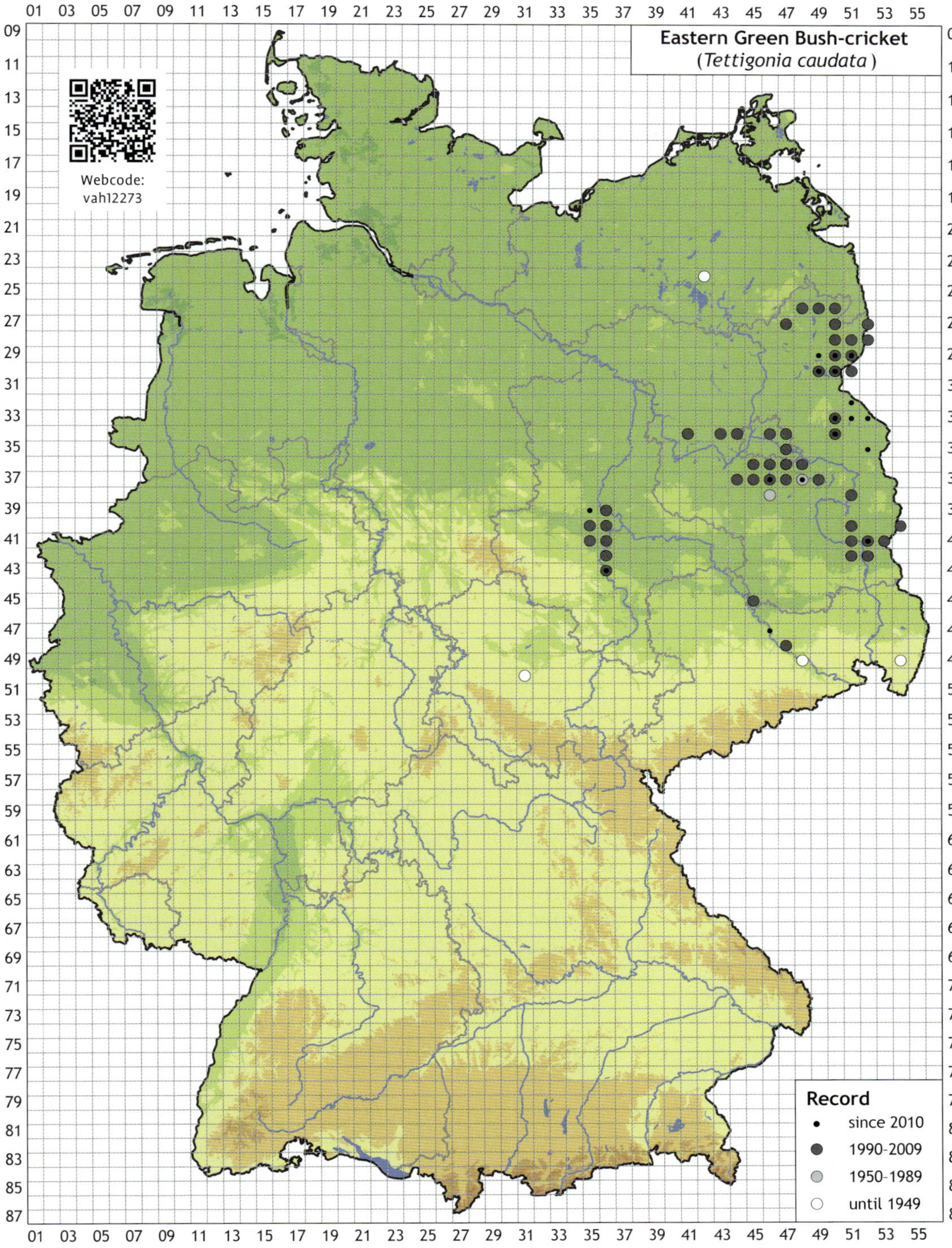
Eastern Green Bush-cricket
(*Tettigonia caudata*)
Webcode:
vah12273
Record
since 2010
1990-2009
1950-1989
until 1949

Abandoned arable fields, in particular on sandy soils, were and are partly still important habitats of the Common Wart-biter (Mecklenburg-Western Pomerania: Mecklenburg Lake District).
(Photo credit: Thomas Fartmann)

Male Common Wart-biter (*Decticus verrucivorus*). (Photo credit: Thomas Fartmann)

Common Wart-biter – *Decticus verrucivorus* (Linnaeus, 1758)

Distribution The Common Wart-biter occurs in all seven major biogeographical regions in Germany; however, regionally, its frequency varies greatly. Strongholds are the Alps, low mountain ranges, the Lüneburg Heath, the Pre-Alpine Hills and Moorland as well as sandy regions of the Northeast German Lowland. By contrast, in landscapes with intensive agriculture, the species is very rare.
Habitat The Common Wart-biter inhabits nutrient-poor open habitats characterised by low land-use intensity, exhibiting heterogeneous swards with some patches of bare ground. Accordingly, the SFI of 28.1 % is above average. Sunlit open soil is necessary for oviposition and is also preferred by the very thermophilic young nymphs. The larger older nymphs and later particularly the adults also depend on taller vegetation as a shelter against vertebrate predators and too high temperatures. Characteristic habitats are heathlands and all kinds of semi-natural grasslands. Moreover, in the 1990s and 2000s, many populations colonised abandoned arable fields. The eggs of the Common Wart-biter are highly sensitive to drought. Due to the usually at least biennial egg development, sufficient precipitation or moist soils are normally required. However, under continental climatic conditions, an annual development is possible so that the eggs are not exposed to periods of summer drought. This might explain why the species regularly occurs on permeable sandy soils in the dry Northeast German Lowland.
Population trend and threats Due to the dependence on semi-natural open habitats, the Common Wart-biter has already suffered for many decades from land-use change in Germany. Recently, the cessation of the mandatory set-aside payments of the EU in 2008 has caused the large-scale loss of abandoned arable fields as a further important habitat. Moreover, the remaining populations outside the strongholds are often small, greatly isolated and threatened by succession.

THOMAS FARTMANN

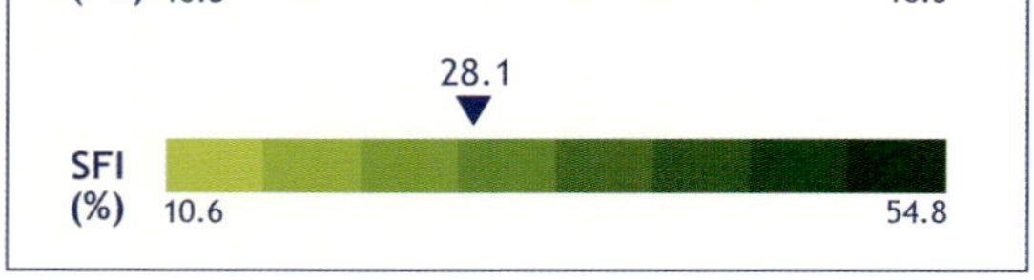

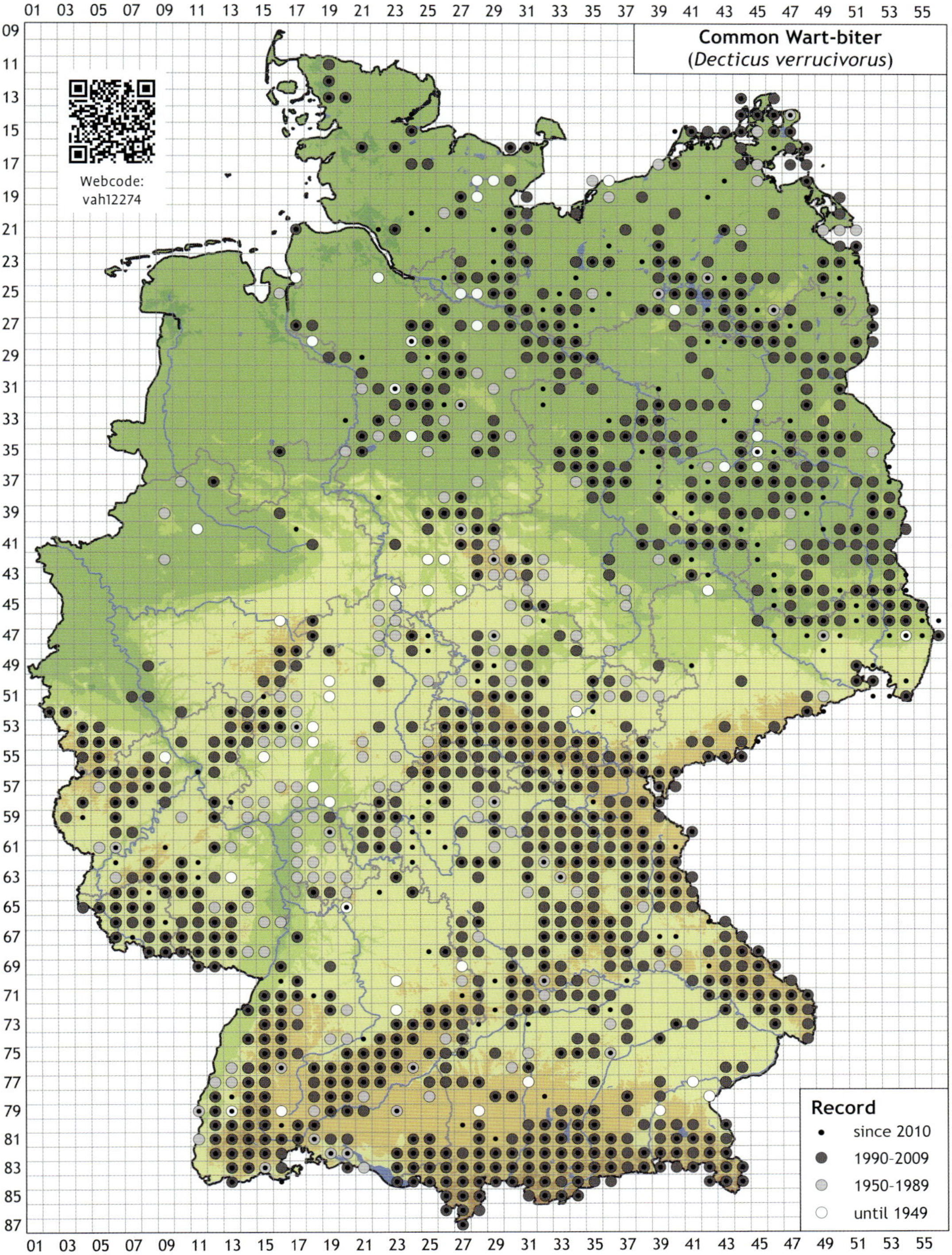
Common Wart-biter
(Decticus verrucivorus)
Webcode:
vah12274
Record
since 2010
1990-2009
1950-1989
until 1949

Habitat of the Heath Bush-cricket at the Munster South Military Training Area in a heterogeneous heathland rich in bare soil (Lower Saxony: Lüneburg Heath). (Photo credit: Thomas Fartmann)

Male Heath Bush-cricket (*Gampsocleis glabra*). (Photo credit: Jonas Brüggeshemke)

Heath Bush-cricket – *Gampsocleis glabra* (Herbst, 1786)

Distribution Within Germany, the Heath Bush-cricket occurs only very locally in the North German Lowland. Populations exist on the active military training areas Munster South, Altmark, Klietz and Altengrabow as well as on the Rheinmetall firing range near Unterlüß. The species was recently reintroduced to the Lüneburg Heath Nature Reserve and the Munster North Military Training Area. A recent record at the former Munich Airport was most likely based on released individuals.
Habitat Nowadays, the Heath Bush-cricket only occurs in open sandy heathlands and their grassy degradation stages as well as on dry sandy grasslands that are regularly exposed to fire. Correspondingly, the SFI is very high at 38.8 %. The fires caused by military training activity or used as heathland management measures ensure the preservation of patches of bare ground, which are important for egg laying and embryonic development. Adults prefer older heathlands (>4 years after fire). Emerging solitary trees are often used as perches by displaying males.
Population trend and threats In the past, the Heath Bush-cricket was more widespread in Germany than today. Besides additional records from the North German Lowland, particularly in the Lüneburg Heath, there were also observations from the northern Upper Rhine Valley (Griesheimer Sand) and the alluvial calcareous grasslands of the Alpine Foreland (Königsbrunn and Garching Heaths). After a strong decline in the course of the 20th century, the distribution appears to have been stable since the 1990s. Although the populations on the military training areas in the Northeast German Lowland – which are difficult to access – were only recently discovered, it is very likely that they have already existed for a long time. Due to the continued military activity, the future prospects of all long-established populations seem to be good.

JONAS BRÜGGESHEMKE

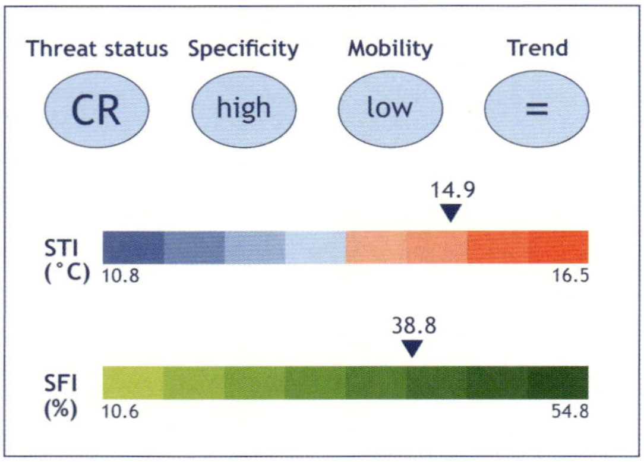

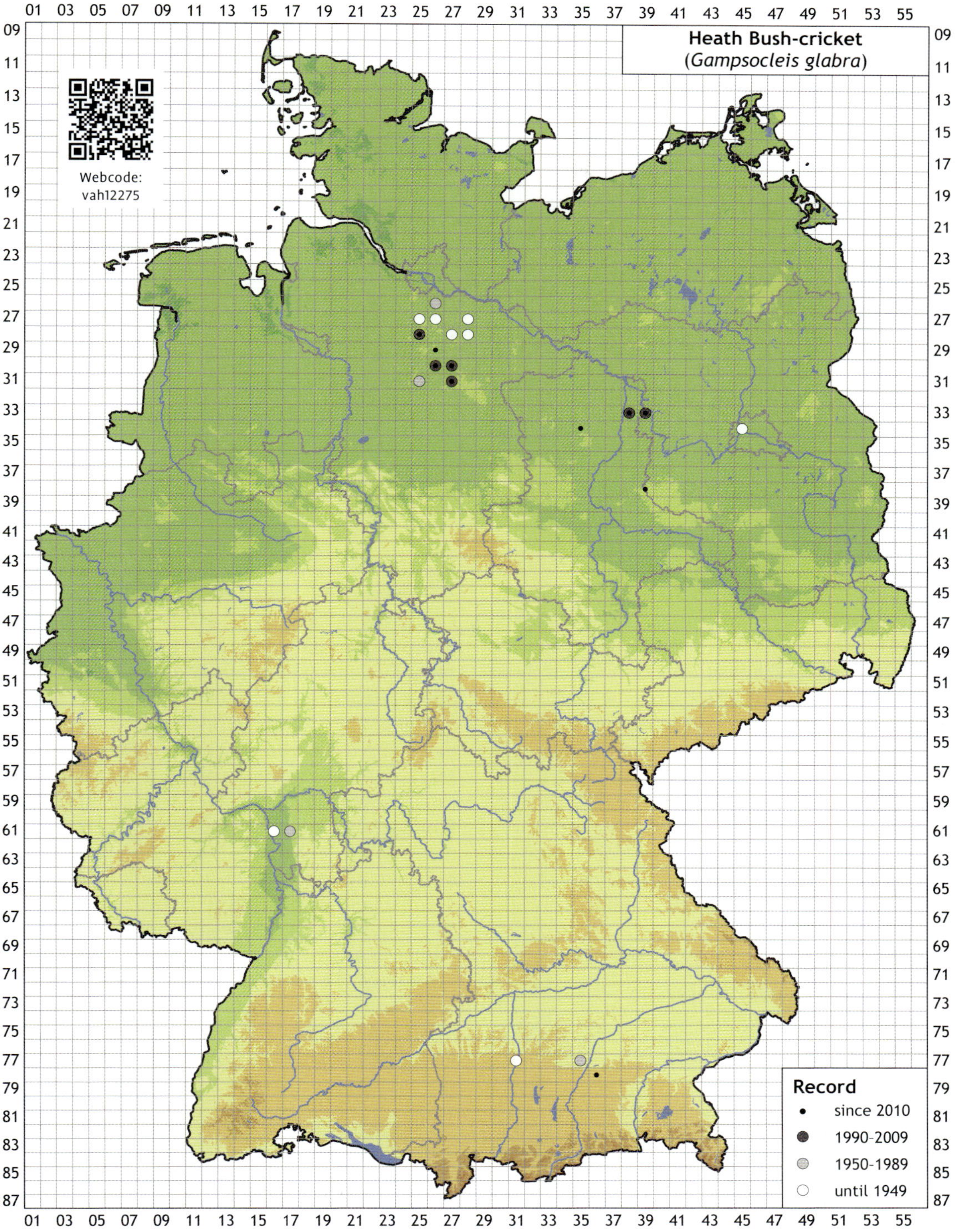
Heath Bush-cricket
(Gampsocleis glabra)
Webcode:
vah12275
Record
since 2010
1990-2009
1950-1989
until 1949

Female Common Grey Bush-cricket (*Platycleis albopunctata*). (Photo credit: Thomas Fartmann)

Sunlit, heterogeneous habitats rich in bare soil, such as this calcareous grassland, are preferred by the Common Grey Bush-cricket (Baden-Württemberg: Upper Donau Valley). (Photo credit: Dominik Poniatowski)

Common Grey Bush-cricket – *Platycleis albopunctata* (Goeze, 1778)

Distribution Within Germany, the Common Grey Bush-cricket mainly occurs in regions that are characterised by warm summers. In line with this, the STI of 15.2 °C is higher than average. The distribution centres are the Northeast German Lowland and the Southwestern Uplands, but the species also occurs at lower elevations of the Western and Eastern Uplands. However, due to climate change, the species is currently expanding its range, particularly into the oceanic northwest of Germany.

Habitat The xero-thermophilic Common Grey Bush-cricket inhabits dry and warm open habitats that exhibit heterogeneous swards. Characteristic habitats are dry and semi-dry grasslands, heathlands or vineyards and also ruderal habitats such as brownfields, gravel pits, quarries or railway systems. The eggs are relatively drought resistant and preferentially deposited near the ground at sparsely vegetated patches. Young nymphs also prefer these open microhabitats. By contrast, older nymphs and adults additionally need patches of more closed and taller vegetation. Such structures serve as shelter against predators and during periods of high temperatures on hot summer days.

Population trend and threats In former times, many habitats of the Common Grey Bush-cricket in Germany were lost due to abandonment of traditional land use and subsequent succession or afforestation. However, since the species has expanded its range in response to global warming, it is considered of least concern. Although the Common Grey Bush-cricket is capable of flight, a faster range expansion is constrained by a limited availability and strong isolation of suitable habitat patches.

WOLFGANG WRANIK

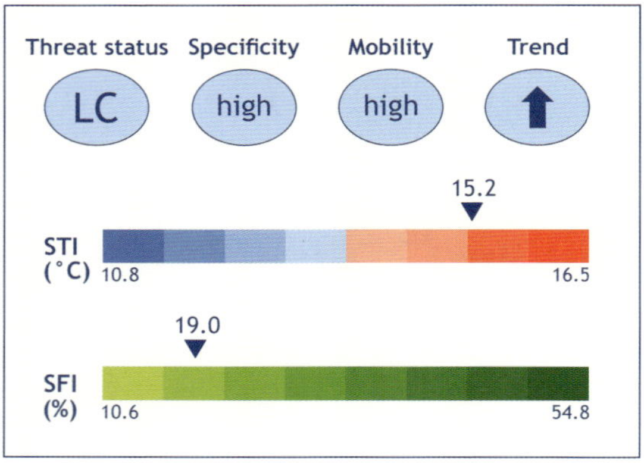

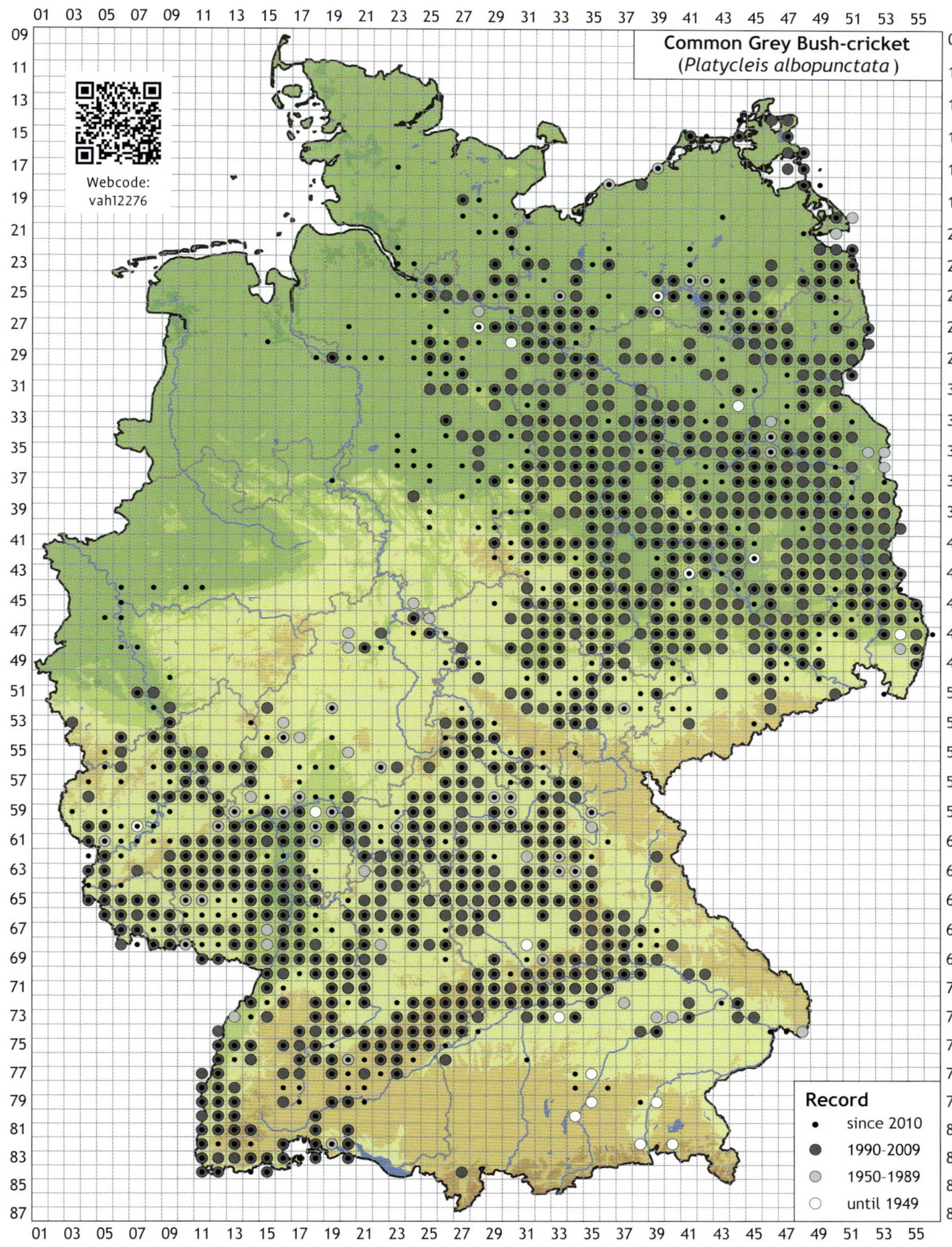
Common Grey Bush-cricket
(Platycleis albopunctata)
Webcode:
vah12276
Record
since 2010
1990-2009
1950-1989
until 1949

The Brown-spotted Bush-cricket is wide-spread in the semi-natural grasslands of this cattle pasture (Baden-Württemberg: Kaiserstuhl).
(Photo credit: Dominik Poniatowski)

Female Brown-spotted Bush-cricket (*Tessellana tessellata*).
(Photo credit: Dominik Poniatowski)

Brown-spotted Bush-cricket – *Tessellana tessellata* (Charpentier, 1825)

Distribution As a western Mediterranean faunal element, the Brown-spotted Bush-cricket reaches the northeastern edge of its distribution in Central Europe. Accordingly, it is rare in Germany. All of the few known records are located in the Upper Rhine Valley, which is characterised by very warm summers (high STI of 16.4 °C). Many of these populations are highly isolated. The largest populations are currently located in and around Freiburg (e. g. the airfields Bremgarten and Freiburg, Müllheim Military Training Area and Schelingen cattle pasture). Further populations, some with a large number of individuals, have recently been discovered in northern Baden and the neighbouring Rhineland-Palatinate.

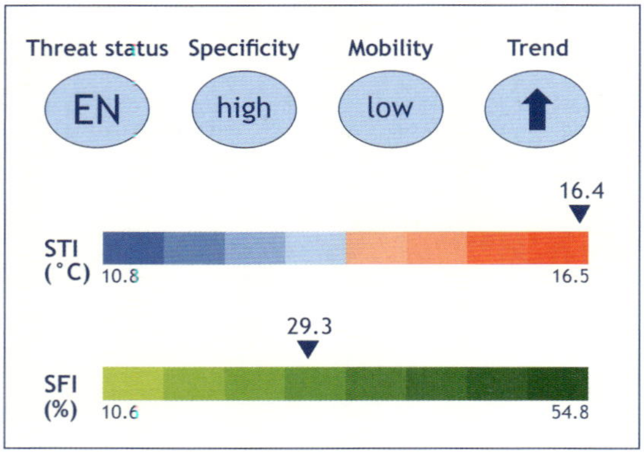

Habitat In Germany, the species prefers dry and warm habitats that are characterised by a largely closed and, at best, medium-high sward. Tall-growing grasses and herbs may be present sporadically. Open bare ground is also essential in the habitats at the Upper Rhine Valley. However, the proportions are often low and fluctuate between 5 and 10%. Typical habitats include fallow arable fields, ruderal vegetation and semi-natural grasslands.

Population trend and threats Due to its high thermal requirements, it can be assumed that the Brown-spotted Bush-cricket will benefit from climate change in Germany. Numerous new observations far away from known populations indicate a positive population development. However, many of the species' habitats are not protected by law. These include, for example, fallow arable fields and ruderal vegetation. The greatest threat to the Brown-spotted Bush-cricket is therefore habitat loss due to ploughing and urbanization. However, the absence of land use over longer time periods is also problematic since the species avoids uniformly tall and dense vegetation.

DOMINIK PONIATOWSKI

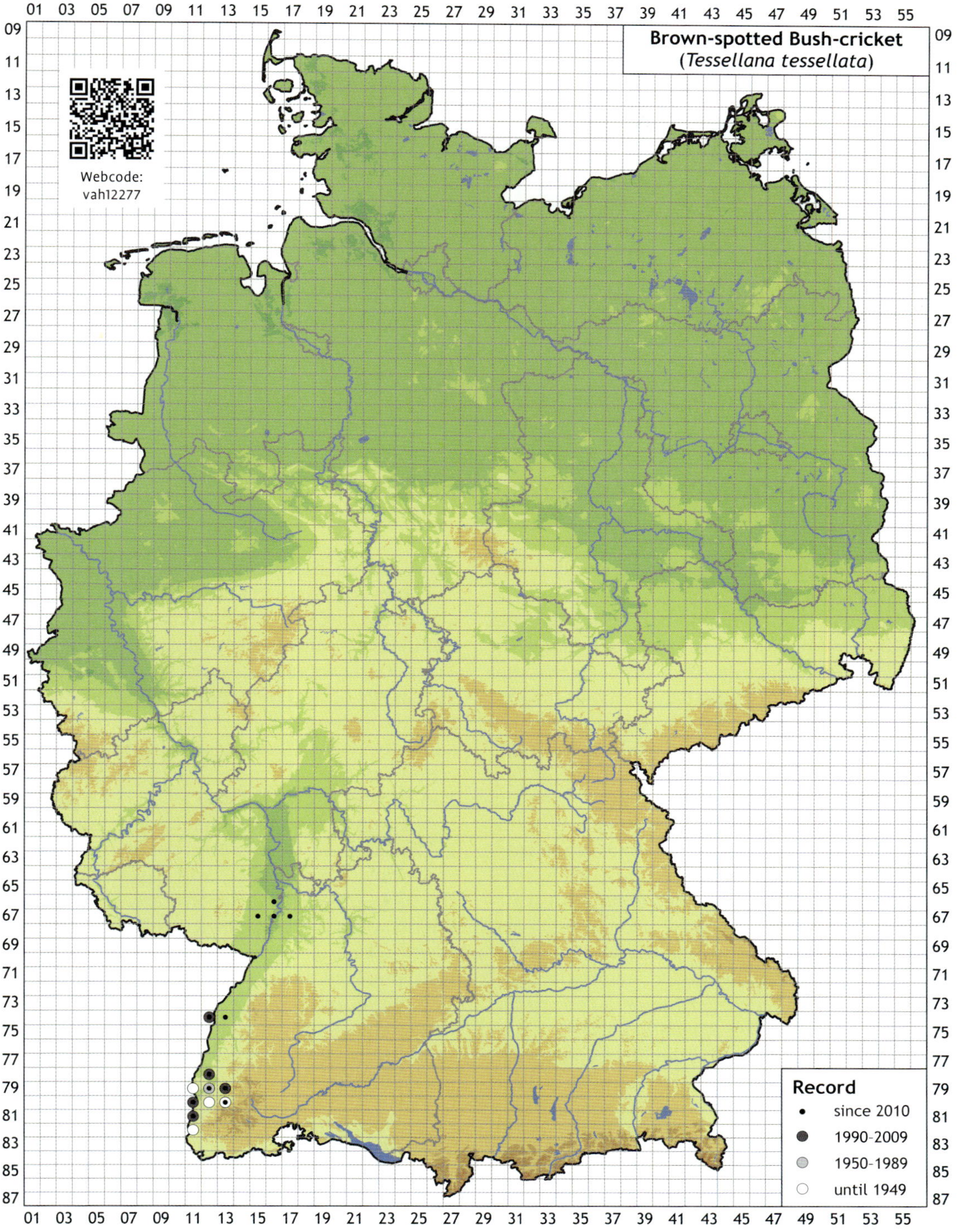
Brown-spotted Bush-cricket
(*Tessellana tessellata*)
Webcode:
vah12277
Record
since 2010
1990-2009
1950-1989
until 1949

Heterogeneous dry tussock grasslands on sandy, calcareous soils were the last habitat of the Steppe Bush-cricket in Germany (Brandenburg: Uckermark). (Photo credit: Thomas Fartmann)

Female Steppe Bush-cricket (*Montana montana*). (Photo credit: Petr Kočárek)

Steppe Bush-cricket – *Montana montana* (Kollar, 1833)

Distribution The few historical records of the Steppe Bush-cricket in Germany originate from the Northeast German Lowland, more precisely from the most continental parts of the federal states of Berlin and Brandenburg. After 60 years without any observation, three small and isolated subpopulations of the species were rediscovered in 1994 in the Gartzer Silberberge Nature Reserve in the Uckermark (NE Brandenburg). Here, the Steppe Bush-cricket was recorded until 2000.

Habitat In the Gartzer Silberberge, subcontinental, dry grasslands on calcareous sands were colonised by the Steppe Bush-cricket. They were sunlit and had a heterogeneous vegetation structure, including patches of bare ground. Bare soil enhanced the warming of the habitats on sunny days and was additionally used for oviposition. Tussock grasses (*Corynephorus canescens*, *Fectuca brevipila*, *F. polesica*, *F. psammophila*, *Koeleria glauca* and *Stipa borysthenica*) and the Field Wormwood (*Artemisia campestris*), which is already heavily branched near the ground, were essential structures within the habitat. The males sang their songs on their bases, and the plants served as a shelter against vertebrate predators.

Population trend and threats When the Steppe Bush-cricket was rediscovered in Germany, the dry grasslands of the Gartzer Silberberge had already lacked any land use for a long time. Accordingly subcontinental grasslands, rich in bare ground, covered only small patches. Until 2000, their area had further decreased due to succession, which had favoured litter accumulation, the invasion of False Oat-grass (*Arrhenatherum elatius*) and the spreading of Scots Pine (*Pinus sylvestris*) in two of the three previously colonised areas. In the mid-2000s, when the species was probably already extinct because of successional processes, pine trees were removed and cattle grazing was introduced as a conservation measure. Today, the extent of heterogeneous tussock grasslands on calcareous sands in the nature reserve is as large as it was in the mid-1990s.

THOMAS FARTMANN

Threat status	Specificity	Mobility	Trend
EX	high	low	† 2000

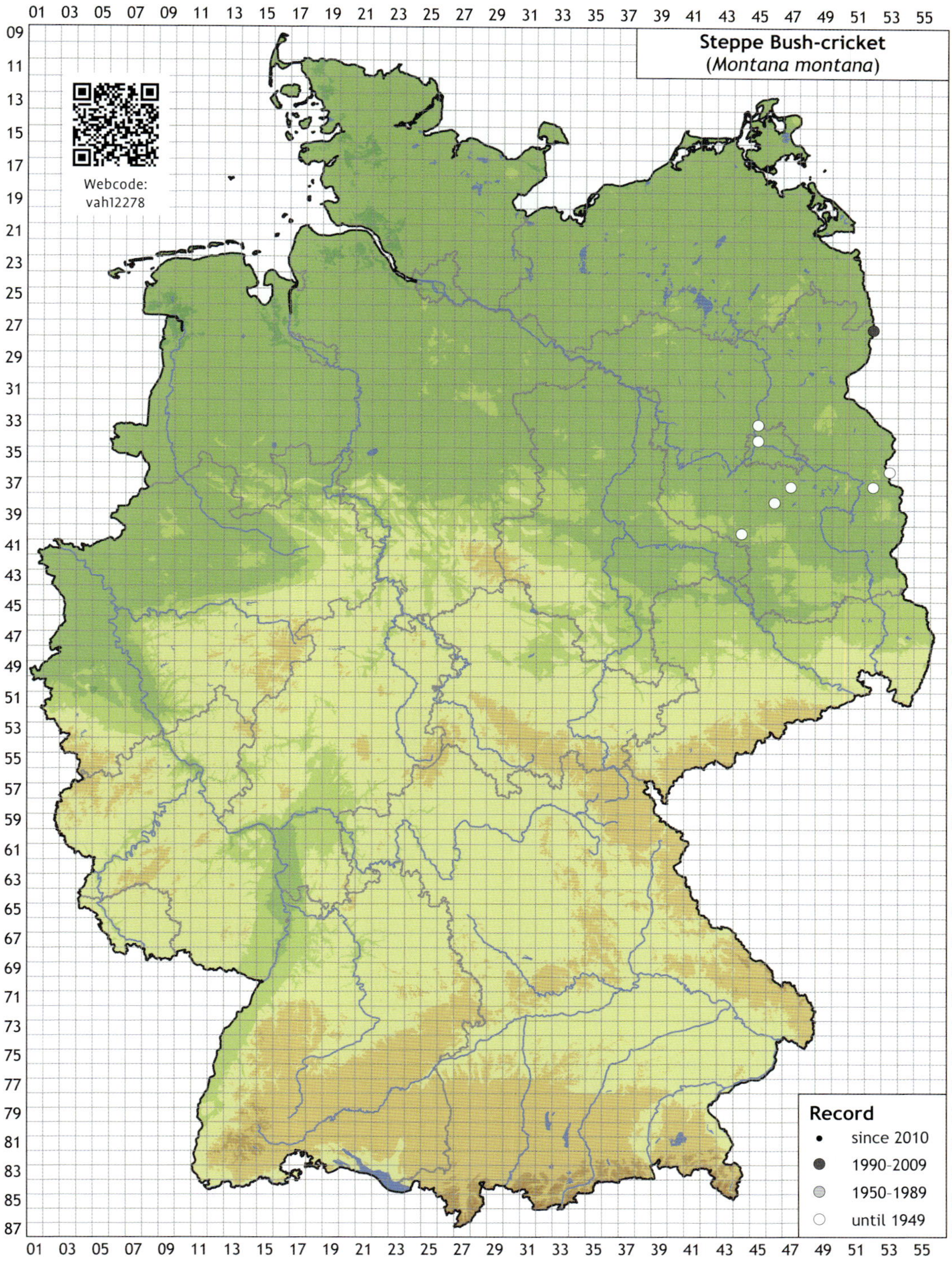
Steppe Bush-cricket
(Montana montana)
Webcode:
vah12278
Record
since 2010
1990-2009
1950-1989
until 1949

Heterogeneous montane heathlands are typical habitats of the Bog Bush-cricket (North Rhine-Westphalia: Rothaar Mountains).
(Photo credit: Dominik Poniatowski)

Female Bog Bush-cricket (*Metrioptera brachyptera*).
(Photo credit: Dominik Poniatowski)

Bog Bush-cricket – *Metrioptera brachyptera* (Linnaeus, 1761)

Distribution The Bog Bush-cricket has a scattered distribution in Germany. Due to the relatively high moisture requirements of the eggs, its occurrence is predominantly restricted to regions with high precipitation. These include the Alps, the foothills of the Alps, the low mountain ranges and the oceanic Northwest German Lowland. In dry parts of the country, such as the Northeast German Lowland and the Upper Rhine Valley, the species is rare and limited to habitats with high soil moisture.
Habitat The Bog Bush-cricket prefers nutrient-poor habitats without land use or that are only infrequently managed. The habitat spectrum ranges from wet heaths and stands with Purple Moor Grass (*Molinia caerulea*) in degraded bogs to *Nardus* grasslands, litter meadows and dwarf shrub heaths to semi-dry, calcareous grasslands. However, the latter are only suitable as a habitat if a moist microclimate ensures successful egg development – for example due to a north-facing slope.
Population trend and threats Although the Bog Bush-cricket mainly inhabits protected habitats in Germany, a decline in populations can currently be assumed. The abandonment of land use on many marginal sites is particularly problematic. As a result, valuable open habitats such as heaths and semi-natural grasslands become overgrown by shrubs and trees. Accordingly, they gradually lose their suitability as a habitat due to excessive shading. Climate change also poses a threat as the eggs are sensitive to drought and have to survive in the soil for at least two years before the nymphs hatch. Therefore, frequent droughts have to be considered as an additional threat to the species – especially on dry sites such as calcareous grasslands and sandy heaths.

DOMINIK PONIATOWSKI

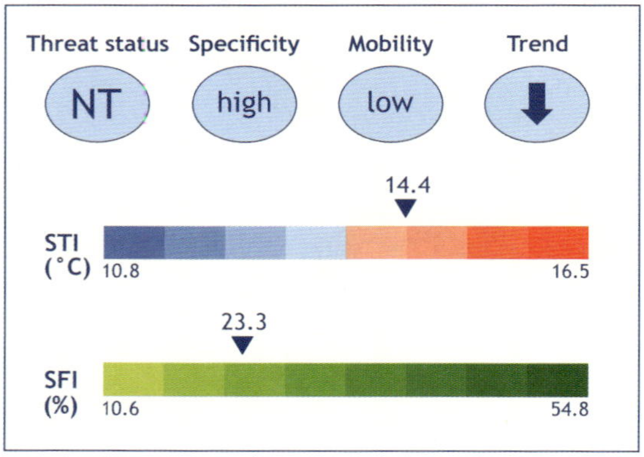

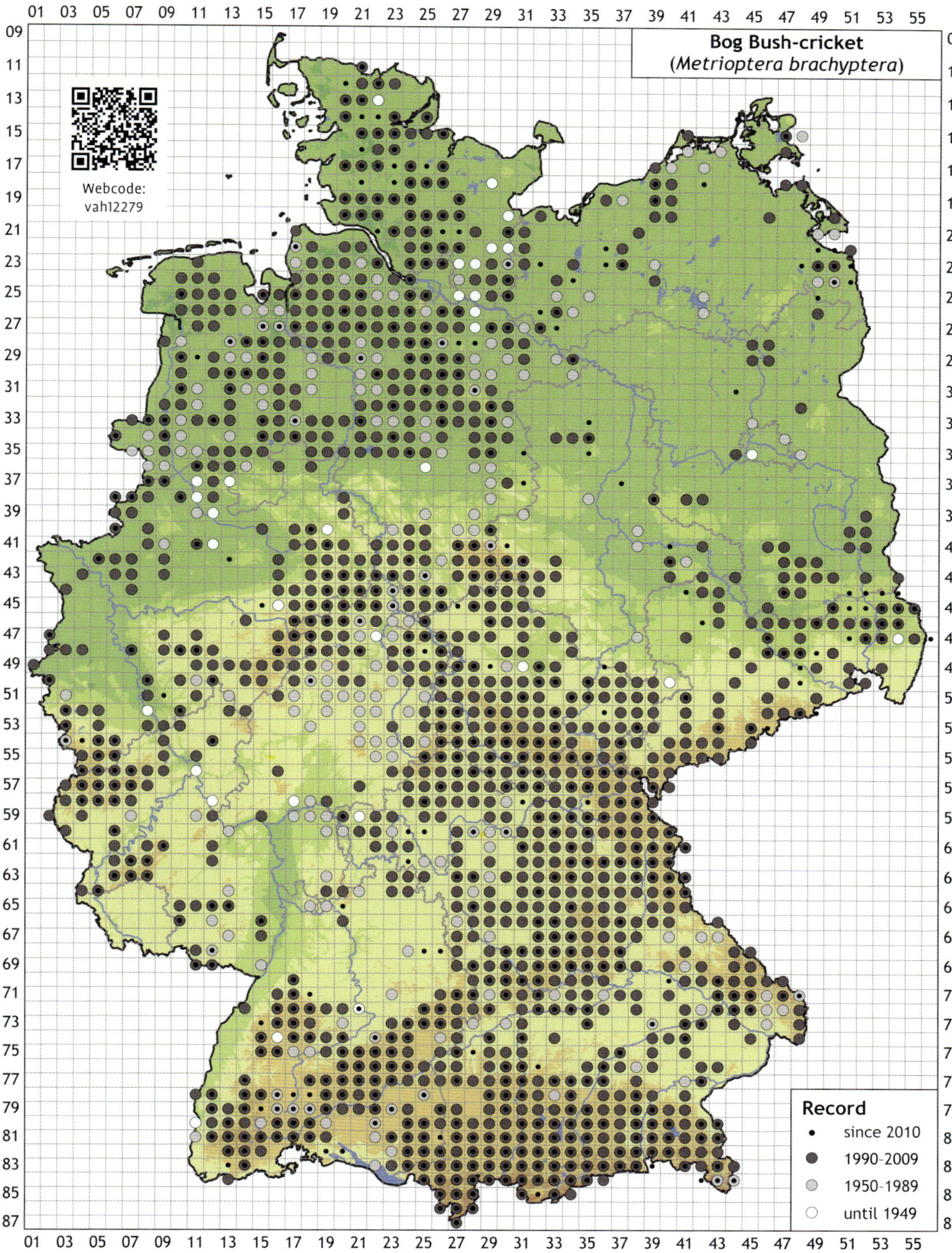
Bog Bush-cricket
(Metrioptera brachyptera)
Webcode:
vah12279
Record
since 2010
1990-2009
1950-1989
until 1949
01 03 05 07 09 11 13 15 17 19 21 23 25 27 29 31 33 35 37 39 41 43 45 47 49 51 53 55
09 11 13 15 17 19 21 23 25 27 29 31 33 35 37 39 41 43 45 47 49 51 53 55 57 59 61 63 65 67 69 71 73 75 77 79 81 83 85 87

Subcontinental, sandy grasslands with tall swards are characteristic habitats of the Two-coloured Bush-cricket (Brandenburg: Uckermark).
(Photo credit: Thomas Fartmann)

Female Two-coloured Bush-cricket (*Bicolorana bicolor*).
(Photo credit: Thomas Fartmann)

Two-coloured Bush-cricket – *Bicolorana bicolor* (Philippi, 1830)

Distribution The Two-coloured Bush-cricket is highly xero-thermophilic and hence, in Germany, restricted to those biogeographic regions that are characterised by warm summers but also a frequent occurrence of dry habitats. Accordingly, the strongholds are the Southwestern Uplands, the west of the Western Uplands, the Thuringian Basin and the warmest parts of the Northeast German Lowland. The highest permanent populations are located at 1,000 m a.s.l. in the southern Black Forest. The strong dependence on warm summers is reflected by an above average STI of 15.0 °C.

Habitat The Two-coloured Bush-cricket is a specialised inhabitant of dry and warm open habitats. The eggs are deposited in plant stems and are highly drought resistant. Accordingly, even during a facultative biannial egg development, they are not harmed by summer drought. Characteristic habitats are calcareous, sandy and steppe grasslands as well as nutrient-poor mesic grasslands or thermophilic fringes. The vegetation cover is usually high, and some taller plants, such as tall forbs or high-growing grasses, are always present.

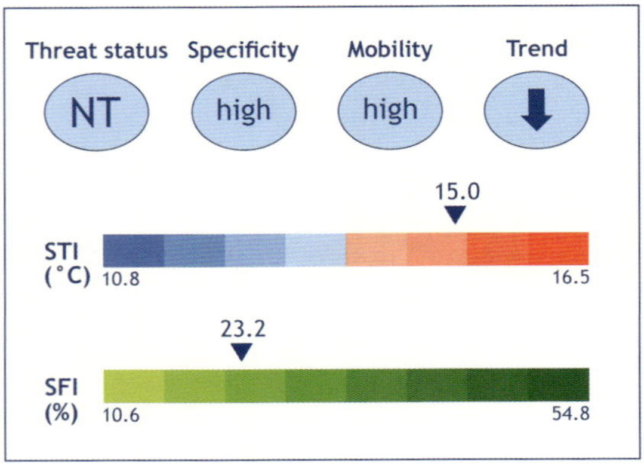

Population trend and threats Recently, due to climate change, the Two-coloured Bush-cricket has locally expanded its distribution at its German range margin. Especially long-winged individuals capable of flight have very likely contributed significantly to this expansion. By contrast, land-use change has caused population declines in the distribution centres. Land-use intensification but also long-term abandonment on marginal sites and subsequent shrub encroachment have been responsible for these losses. As a result, in the short term, a moderate decline has been assumed and the species is now considered near threatened.

CARSTEN RENKER

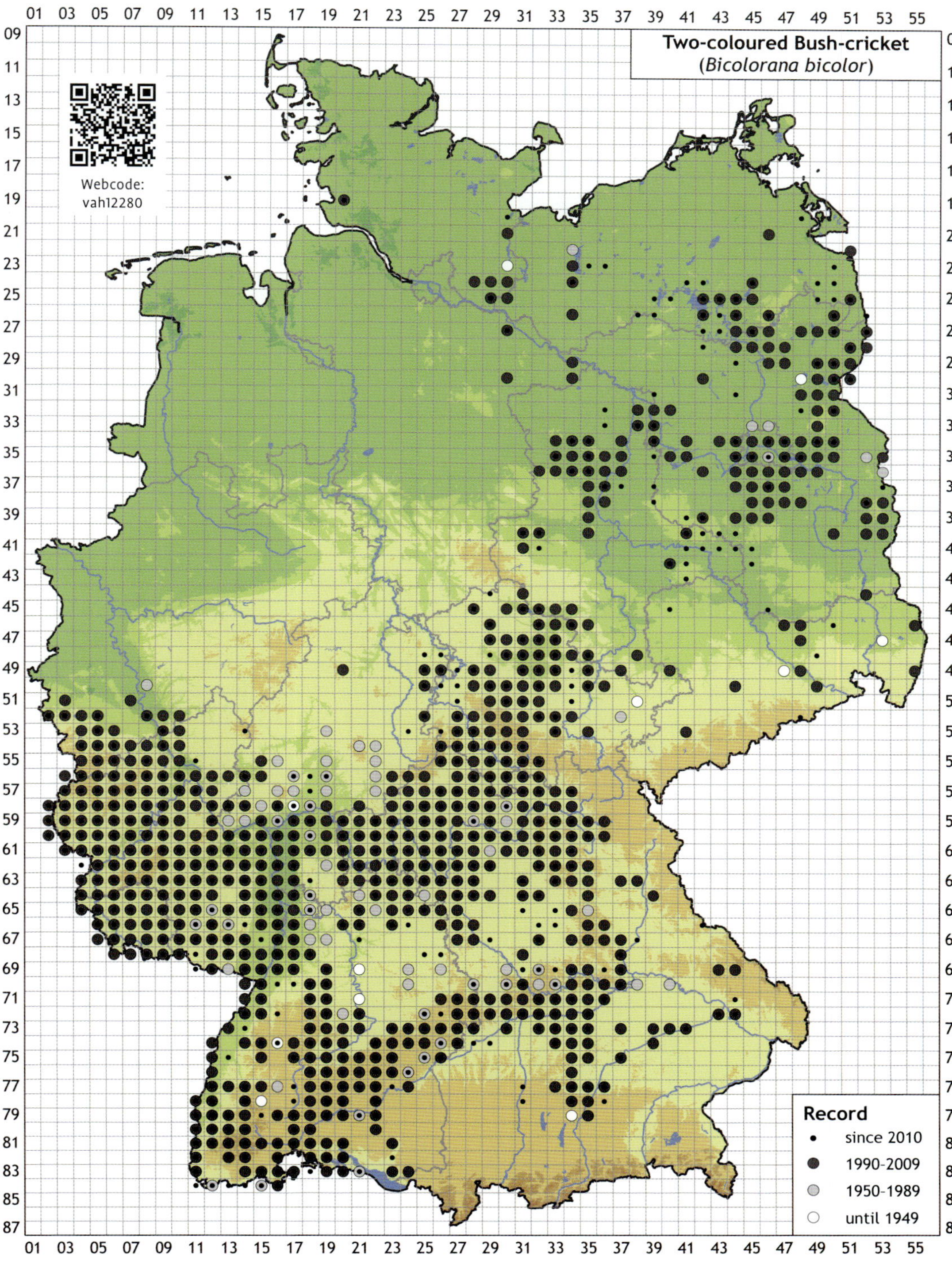
Two-coloured Bush-cricket
(Bicolorana bicolor)
Webcode:
vah12280
Record
since 2010
1990-2009
1950-1989
until 1949

Roesel's Bush-cricket prefers mesic to moist grasslands (Hesse: Diemel Valley). (Photo credit: Dominik Poniatowski)

Male of Roesel's Bush-cricket (*Roeseliana roeselii*). (Photo credit: Dominik Poniatowski)

Roesel's Bush-cricket – *Roeseliana roeselii* (Hagenbach, 1822)

Distribution Roesel's Bush-cricket was already widespread in Germany 20 to 30 years ago. However, at that time, there were still larger distribution gaps in the Northwest German Lowland and at higher elevations of the low mountain ranges and within the Alps. Due to climate change, Roesel's Bush-cricket has managed to largely close these gaps. The long-winged individuals of this predominantly short-winged species have particularly contributed to this development. It is now the most common bush-cricket in Germany.

Habitat Roesel's Bush-cricket is a habitat generalist. The habitat spectrum comprises a large variety of different types of grasslands and moisture levels. Only very wet and very dry locations are avoided. Comparatively high and dense vegetation is fundamentally important for an occurrence. Optimal habitats are mesic to moist meadows and pastures that are temporarily subject to no or only very low land-use intensity. In intensively used agricultural landscapes, linear structures such as field margins and roadside verges are also colonised if their vegetation is rarely mown. In semi-dry grasslands, the occurrence of Roesel's Bush-cricket is limited to tall-growing fringes.

Population trend and threats Due to its widespread distribution in Germany, range expansions of Roesel's Bush-cricket can only be observed very locally. The populations are considered stable on a national level. Roesel's Bush-cricket is one of the few grasshopper species in Germany that is able to colonise intensively used agricultural landscapes – given that grassland-like habitats are at least locally available. In the current red list, the species is classified as of least concern.

DOMINIK PONIATOWSKI

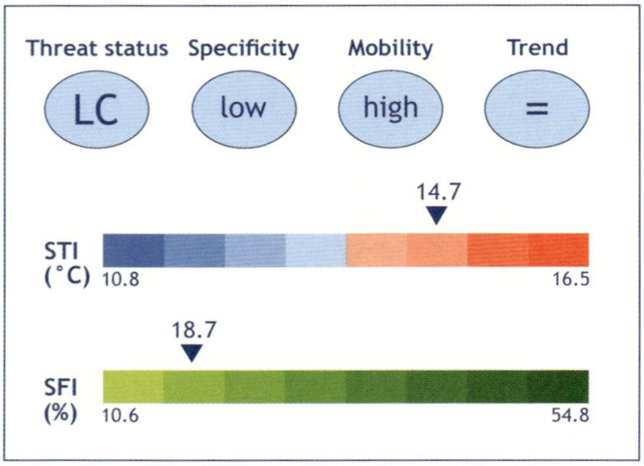

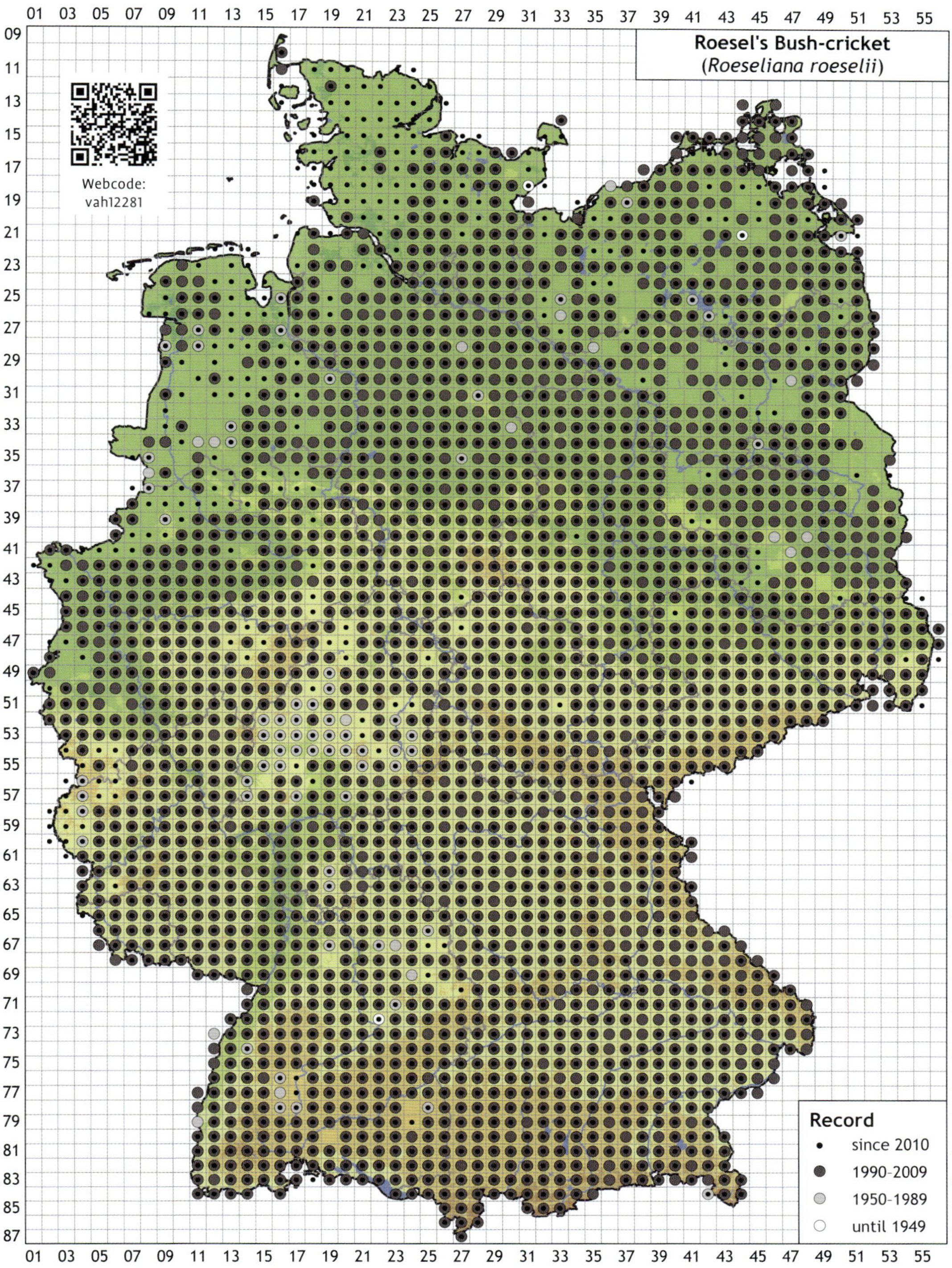
Roesel's Bush-cricket
(Roeseliana roeselii)
Webcode:
vah12281
Record
since 2010
1990-2009
1950-1989
until 1949

Sunlit vegetation mosaics on gravel with tall forbs and some woody plants are typical habitats of the Alpine Dark Bush-cricket in the Bavarian Alps (Bavaria: Ammer Mountains). (Photo credit: Andreas Zehm)

Male Alpine Dark Bush-cricket (*Pholidoptera aptera*). (Photo credit: Thomas Fartmann)

Alpine Dark Bush-cricket – *Pholidoptera aptera* (Fabricius, 1793)

Distribution In Germany, the Alpine Dark Bush-cricket is characteristic of the Bavarian Alps. Approximately 90 % of all observations are from this biogeographic region; the species is particularly widespread in elevations of 600–1,800 m a.s.l. Further north, there are also some populations at lower elevations in the Bavarian Alpine Foreland, especially between the rivers Lech and Isar up until Lake Starnberg, in the Bavarian Forest near Passau and in the Saxon Elbe Sandstone Mountains. Due to its strong affinity towards the Alps, the STI of 12.2 °C is one of the lowest among the native grasshopper species.

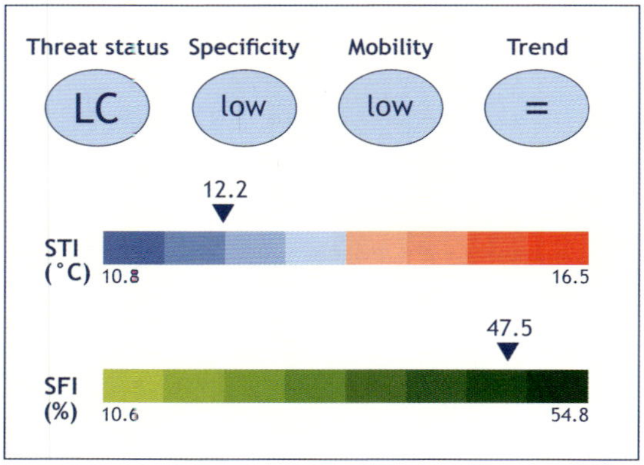

Habitat The semi-open and open habitats that are colonised by the Alpine Dark Bush-cricket are quite diverse. However, common features are a high air humidity, strong solar irradiation, a certain amount of forbs/shrubs and no or at best a low intensity of land use (e. g. grazing). As a result, the SFI of 47.5 % is one of the highest among all German grasshopper species. Characteristic habitats of the species are clear-cuts, forest edges, glades or light slope forests. Beyond woodlands, stands of Bracken, Butterbur, Dwarf Mountain Pine, Raspberry and *Rhododendron,* semi-natural grassland fallows or overgrown screes are colonised. With increasing elevation (> 1,300 m a.s.l.), the occurrence of the Alpine Dark Bush-cricket is more or less restricted to south-facing slopes rich in gravel and rocks that warm up well on sunny days. In the Alpine Foreland, moreover, abandoned wet grasslands rich in tall forbs function as a habitat.

Population trend and threats The German populations of the Alpine Dark Bush-cricket are considered relatively stable both historically and today. Local extinction seems to be remotely possible solely for the outposts north of the Alps, for example if the habitats become overgrown by woody plants.

ANDREAS ZEHM

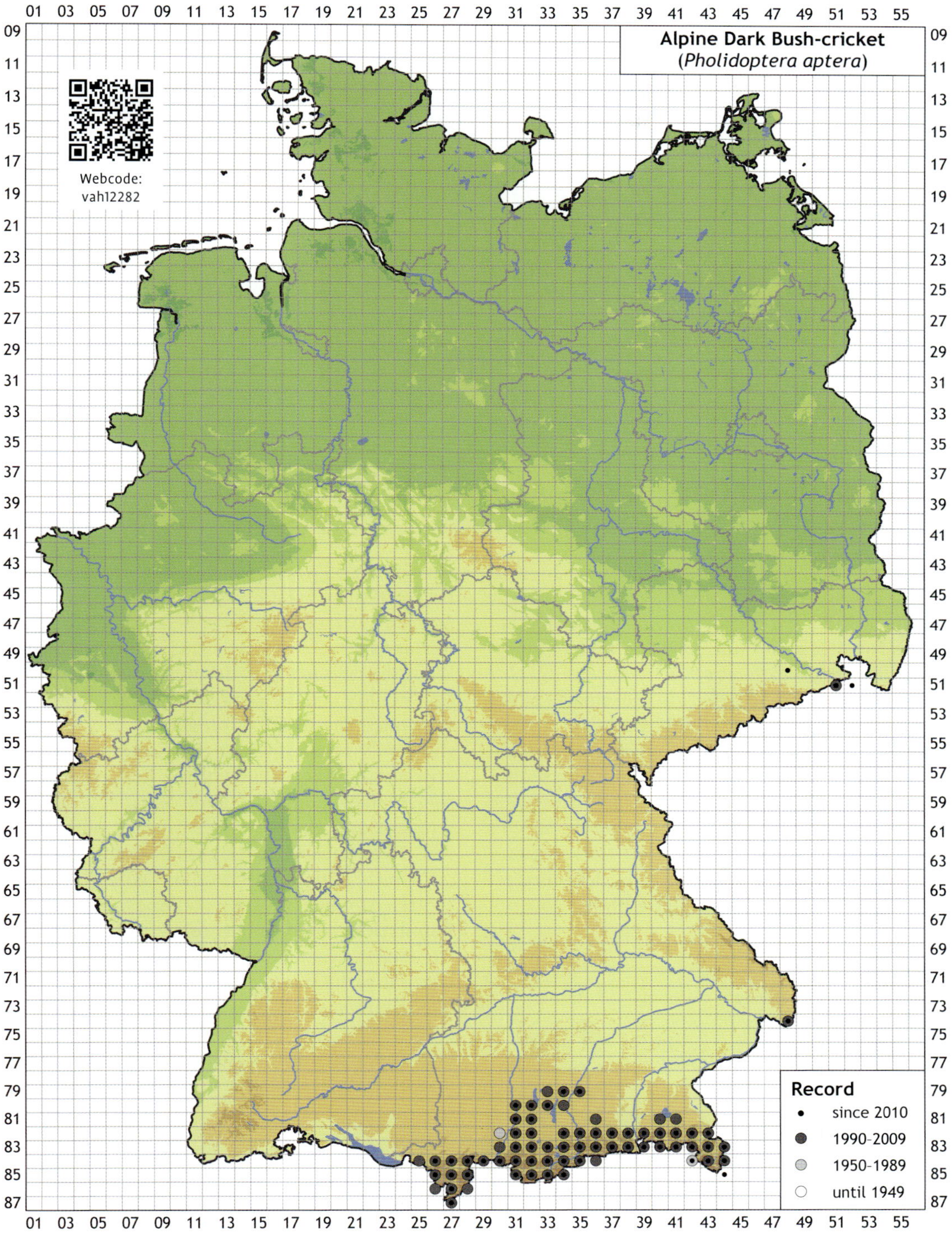
Alpine Dark Bush-cricket
(*Pholidoptera aptera*)
Webcode:
vah12282
Record
since 2010
1990-2009
1950-1989
until 1949
01 03 05 07 09 11 13 15 17 19 21 23 25 27 29 31 33 35 37 39 41 43 45 47 49 51 53 55
09 11 13 15 17 19 21 23 25 27 29 31 33 35 37 39 41 43 45 47 49 51 53 55 57 59 61 63 65 67 69 71 73 75 77 79 81 83 85 87

Woodland clearings are also an important habitat for the Common Dark Bush-cricket (Lower Saxony: Weser Uplands). (Photo credit: Marcel Kettermann)

Male Common Dark Bush-cricket (*Pholidoptera griseoaptera*). (Photo credit: Thomas Fartmann)

Common Dark Bush-cricket – *Pholidoptera griseoaptera* (De Geer, 1773)

Distribution The Common Dark Bush-cricket is widespread throughout Germany. It is only absent in the most continental sandy landscapes of the Northeast German Lowland and in biogeographical regions that are characterised by cool summers. Among the latter are those parts of the Northwest German Lowland that exhibit the most oceanic climate as well as the highest elevations of the uplands (Rothaar Mountains, Harz Mountains, Ore Mountains, Black Forest, Bavarian Forest) and Alps. However, due to climate change, the species is now even expanding its range into these regions.
Habitat The Common Dark Bush-cricket prefers semi-open habitats that exhibit a higher cover of shrubs as well as closed and tall herbaceous vegetation. Typical habitats are abandoned grasslands, forest clearings and edges, hedges, shrub fringes or stands of tall forbs as well as gardens and parks. Since the eggs are very drought sensitive, oviposition mainly occurs in moist soil and plants. Consequently, the distribution gaps of the species in the sandy landscapes of the Northeast German Lowland can be explained by the low precipitation, high evapotranspiration due to high summer temperatures and low water capacity of the sandy soils. Even on more cohesive soils within this biogeographical region, the Common Dark Bush-cricket is often common only in woodlands.
Population trend and threats The short-term trend of the Common Dark Bush-cricket is stable. Currently, the species is even expanding its range due to climate change. Consequently, it is considered of least concern. However, on sandy soils in very dry and warm regions, climate change may lead to declines since the eggs are hygrophilic.

WOLFGANG WRANIK

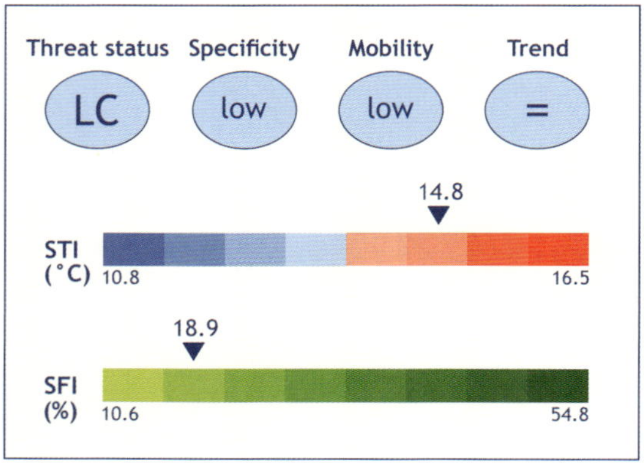

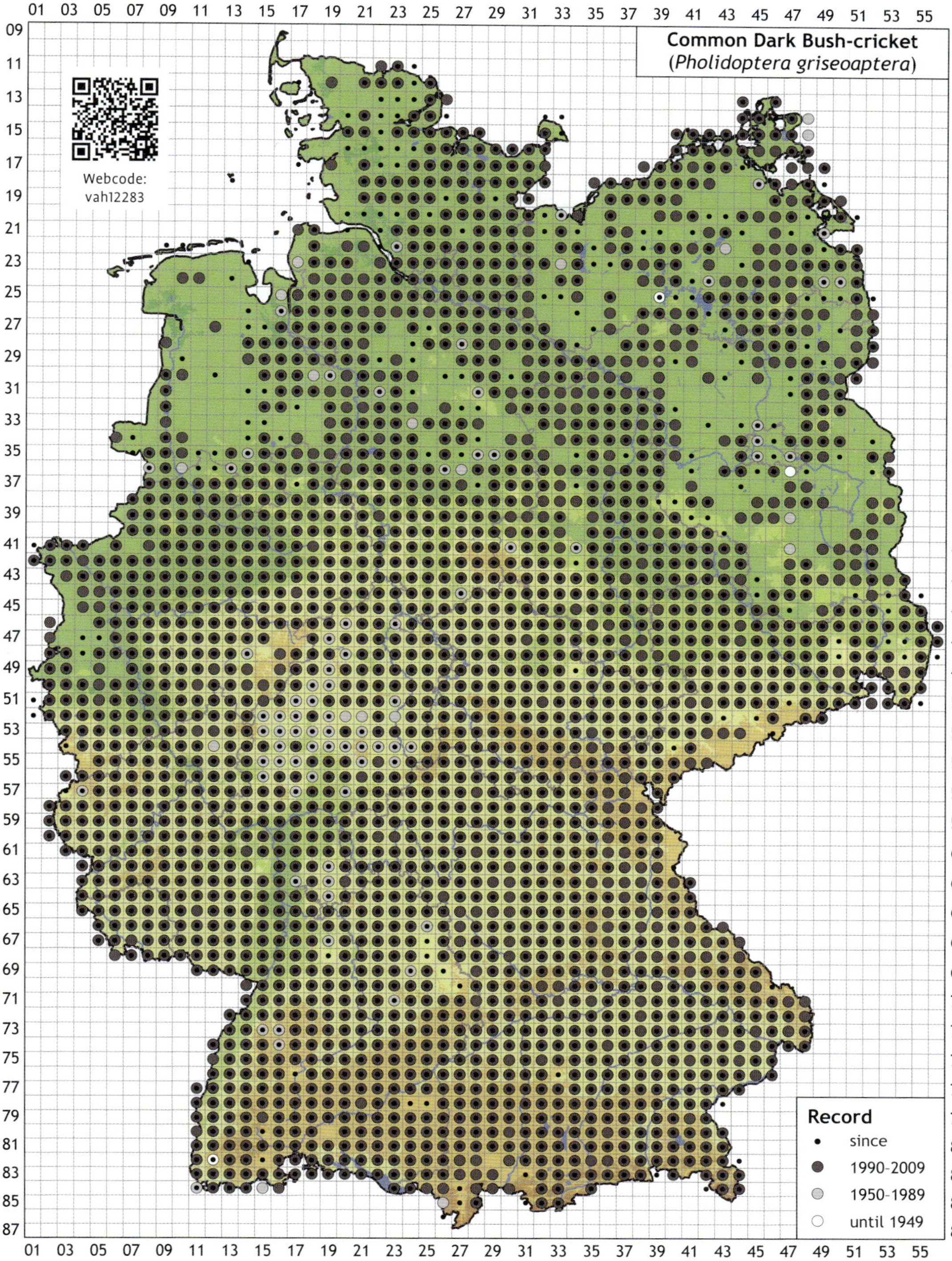

Common Dark Bush-cricket
(*Pholidoptera griseoaptera*)
Webcode:
vah12283
Record
since
1990-2009
1950-1989
until 1949

Sunlit, xerothermic and steep slopes exhibiting some scrub are preferred habitats of the Western Saddle Bush-cricket (Rhineland-Palatinate: Saar Nahe Uplands). (Photo credit: Marcel Kettermann)

Male Western Saddle Bush-cricket (*Ephippiger diurnus*). (Photo credit: Marcel Kettermann)

Western Saddle Bush-cricket – *Ephippiger diurnus* Dufour, 1841

Distribution The Rhine river constitutes the eastern range margin of the Western Saddle Bush-cricket in Germany. Almost all populations are located in the wine regions of Rhineland-Palatinate on the west bank of the Rhine. Correspondingly, the STI of 15.5 °C is very high. Only in the Middle Rhine Valley do there remain some populations on the east bank of the Rhine. The northernmost outposts are situated in the Lower Rhine Lowland at the border to the Netherlands.

Habitat The Western Saddle Bush-cricket prefers sun-exposed, dry and warm open habitats, mostly on steep slopes that exhibit some shrubs. Such conditions are often found in wine regions, where especially fallows between managed vineyards are inhabited. The main habitats are abandoned dry grasslands, orchards or vineyards that may contain some patches of bramble shrubberies or thermophilic fringe vegetation. Outside the wine regions, the species occurs in heterogeneous nutrient-poor grasslands and heathlands.

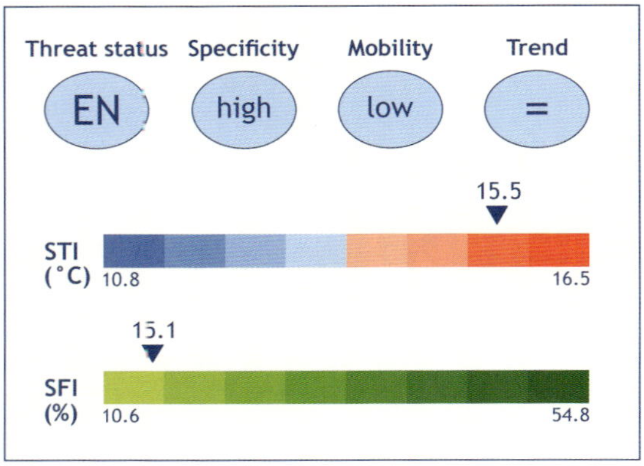

Population trend and threats In former times, the German populations of the Western Saddle Bush-cricket strongly declined. An important driver of decrease was land consolidation. This resulted in a loss of many habitat patches and, hence, the disappearance of many populations. Detrimental was also the intensification of viticulture through the application of insecticides and mechanisation of grape harvests. On marginal land, abandonment of land use in the vineyard slopes and the subsequent encroachment of shrubs deteriorated the habitats. Partly, former vineyard slopes have even been afforested. Currently, however, the remaining populations are considered stable.

MANFRED ALBAN PFEIFER

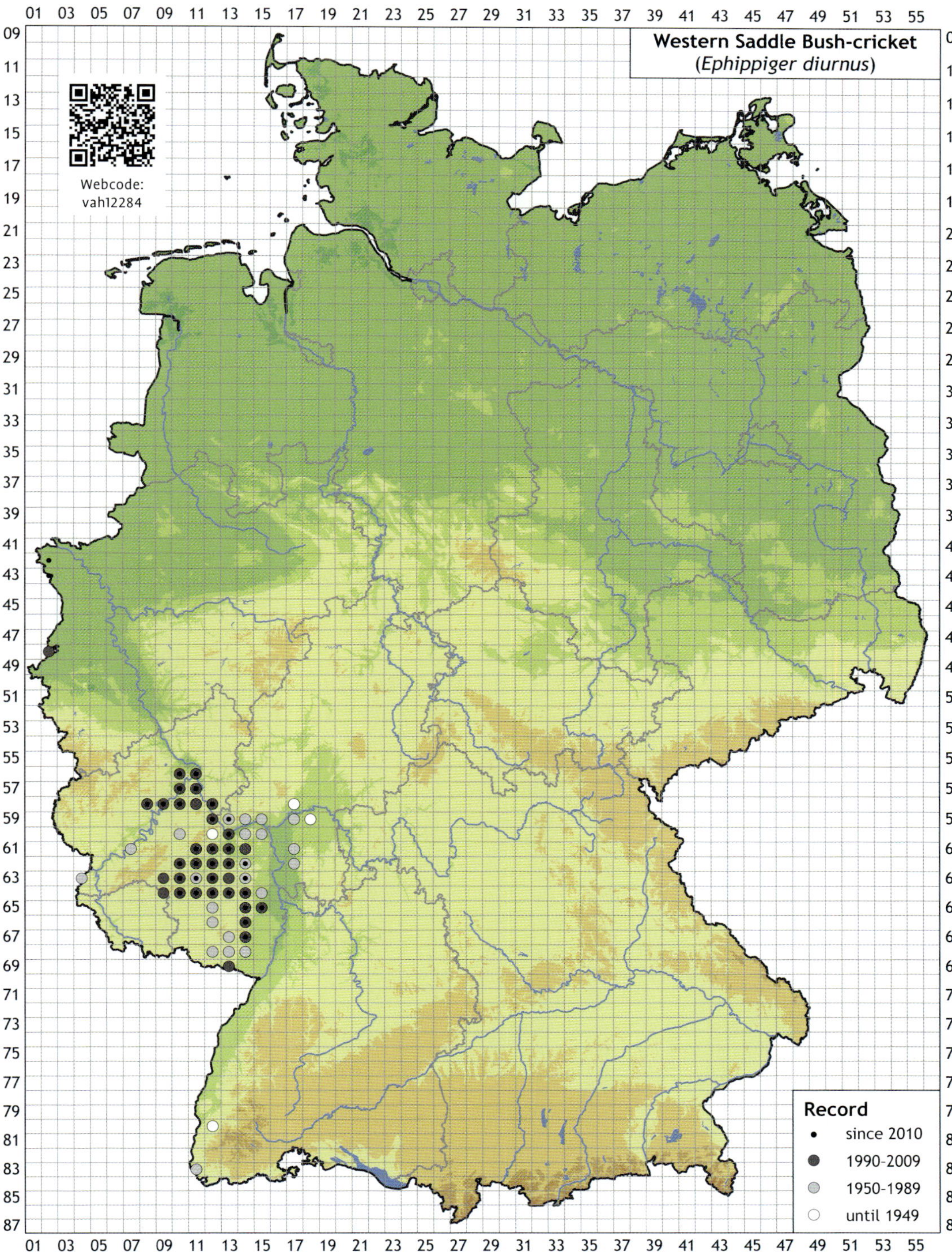
Western Saddle Bush-cricket
(Ephippiger diurnus)
Webcode:
vah12284
Record
since 2010
1990-2009
1950-1989
until 1949

Cave system with sinter formations that is inhabited by the Neglected Cave-cricket (Bavaria: Fichtel Mountains). (Photo credit: Jürgen Fischer)

Female Neglected Cave-cricket (*Troglophilus neglectus*). (Photo credit: Jürgen Fischer)

Neglected Cave-cricket – *Troglophilus neglectus*
Krauss, 1879

Distribution The main distribution area of the Neglected Cave-cricket reaches from Greece across the Balkans to Northeast Italy and Southeast Austria. Only a very few records from the low mountain ranges exist in Germany, namely in the Eifel Mountains (Mayener Grubenfeld), Fichtel Mountains (marble quarry near Sinatengrün), southern Franconian Jura and Elbe Sandstone Mountains. These mostly small and strongly isolated populations represent the northernmost records of the species in Europe. However, it is very likely that all populations north of the Alps originate from introduced or passively dispersed individuals.

Habitat The crepuscular and nocturnal Neglected Cave-cricket occurs in habitats with a year-round balanced microclimate (relative humidity > 80 %, temperature 5–10 °C). Typical habitats include bunkers, caves, shafts and tunnels. Between May and September, individuals can also be found outside these habitats in nearby forests, screes and spoil heaps. During the daytime, they hide under stones or in foliage.

Threat status	Specificity	Mobility	Trend
R	high	low	=

Population trend and threats Despite their small size and strong isolation, the German populations of the Neglected Cave-cricket are considered stable. The main threats are the destruction of the habitats or an impairment of the specific microclimate of the cave system, for example by quarrying.

CINJA SCHWARZ

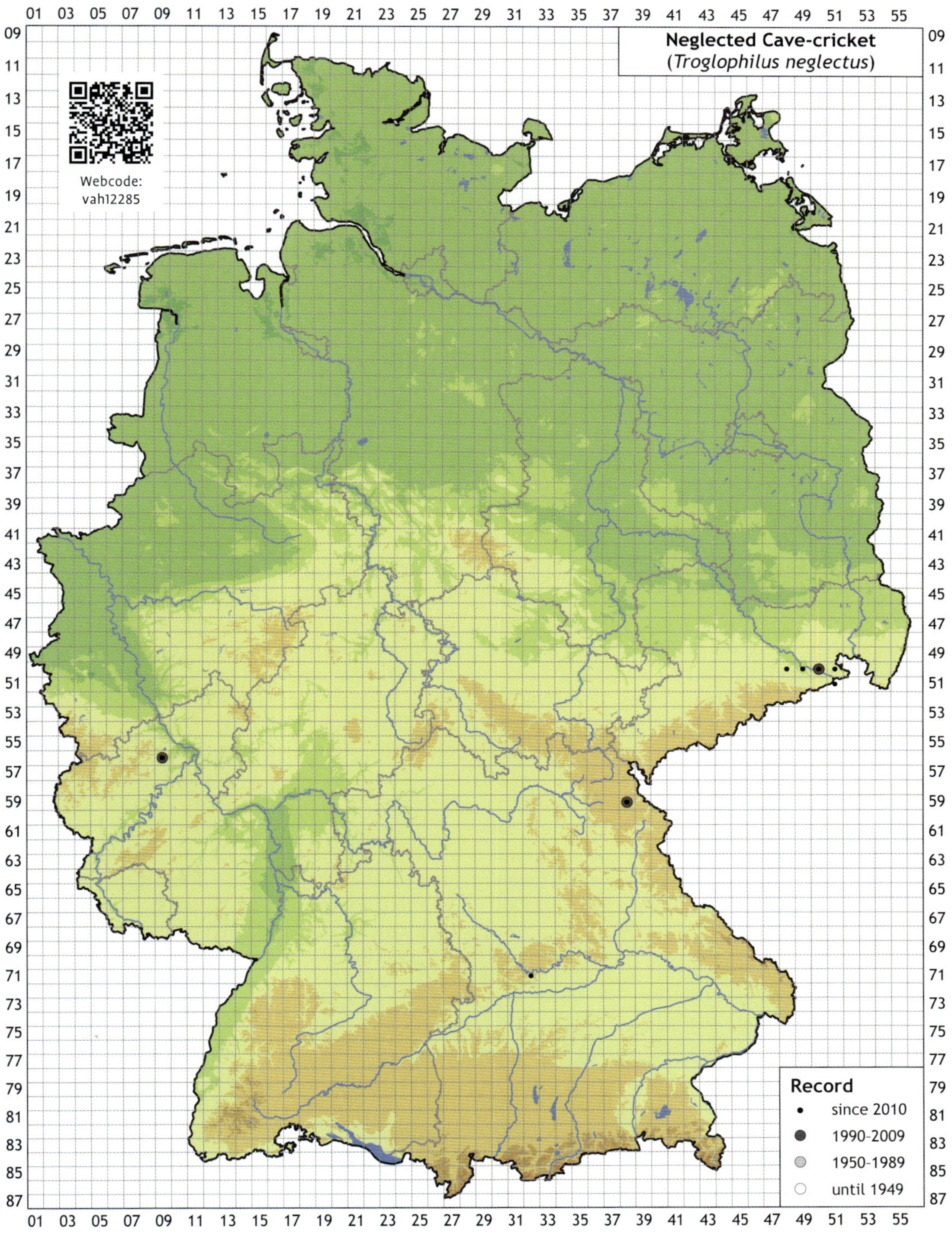
Neglected Cave-cricket
(Troglophilus neglectus)
Webcode:
vah12285
Record
since 2010
1990-2009
1950-1989
until 1949

Female European Mole Cricket (*Gryllotalpa gryllotalpa*). (Photo credit: Jürgen Fischer)

In this sand quarry, the European Mole Cricket colonises the moist, sparsely vegetated and sunny riparian zones of small water bodies (Lower Saxony: Ems-Hunte old moraine landscape). (Photo credit: Dominik Poniatowski)

European Mole Cricket – *Gryllotalpa gryllotalpa* (Linnaeus, 1758)

Distribution The European Mole Cricket has a scattered distribution across all seven major biogeographical regions in Germany. However, the strongholds are the eastern and southern parts of the country below 800 m a.s.l., which are characterised by sunny and warm summers. The weak dependence on a warm climate is reflected by an STI of 15.0 °C, which is slightly above average. Accordingly, the species has recently expanded its range in some regions as a response to climate change.

Habitat Due to the subterranean way of life, the European Mole Cricket needs loose soil substrates that allow for digging. Moreover, a sufficient humidity and warming of the soils are prerequisites for a successful development of the eggs and nymphs. Therefore, the species colonises sunny soils that are covered by sparse vegetation or at least short swards. To secure a warm microclimate, the animals may even crop all plants above the nests. Fens and edges of bogs are the natural habitats of the species. Today, however, gardens are often the main secondary habitats. Due to regular irrigation, their soils remain moist even during periods of summer drought. Further important habitats are the riparian zones of water bodies, for example in sand quarries, as well as wet grasslands that exhibit a low land use intensity.

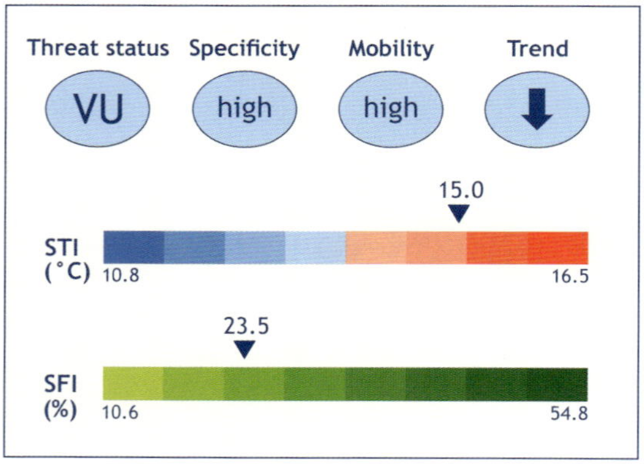

Population trend and threats The European Mole Cricket is considered vulnerable in Germany. The decline of the species during recent decades has mainly been driven by habitat loss due to drainage and intensification of agricultural management but also by natural succession after abandonment of land use. Moreover, in gardens, the species is still treated as a pest. All these losses have hardly been compensated for by the local expansion due to climate change.

INGA HUNDERTMARK

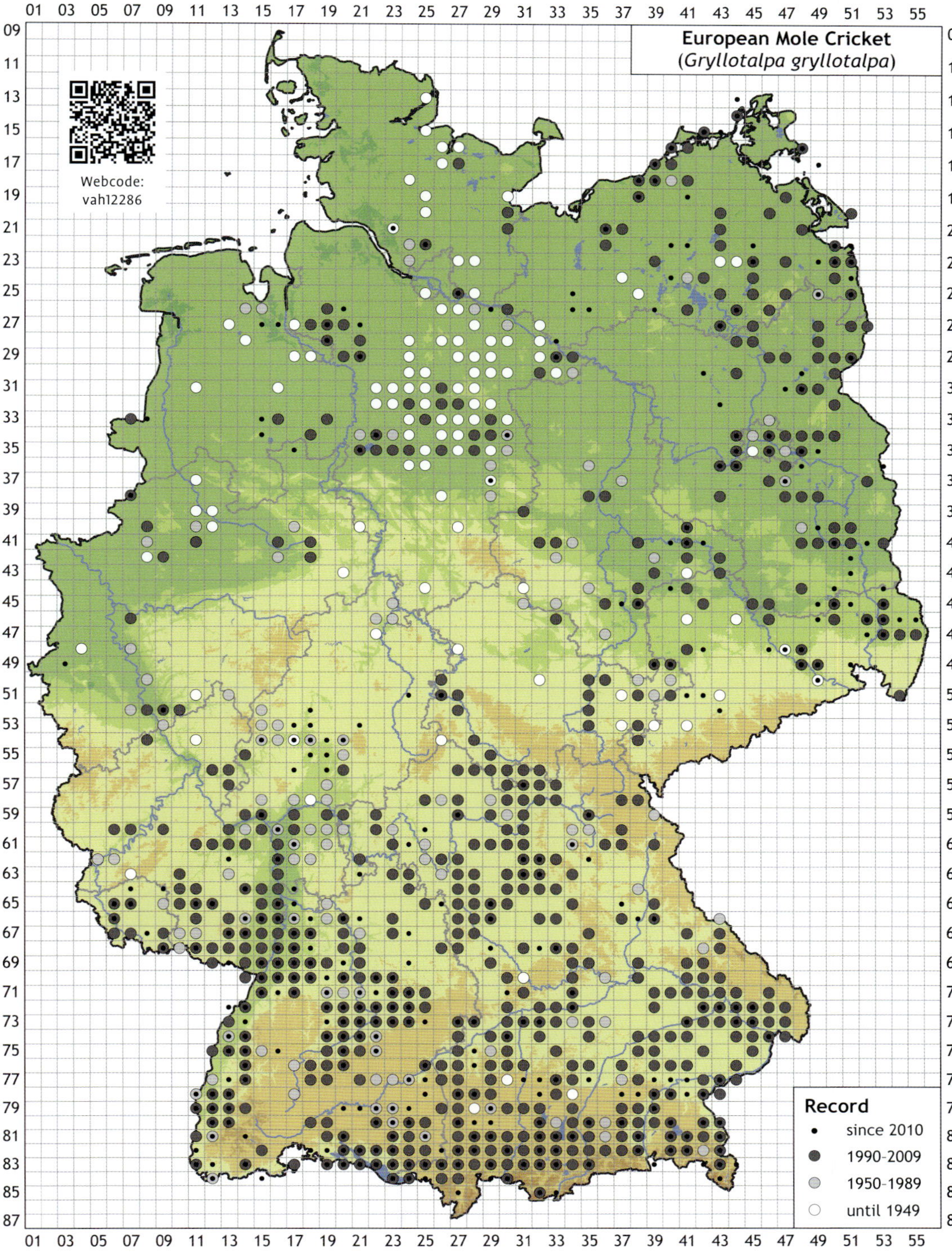
European Mole Cricket
(*Gryllotalpa gryllotalpa*)
Webcode:
vah12286
Record
since 2010
1990-2009
1950-1989
until 1949

Quarries are a typical habitat of the Common Ant-cricket (North Rhine-Westphalia: Westphalian Bay).
(Photo credit: Marcel Kettermann)

Female Common Ant-cricket (*Myrmecophilus acervorum*).
(Photo credit: Marcel Kettermann)

Common Ant-cricket – *Myrmecophilus acervorum* (Panzer, [1799])

Distribution Due to its cryptic way of life in ant nests, the precise distribution of the Common Ant-cricket in Germany has not yet been adequately surveyed. Up until now, the species has been recorded in all seven major biogeographical regions except the Alps. However, the accumulation of observations in regions exhibiting relatively warm summers, such as the Ruhr area, the Thuringian Basin and its vicinity, the Northeast German Lowland or the Nördlinger Ries together with the neighbouring southern Franconian Jura, is striking.

Habitat The Common Ant-cricket is considered thermo- and hygrophilic. It lives as a parasite in the nests of different ant species. A specialisation for a certain species of host ant does not exist. The ants maintain a constant warm and moist microclimate for rearing their brood; the Common Ant-cricket benefits from this microclimate. The species inhabits a wide range of dry and warm habitats such as urban brownfields, dry grasslands, thermophilic fringes, gardens, parks, quarries, ruderal areas or light woodlands. In highly xerothermic habitats, the colonised nests are often located within taller vegetation.

Population trend and threats Given the poor data basis, it is difficult to assess the population trend of the Common Ant-cricket in Germany. Due to its hidden way of life, it can be assumed that the species is much more widespread than the few records show so far. Based on current knowledge, the Common Ant-cricket is considered of least concern. However, locally, populations may disappear due to habitat destruction, for example by building development in urban areas.

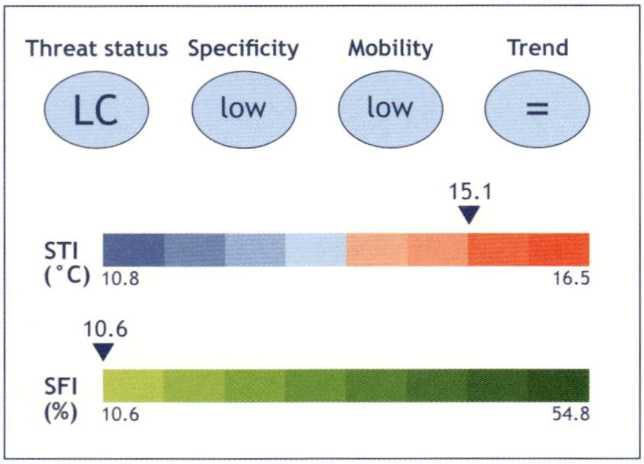

MARCEL KETTERMANN

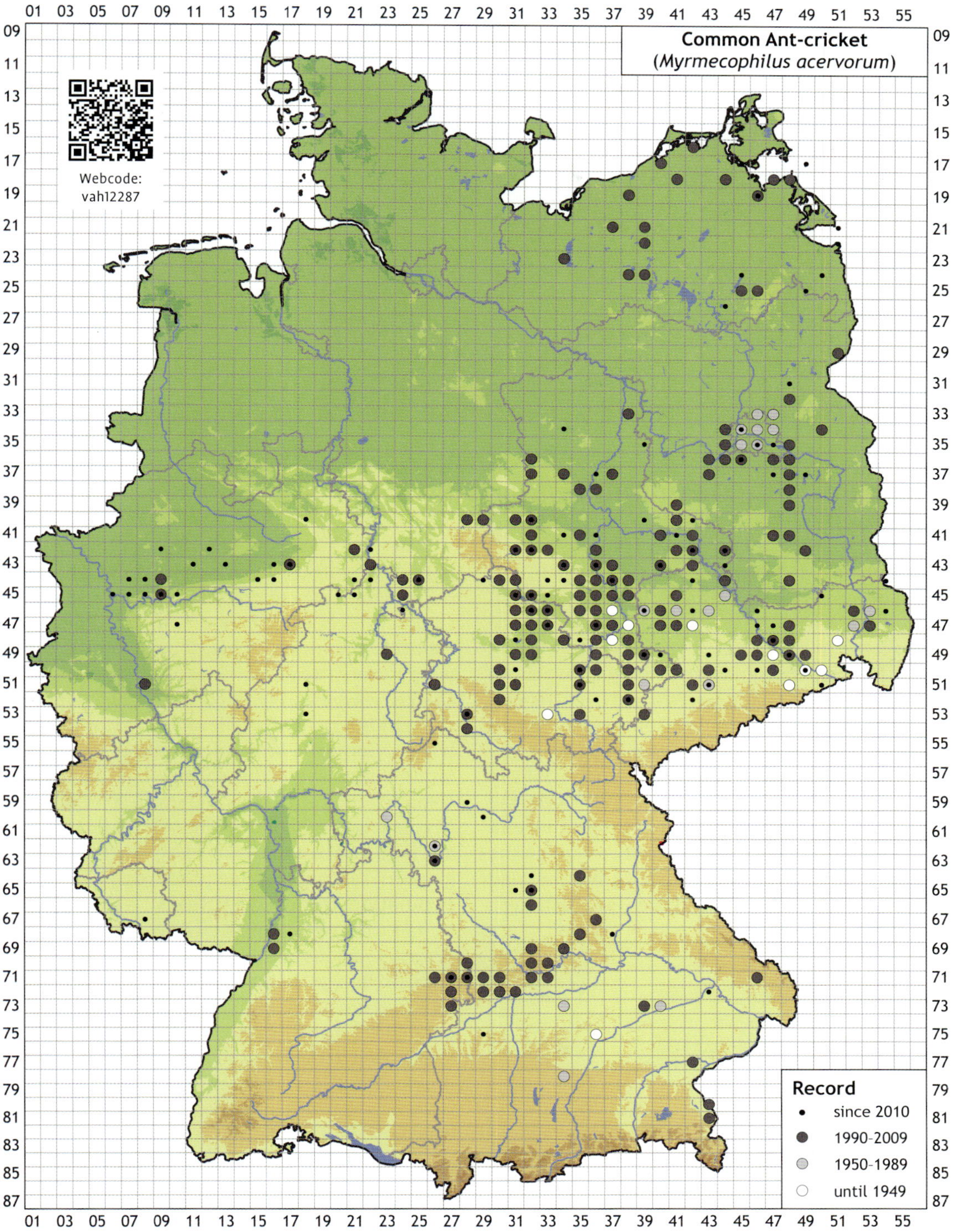
Common Ant-cricket
(Myrmecophilus acervorum)
Webcode:
vah12287
Record
since 2010
1990-2009
1950-1989
until 1949

In times of climate change, quarries exhibiting short, sparsely vegetated swards and a warm microclimate enable the occurrence of the Italian Cricket in the low mountain ranges (Hesse: Wetterau). (Photo credit: Stefan Stübing)

Female Italian Cricket (*Oecanthus pellucens*). (Photo credit: Dominik Poniatowski)

Italian Cricket – *Oecanthus pellucens* (Scopoli, 1763)

Distribution In Germany, until about 2010, the distribution of the Italian Cricket was restricted to those regions that are characterised by the warmest summers, namely the valleys of the rivers Rhine, Neckar, Main and Moselle as well as the adjacent areas. However, a few isolated occurrences, which at least partly stemmed from passive dispersal, were already recorded at that time far away from these strongholds. Subsequently, especially in the dry and hot summers from 2018 onwards, the species has strikingly expanded its range. Currently, the Italian Cricket even regularly occurs in the warmest zones of the Southwestern, Western and Eastern Uplands as well as in the warm southern part of the Northeast German Lowland. The strong dependence on regions that exhibit very warm summers is reflected by a very high STI of 15.6 °C.

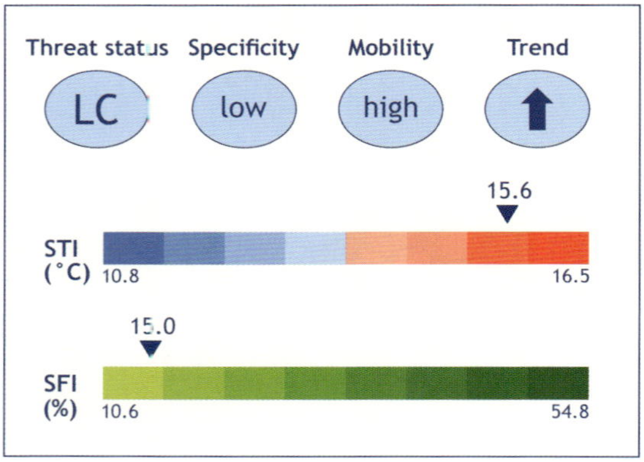

Habitat Until the 1990s, thus before the strong temperature rise due to climate change, the Italian Cricket was also restricted within heat islands to dry and hot habitats such as fallows or meadows within vineyards or calcareous and sandy grasslands. Since then, the species has increasingly colonised the herbaceous and shrub layer of further dry and warm habitats. Characteristic examples are industrial and railway system brownfields or gardens, road embankments and waysides.

Population trend and threats Originally, the Italian Cricket only occurred in the warm wine-growing regions within Germany. However, due to climate change, the species has not only broadened its spectrum of colonised habitats but has also expanded its distribution area at an extraordinary speed. Due to the observed range expansion and a predicted further spread as a consequence of global warming, the Italian Cricket is considered of least concern.

STEFAN STÜBING

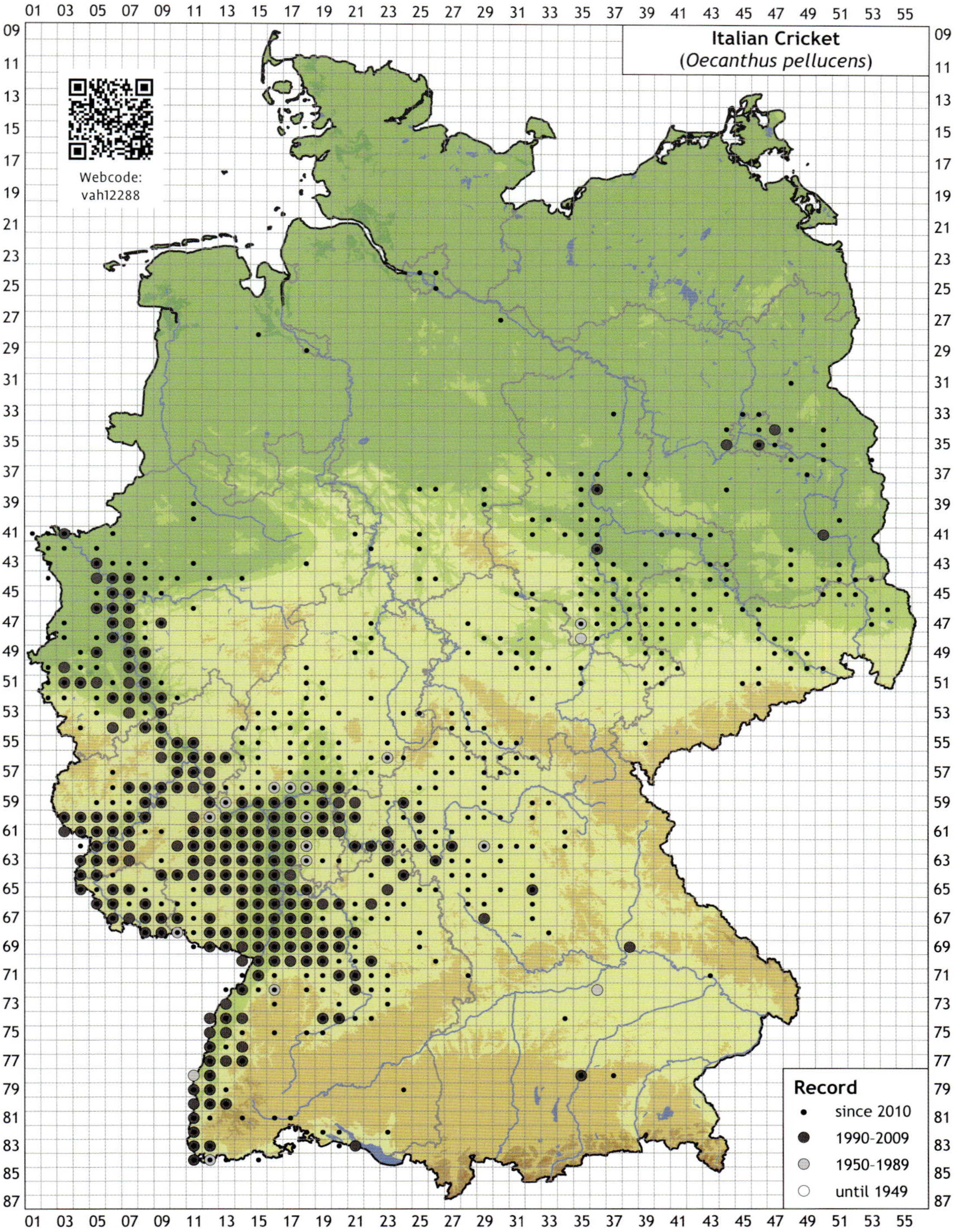
Italian Cricket
(*Oecanthus pellucens*)
Webcode:
vah12288
Record
since 2010
1990-2009
1950-1989
until 1949

Sunlit edge of an oak-beech forest that is colonised by a strong population of the Wood Cricket (North Rhine-Westphalia: Münsterland).
(Photo credit: Thomas Fartmann)

Female Wood Cricket (*Nemobius sylvestris*).
(Photo credit: Thomas Fartmann)

Wood Cricket – *Nemobius sylvestris* (Bosc, 1792)

Distribution The Wood Cricket is a west Mediterranean species that, in Germany, also depends on those biogeographical regions that are sufficiently warm and where the winters are not too cold. Accordingly, it is widespread in the Southwestern and Western Uplands, except the coolest parts of the higher elevations. Already in the Eastern Uplands, the species is rarer. In the North German Lowland, it occurs only locally in the southern third, while in the Alpine Foreland, it is found only in the lower elevations in the north and in the Lake Constance Basin. The dependence of the Wood Cricket on a warm climate is reflected by an above average STI of 14.9 °C.

Habitat The Wood Cricket inhabits sun-exposed forest edges, open deciduous and mixed forests as well as forest clearings. Shrubby or nutrient-poor grasslands adjacent to forests are also typical habitats of the species. The Wood Cricket can also be found in warm quarries, screes or railroad embankments. The crickets are thermophilic but sensitive to dehydration and prefer to dwell in a pronounced layer of foliage litter. It provides both protection from predators and higher humidity compared to the surroundings. Furthermore, the insulating effect of fallen leaves plays an important role for overwintering. The Wood Cricket overwinters in the egg or nymphal stage; however, adults also regularly appear until winter.

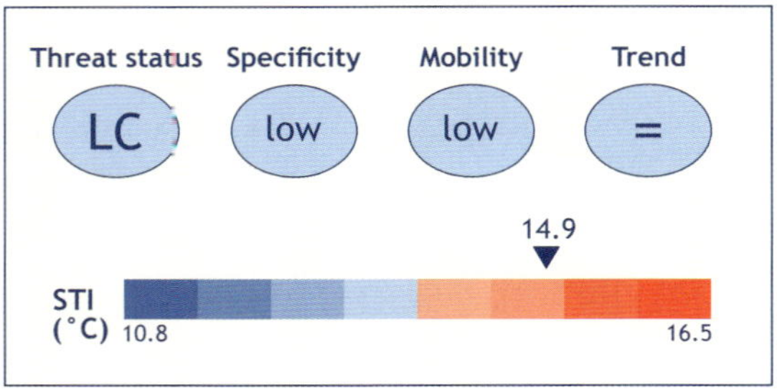

Population trend and threats In Germany, the Wood Cricket is mostly common throughout its distribution area, with stable populations in the long term. Therefore, it is considered of least concern. Recent observations have shown local spreads; however, on the national scale, these are only incompletely reflected by changes in grid occupancy. It is very likely that climate change was the driving force behind the spreading. Further range expansions and population densifications due to warmer summers and milder winters can be assumed.

FELIX HELBING

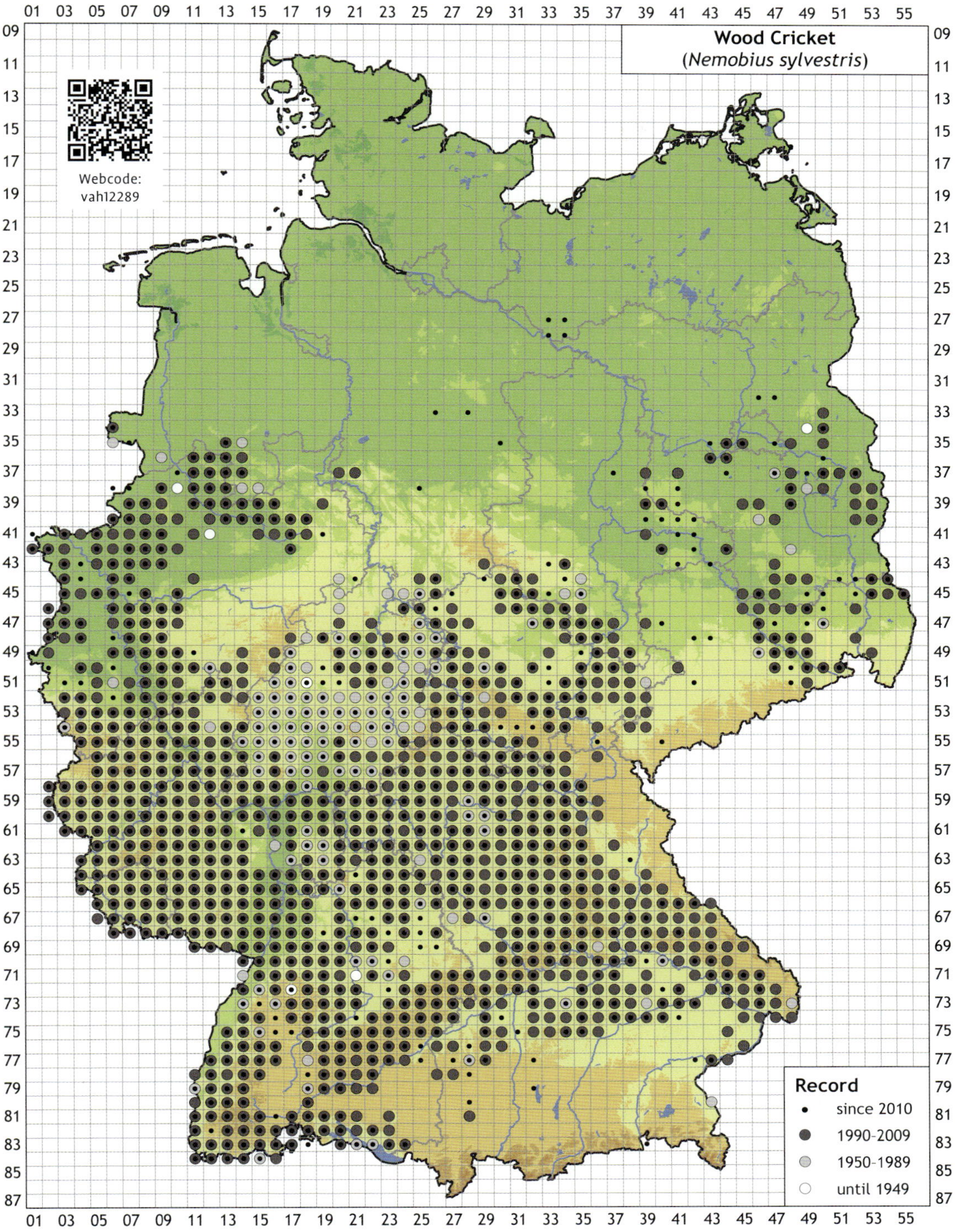
Wood Cricket
(*Nemobius sylvestris*)
Webcode:
vah12289
Record
since 2010
1990-2009
1950-1989
until 1949
01 03 05 07 09 11 13 15 17 19 21 23 25 27 29 31 33 35 37 39 41 43 45 47 49 51 53 55
09 11 13 15 17 19 21 23 25 27 29 31 33 35 37 39 41 43 45 47 49 51 53 55 57 59 61 63 65 67 69 71 73 75 77 79 81 83 85 87

Grasslands and mires exhibiting permanently high water levels are among the preferred habitats of the Marsh Cricket (Baden-Württemberg: southern Upper Rhine Valley).
(Photo credit: Marcel Kettermann)

Female Marsh Cricket (*Pteronemobius heydenii*). (Photo credit: Jürgen Fischer)

Marsh Cricket – *Pteronemobius heydenii* (Fischer, 1853)

Distribution First records of the Marsh Cricket in Germany date back to the beginning of the 20th century. For a long time, the species only occurred locally in the southwesternmost parts of the Southwestern Uplands (Upper Rhine Valley) and Alpine Foreland (Lake Constance Basin). Recently, however, many further populations have been discovered not only in the vicinity of the two known strongholds but also in warmer, southern parts of the Alpine Foreland and even the Alps. The German populations are located at the northern range margin of the Mediterranean species. The dependence of the Marsh Cricket on a warm climate is reflected by a disproportionally high STI of 15.0 °C.

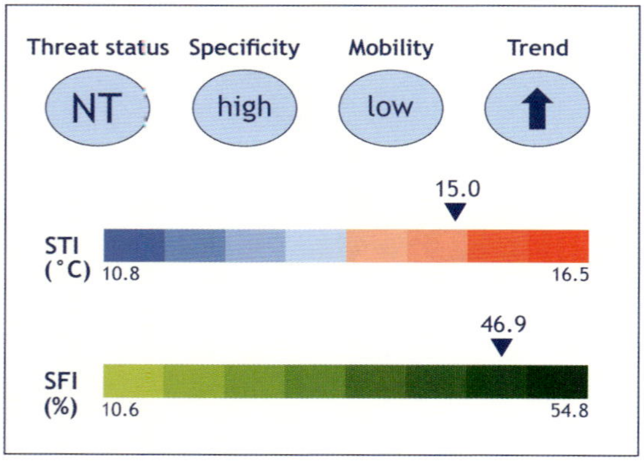

Habitat The occurrence of the thermophilic Marsh Cricket is restricted to open habitats on permanently wet or soaked soils. The presence of small-scale patches of bare ground and a not-to-dense vegetation are further prerequisites. The main habitats are wet grasslands as well as fens and spring mires. Additionally, the species colonises the riparian zone of larger water bodies and drenched slopes of grasslands.

Population trend and threats The Marsh Cricket has always been rare in Germany. Many new records in recent times are a result of systematic surveys of grasshopper communities or targeted searches for the species. Since the Marsh Cricket is very thermophilic, its populations may have benefited from climate change and may do so in future. Nevertheless, the species is threatened due to its very specific habitat requirements. The main threats are habitat deterioration due to drainage and abandonment of land use as well as habitat loss due to overbuilding.

JENS SACHTELEBEN

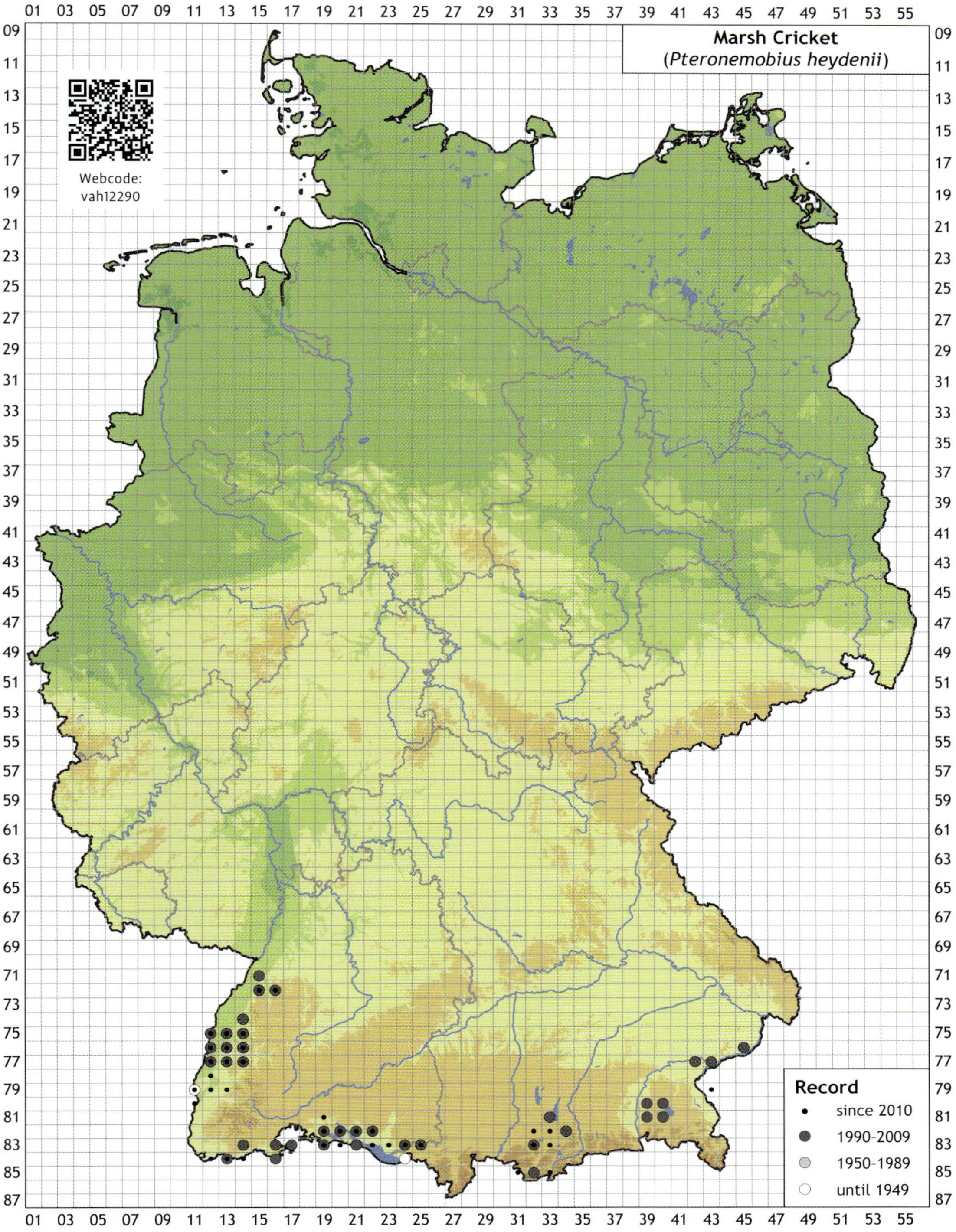
Marsh Cricket
(*Pteronemobius heydenii*)
Webcode:
vah12290
Record
since 2010
1990-2009
1950-1989
until 1949
01 03 05 07 09 11 13 15 17 19 21 23 25 27 29 31 33 35 37 39 41 43 45 47 49 51 53 55
09 11 13 15 17 19 21 23 25 27 29 31 33 35 37 39 41 43 45 47 49 51 53 55 57 59 61 63 65 67 69 71 73 75 77 79 81 83 85 87

Abandoned arable field on a sandy ground moraine, which holds a high abundance of the Field Cricket (Brandenburg: Märkische Schweiz).
(Photo credit: Thomas Fartmann).

Male Field Cricket (*Gryllus campestris*).
(Photo credit: Jürgen Fischer).

Field Cricket – *Gryllus campestris*
Linnaeus, 1758

Distribution The Field Cricket occurs in all seven major biogeographical regions of Germany. However, its frequency strongly differs regionally. Within the Alps, Alpine Foreland, Southwestern Uplands as well as the southern part of the Western Uplands and Northeast German Lowland, it has always been common. Due to climate change, it is now expanding its range northwards and to previously unoccupied elevations of the uplands. The above average STI of 14.9 °C reflects the dependence on a warm climate.

Habitat The Field Cricket inhabits dry and warm, mostly flat or south-facing habitats usually with sparse swards. Characteristic habitats are heathlands and dry grasslands but also field margins, nutrient-poor mesic grasslands, ruderal areas and woodlands edges. In the 1990s and 2000s, abandoned arable fields were also an important habitat in the Northeast German Lowland. Generally, in the warm regions of southern and eastern Germany, the range of colonised habitats is much broader than under the oceanic climate in the northwest. In the latter, the Field Cricket is much more dependent on microclimatically favourable habitats, often located on sandy soils. Due to territorial behaviour and strong weather-related population fluctuations, the minimum area requirements of the species are very large. For long-term viable populations, a habitat size of 3 ha is needed.

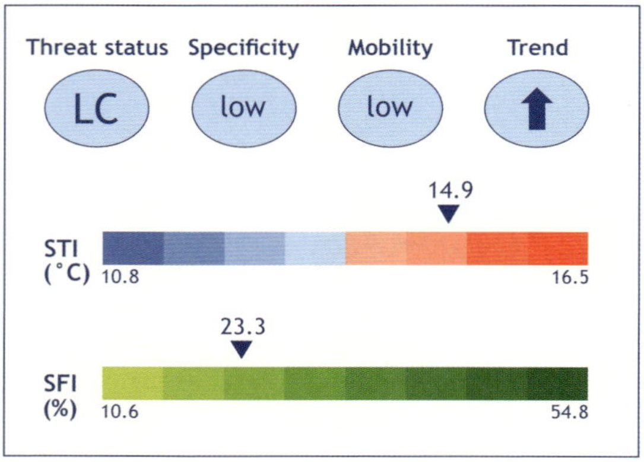

Population trend and threats The Field Cricket has suffered from strong habitat losses due to land-use intensification and afforestation for many decades. Since it disappeared a long time ago from arable landscapes, the species´ common name is no longer appropriate. The cessation of the mandatory set-aside payments by the EU in 2008 has caused a large-scale loss of abandoned arable land – a further important habitat. These declines are contrasted with climate-driven range expansions of the species. Consequently, the Field Cricket is now considered of least concern.

FRANZ LÖFFLER

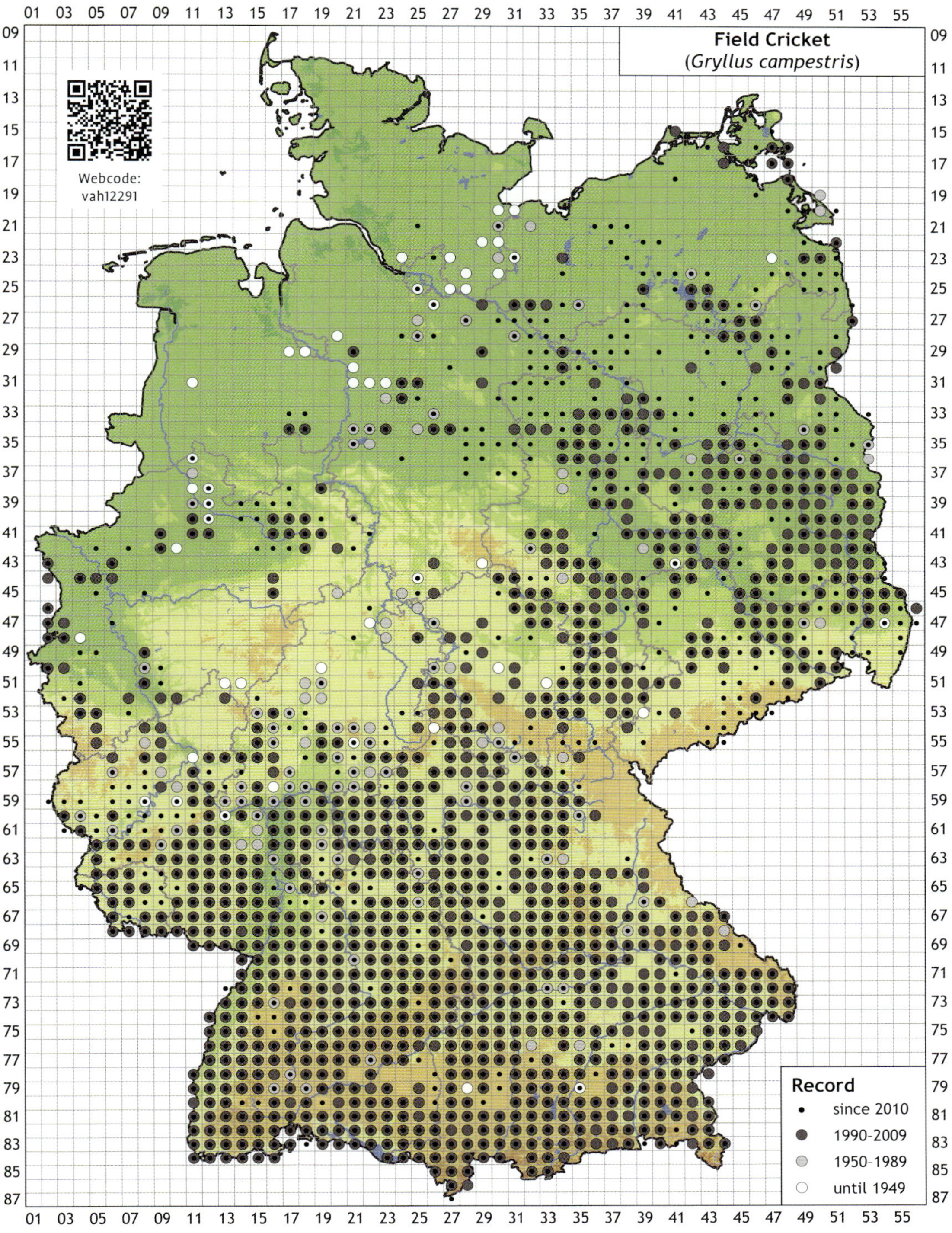
Field Cricket
(*Gryllus campestris*)
Webcode: vah12291
Record
since 2010
1990-2009
1950-1989
until 1949

Even saunas are a habitat of the House Cricket, since they are characterised by high temperatures and air humidity. (Photo credit: Marcel Kettermann)

Male House Cricket (*Acheta domesticus*). (Photo credit: Jürgen Fischer)

House Cricket – *Acheta domesticus* (Linnaeus, 1758)

Distribution Today, the House Cricket is globally distributed, but its origin was most likely in North Africa. In Germany, it was introduced by the Romans. Here, it occurs in all seven major biogeographical regions. However, striking strongholds are located in urban agglomerations and warm regions. The species is widely reared as an important food source for terrarium pets but has also served as a novel food for humans in the European Union since 2022. As a result, both intended and accidental releases of individuals regularly occur, favouring the colonisation of new habitats.

Habitat The House Cricket exhibits a strong dependence on human settlements. For optimal development, it requires a high air humidity and high temperatures. Besides buildings, sewerage, subway and ventilation systems are used as habitats within urban areas. In summer, House Crickets colonise manure heaps, vineyards, wasteland and many other habitats characterised by a warm microclimate in the surrounding landscape. Gradations have been shown to occur under very warm conditions and if air humidity is high enough, for example in waste landfills and sewage plants. Outside buildings, the species can only overwinter if temperatures do not drop too much. Such conditions can exist, if for example sufficient fermentation heat is released in organic waste fermentation and composting plants or in waste landfills.

Population trend and threats Reliable data on the short-term trend of the House Cricket in Germany are missing. Nevertheless, since it is a synanthropic species that additionally benefits from a more Mediterranean-like climate due to climate change, there are no indications for a current threat. Partly a potential decline in the House Cricket has been assumed due to the modern-day construction of houses. Moreover, in waste landfills, consistent waste separation but also compaction and covering of the remaining waste cause declines. However, due to the high reproductive potential and short generation cycles, the species can rapidly compensate for such losses.

CARSTEN RENKER

Threat status	Specificity	Mobility	Trend
DD	low	high	?

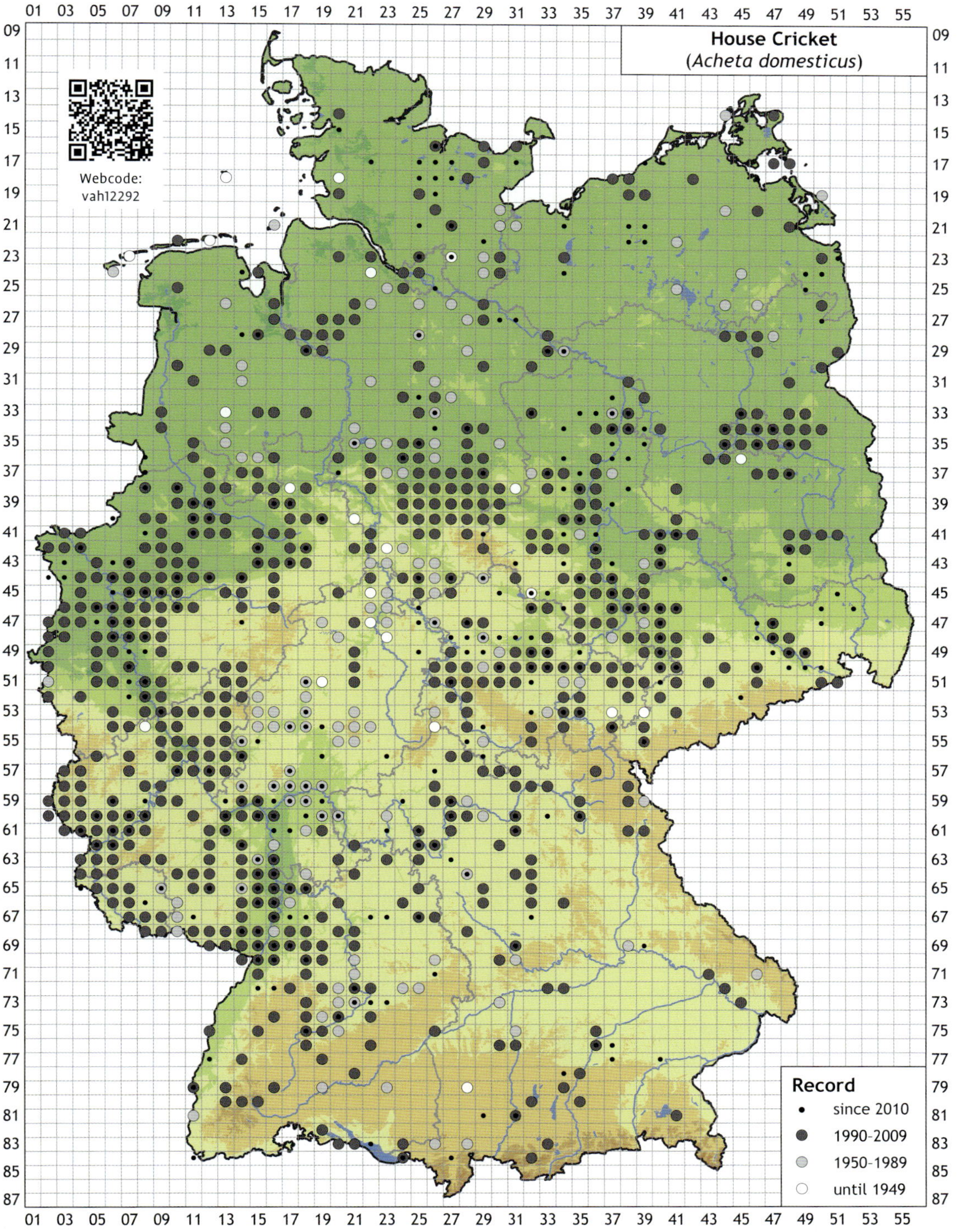
House Cricket
(*Acheta domesticus*)
Webcode:
vah12292
Record
since 2010
1990-2009
1950-1989
until 1949

The ballast of railway tracks is a good heat accumulator and important habitat of the Bordeaux Cricket (Rhineland-Palatinate: northern Upper Rhine Valley). (Photo credit: Ansgar van Elst)

Female Bordeaux Cricket (*Eumodicogryllus bordigalensis*). (Photo credit: Ansgar van Elst).

Bordeaux Cricket – *Eumodicogryllus bordigalensis* (Latreille, 1804)

Distribution The Bordeaux Cricket has a Mediterranean range and its current distribution centre in Germany is located in the outermost southwest along the Upper Rhine Valley. This region is a pronounced heat island characterised by mild winters and very hot summers. Correspondingly, the STI of the Bordeaux Cricket is very high at 16.3 °C. The species was first recorded in Germany in 1995 in the Upper Rhine Valley. Since then, new populations have been discovered continuously. Particularly during recent years, the Bordeaux Cricket has also been detected in further German heat islands.
Habitat The Bordeaux Cricket is very thermo- and xerophilic. Its habitats are characterised by strong solar irradiation and heat storage. Many records are from train stations. Here, the species occurs within the railway tracks that absorb heat and warm up the ground surface. Now, the Bordeaux Cricket can even be found on arable fields since soil crevices are available that can be used as a hideout. Ruderal brownfields are another habitat, where the species stays under stones or wood.
Population trend and threats Since the first observation in Germany almost thirty years ago, the Bordeaux Cricket has rapidly expanded its range. Usually, the species is flightless. However, long-winged morphs capable of flight also occur. Short-distance dispersal is assumed to be the result of active movements on the ground but partly also of flying individuals. By contrast, colonisation of new habitats over long distances are mainly attributed to passive dispersal due to human activities. The Bordeaux Cricket strongly benefits from climate change and is considered of least concern.

MANFRED ALBAN PFEIFER

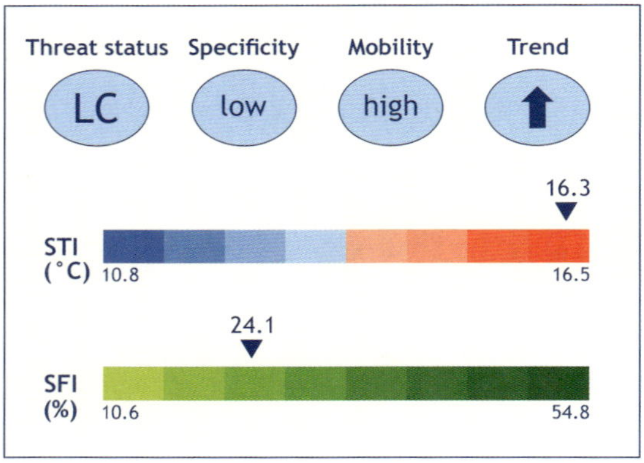

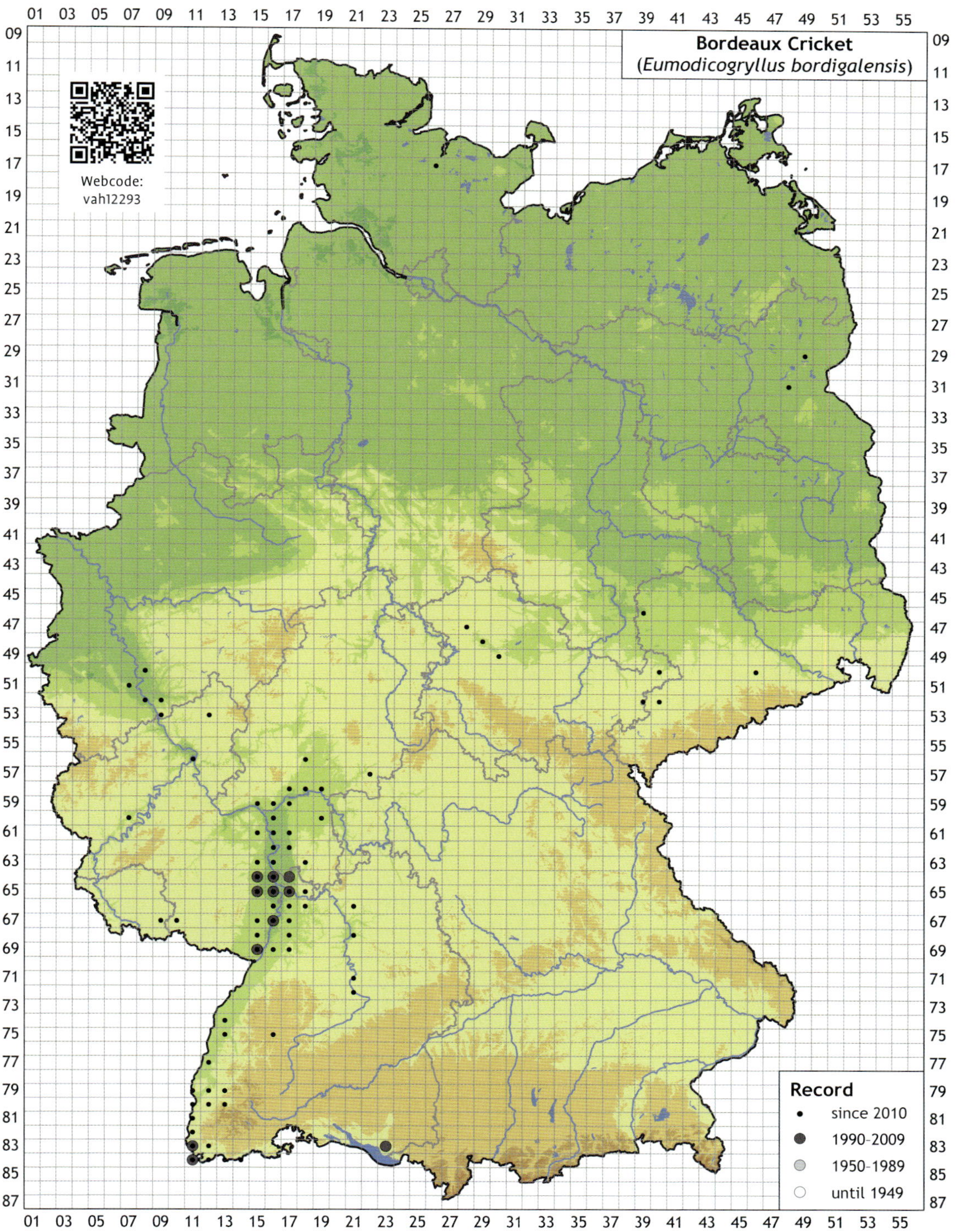
Bordeaux Cricket
(Eumodicogryllus bordigalensis)
Webcode:
vah12293
Record
since 2010
1990-2009
1950-1989
until 1949

Habitat of the Eastern Stripe-headed Cricket in the Jagst Valley. Besides managed vineyards, slopes are colonised that are chisel ploughed every year to foster the species (Baden-Württemberg).
(Photo credit: Jürgen Trautner)

Female Eastern Stripe-headed Cricket (*Modicogryllus frontalis*).
(Photo credit: Jennifer Theobald)

Eastern Stripe-headed Cricket – *Modicogryllus frontalis* (Fieber, 1844)

Distribution The German records of the Eastern Stripe-headed Cricket are outposts advanced far northwest of its range. Only one of the populations has remained in the Jagst Valley in the Southwestern Uplands. In the southern Upper Rhine Valley, the second former stronghold in Germany, the species is considered to be extinct as of 2020. All other former populations in the federal states of Baden-Württemberg and Bavaria became extinct a long time ago.

Habitat The habitats of the Eastern Stripe-headed Cricket are sunny, skeletal soils with sparse or no ground vegetation. The soils are rich in gravel and have a cavity system but also always contain some cohesive substrates. The nymphs and adults partly stay within these cavities. Such environmental conditions sporadically occur in gravel pits in the southern Upper Rhine Valley and, on a larger scale, on a vineyard slope in the Jagst Valley. In the latter region, primarily unvegetated vineyard slopes with vertical vine rows and a high proportion of limestone gravel in the topsoil are colonised. Moreover, there, the species inhabits slopes that are chisel ploughed every year and exhibit comparable habitat structures to the slopes that are used for viticulture.

Population trend and threats Due to the favourable management of the habitats, the long-term viability of the population of the Eastern Stripe-headed Cricket in the Jagst Valley seems to be high. However, the specific vineyard management that fosters the species is increasingly threatened by abandonment. By contrast, the future prospects for the species in the southern Upper Rhine Valley are very poor: two of the populations were confirmed for the last time in 2018 – in each case on the basis of only one individual. In the third gravel pit, which contained the largest population in 2018, the last individual was observed in 2020. In the following years, no individuals have been detected. The main causes of threat in the gravel pits are the abandonment of quarrying and subsequent succession. Moreover, active revegetation and local deposition of material, such as soil or green waste, has also lead there to habitat loss.

JÜRGEN TRAUTNER & JÖRG RIETZE

Threat status	Specificity	Mobility	Trend
CR	high	low	⬇⬇⬇

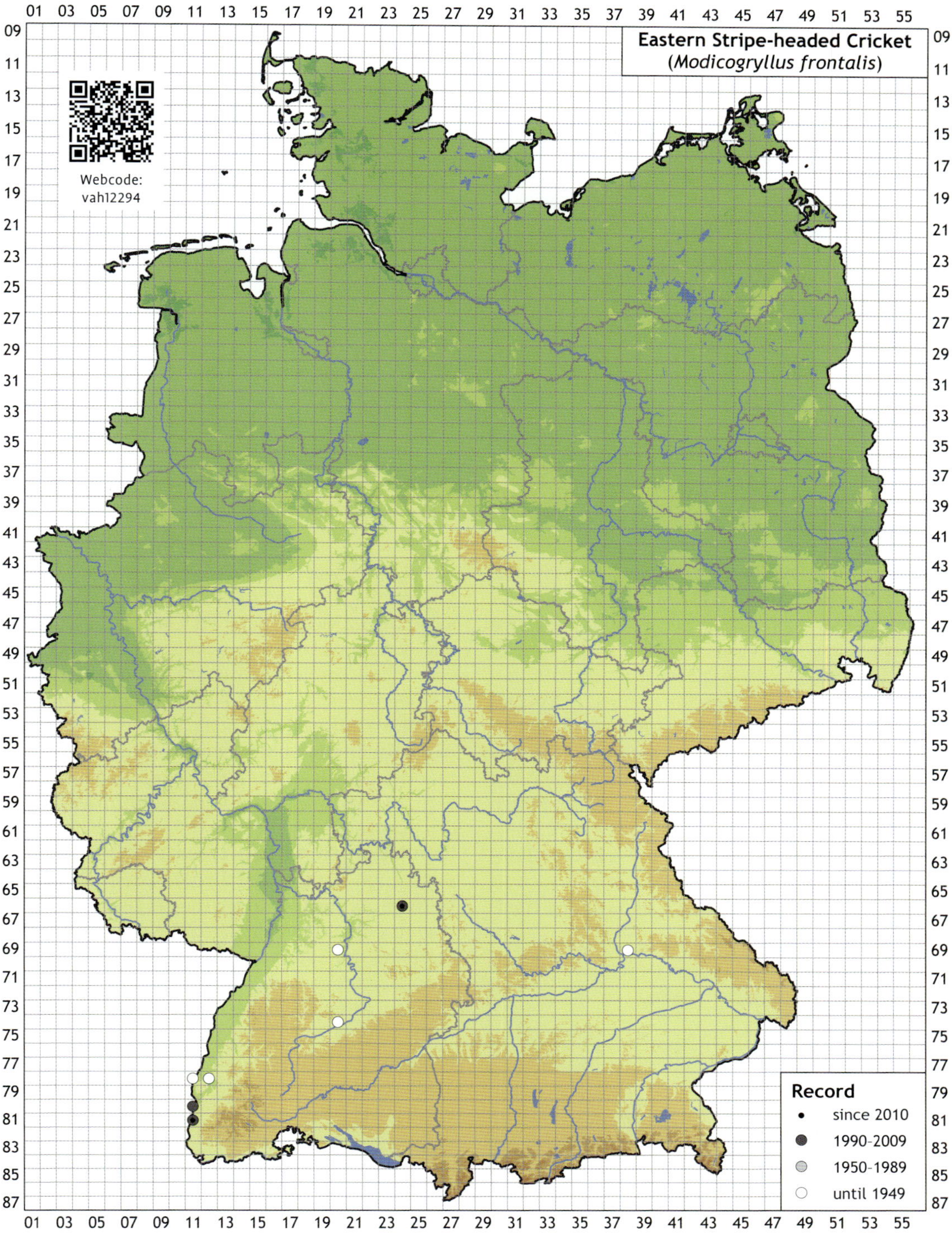
Eastern Stripe-headed Cricket
(Modicogryllus frontalis)
Webcode:
vah12294
Record
since 2010
1990-2009
1950-1989
until 1949

In the lowlands, wet grasslands with small ponds are regularly colonised by the Slender Groundhopper (North Rhine-Westphalia: Westphalian Bay). (Photo credit: Felix Helbing)

Female Slender Groundhopper (*Tetrix subulata*). (Photo credit: Dominik Poniatowski)

Slender Groundhopper – *Tetrix subulata* (Linnaeus, 1758)

Distribution For a long time in Germany, the slightly thermophilic (STI = 14.9 °C) Slender Groundhopper was widespread only in the warm lowlands. However, since the 1990s, the species has at least regionally expanded its distribution due to climate change. This has resulted in a disappearance of distribution gaps in the oceanic northwest and north of Germany as well as in the low mountain ranges. However, in the cool Sauerland low mountain range and generally above 900 m a.s.l., the Slender Groundhopper is still rare.

Habitat The hygrophilic Slender Groundhopper inhabits moist to wet open habitats, such as wet grasslands, mires and reed beds or the riparian zone of water bodies. Sunlit patches of bare ground are of prime importance since they serve as an oviposition habitat. As long as the humidity and temperature requirements are fulfilled, even very small patches can contain a population. As a pioneer species, the Slender Groundhopper can rapidly colonise new habitats. In mid- and late summer, dispersing adults may be found in habitats that are usually unsuitable for reproduction, such as arable fields, clear-cuts and windthrows or even dry grasslands.

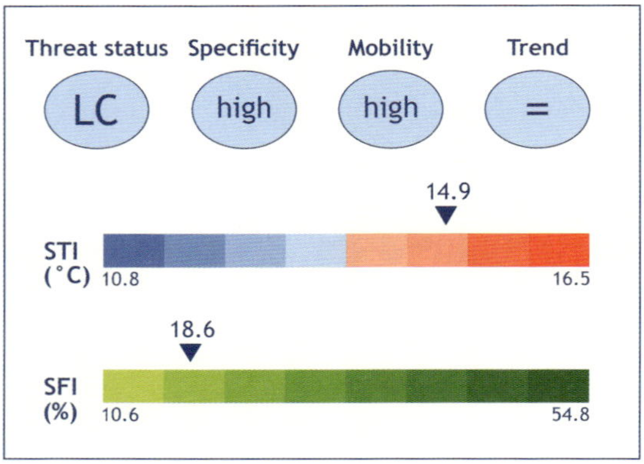

Population trend and threats The long-term population trend of the Slender Groundhopper in Germany is positive due to the range expansions in response to climate change. The short-term trend is stable; the species is currently considered not threatened. The future prospects, however, are uncertain in light of climate change. On the one hand, with ongoing global warming, a further range expansion seems to be possible. On the other hand, an increase in drought periods may lead to habitat loss in the future.

FELIX MAXIMILIAN FREIENSTEIN

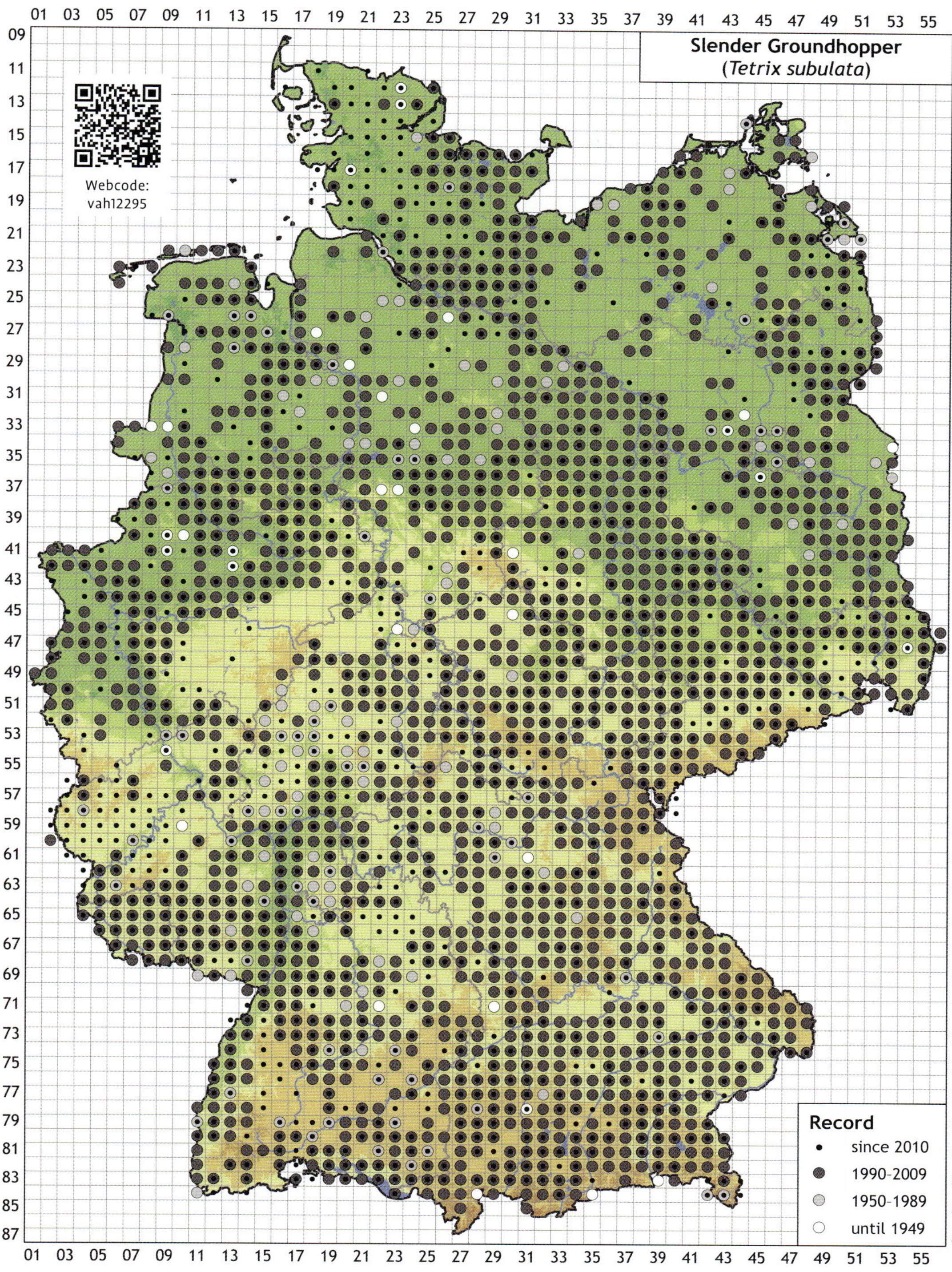
Slender Groundhopper
(*Tetrix subulata*)
Webcode:
vah12295
Record
since 2010
1990-2009
1950-1989
until 1949

Female Cepero's Groundhopper
(*Tetrix ceperoi*).
(Photo credit: Dominik Poniatowski)

A typical habitat of Cepero's Groundhopper is the riparian zone along young ponds – like here in this sand quarry (Lower Saxony: Ems-Hunte-Geest).
(Photo credit: Dominik Poniatowski)

Cepero's Groundhopper – *Tetrix ceperoi* (Bolivar, 1887)

Distribution Cepero's Groundhopper is a western Mediterranean faunal element that reaches the northeastern edge of its distribution in Central Europe. The species is correspondingly rare in Germany. It can primarily be found in sandy areas of the lowlands. The strongholds are the northern Upper Rhine Valley, the Cologne Bay, the Lower Rhine Lowland, the Ems-Hunte-Geest, the East Frisian Islands, the eastern foothills of the Harz Mountains as well as the lowlands around Leipzig and Lower Lusatia.

Habitat As a thermophilic pioneer species, Cepero's Groundhopper has a high habitat specificity. Its occurrence is mostly limited to sunlit locations with sparse vegetation and sandy soils. The species also requires a high soil moisture. Typical habitats are moist dune valleys, sand and mud banks along rivers and the riparian zone of young standing waters – for example, in sand quarry and marl pits. The species also occurs in the post-mining landscapes of western and eastern Germany if moist areas with pioneer vegetation are present.

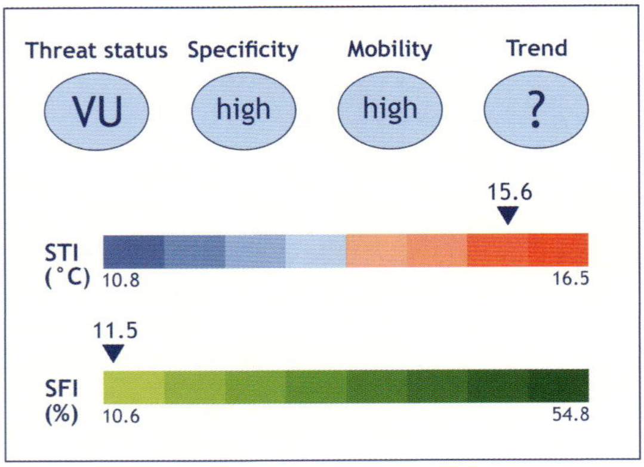

Population trend and threats Cepero's Groundhopper is currently expanding its range in northwestern Germany. For example, initial observations are available for newly created ponds in the Westphalian Bay and numerous open sand quarries in western Lower Saxony. By contrast, for southwestern Germany and in the post-mining areas of eastern Germany, declines have been documented. Natural succession can be seen as the main cause of threat. Due to the loss of open bare ground and the establishment of woody plants, habitat quality deteriorates rapidly. Further threats include recultivation measures in post-mining landscapes as well as the filling or flooding of sand quarries and opencast mines.

DOMINIK PONIATOWSKI

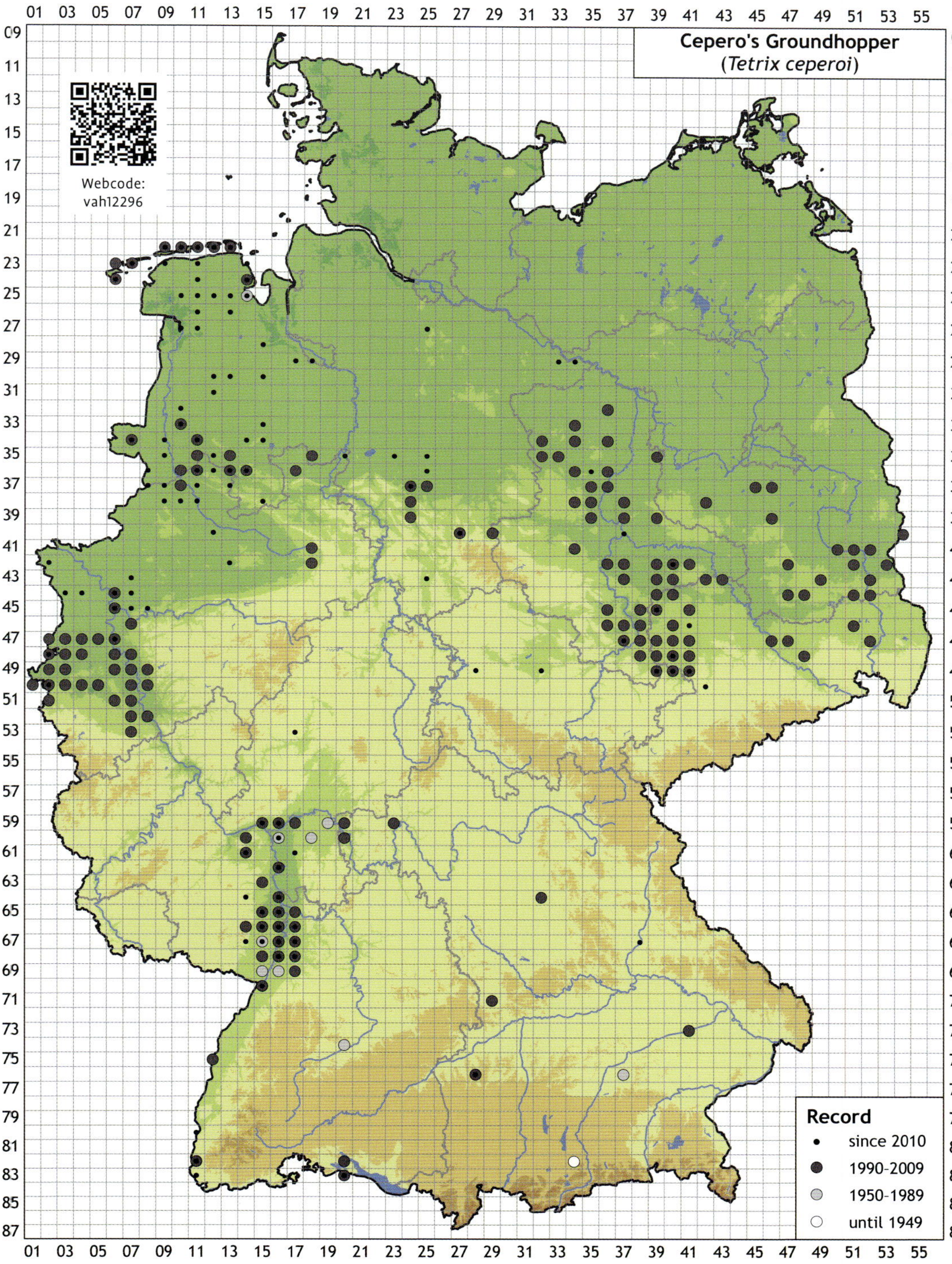
Cepero's Groundhopper
(*Tetrix ceperoi*)
Webcode:
vah12296
Record
since 2010
1990-2009
1950-1989
until 1949

Periodically water-bearing river branches with deposits of fine sediments are a typical habitat of the Alpine Groundhopper (Bavaria: Upper Isar Valley). (Photo credit: Michael Reich)

Female Alpine Groundhopper (*Tetrix tuerki*). (Photo credit: Jürgen Fischer)

Alpine Groundhopper – *Tetrix tuerki* (Krauss, 1876)

Distribution The Alpine Groundhopper is restricted in Germany to the northern edge of the Bavarian Alps. Isolated, particularly historic records stretch along the rivers Lech and Isar into the Alpine Foreland. Today, the strongholds are the tributaries or headwaters of the rivers Ammer, Loisach and Isar. However, a recently discovered population in 2014 at the river Ammer south of the city of Peißenberg shows that there might be further populations at suitable river sections in the Alpine Foreland.

Habitat Characteristic habitats of the Alpine Groundhopper are sparsely vegetated gravel banks along near-natural sections of Alpine rivers. Here, the species occurs together with the Speckled Buzzing Grasshopper and the Gravel Grasshopper. However, in comparison to these two species, the Alpine Groundhopper avoids the drier, elevated parts of the gravel banks. By contrast, it prefers periodically wet, very sparsely vegetated habitats rich in fine-grained, silty sediments along the riparian zones of channels and pools.

Population trend and threats Similar to the Speckled Buzzing Grasshopper, the Alpine Groundhopper had very likely already experienced its strongest losses in Germany before 1950. During that period, hydraulic engineering at the river Lech had largely destroyed its habitats. Generally, reduced discharge dynamics due to reservoirs or water diversion has strong negative effects on the species. Downstream of dams, the bed-load deficit and the discharge of fine-grained sediment causes habitat loss. If water is diverted, the occurrence of suitable periodically wet parts at pools depends on the design of a minimum flow of the river. The Alpine Groundhopper has much smaller area requirements than the Speckled Buzzing Grasshopper but is nevertheless considered threatened with extinction due to habitat loss and habitat deterioration.

MICHAEL REICH

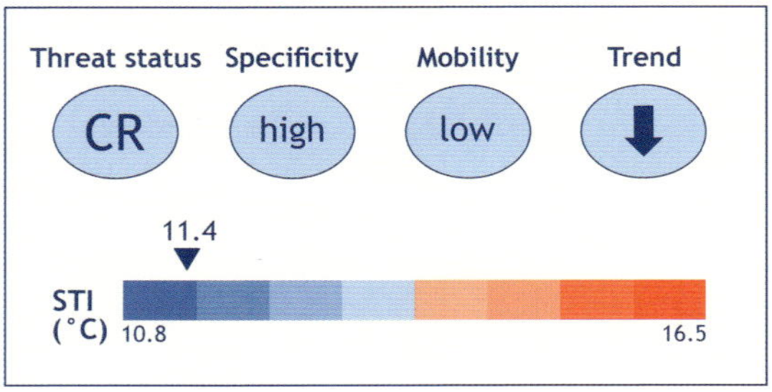

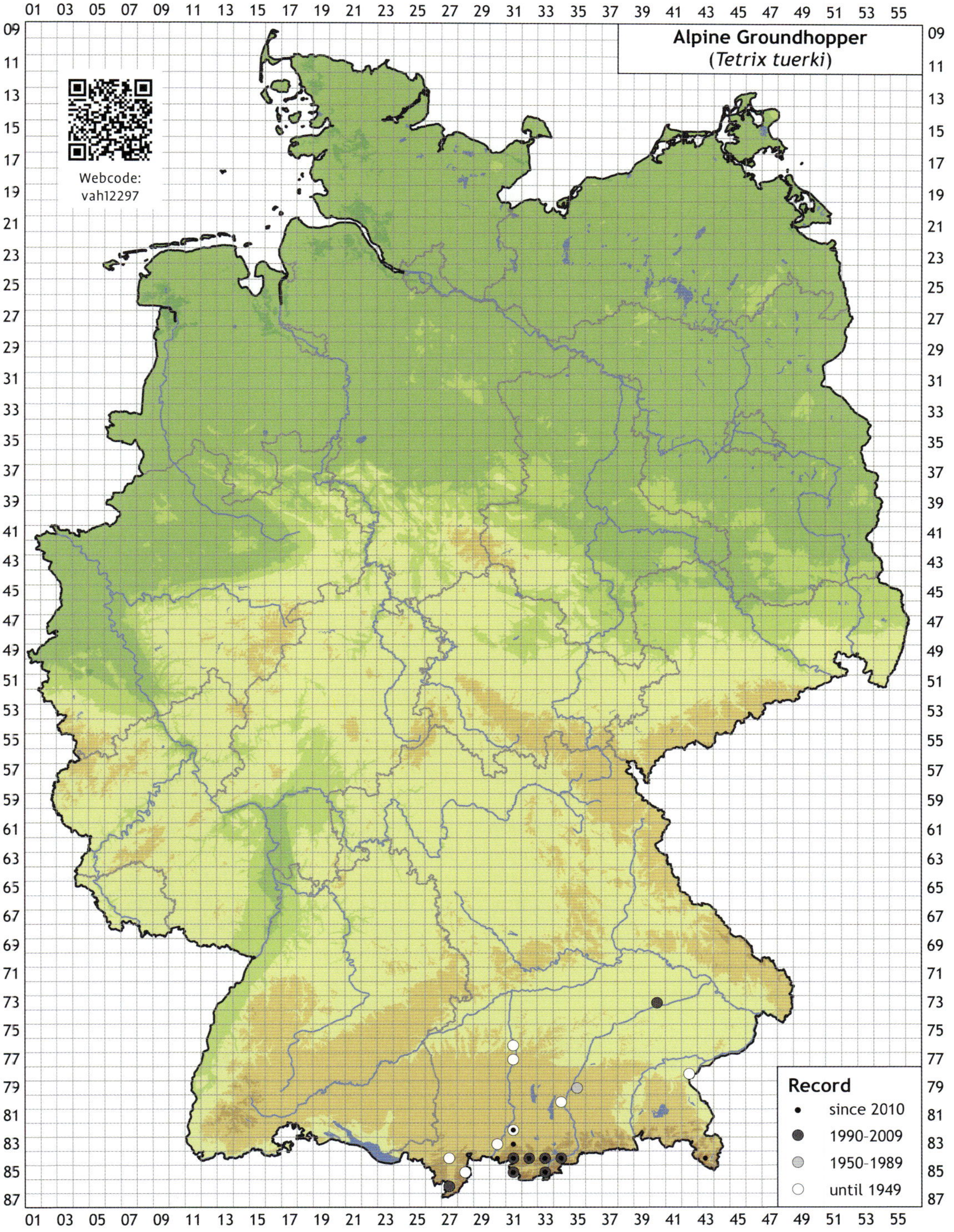
Alpine Groundhopper
(*Tetrix tuerki*)
Webcode:
vah12297
Record
since 2010
1990-2009
1950-1989
until 1949

Female Common Groundhopper (*Tetrix undulata*).
(Photo credit: Thomas Fartmann)

Within wet grasslands, the Common Groundhopper inhabits microhabitats rich in bare ground (North Rhine-Westphalia: Medebach Bay).
(Photo credit: Dominik Poniatowski)

Common Groundhopper – *Tetrix undulata* (Sowerby, 1806)

Distribution The Common Groundhopper is endemic to Europe and is mainly distributed throughout Western Europe and temperate Central Europe. Consequently, the species is common across large parts of Germany. Larger distribution gaps mainly exist in the more continental, low-precipitation areas of the Northeast German Lowland. At the cool, higher elevations of the uplands and Alps from 1,000 m a.s.l. upwards, the Common Groundhopper is entirely absent.

Habitat The slightly hygrophilic Common Groundhopper occurs in a wide range of habitats. It colonises forest edges and light forests, fresh to moist grasslands, wet heathlands, mires, quarries or riparian zones of water bodies. Within these habitats, the species often inhabits moist microhabitats rich in bare soil. In biogeographical regions with low precipitation, the dependence on moist habitats is most pronounced. By contrast, in rainy regions or years, the species even inhabits dry heathlands or grasslands.

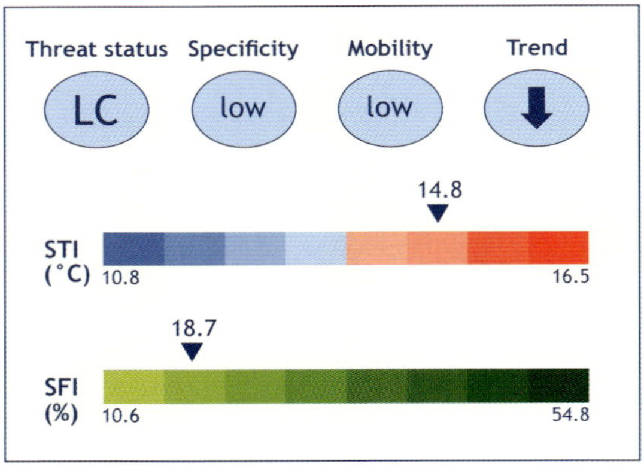

Population trend and threats Recently, a slight decline of the Common Groundhopper has been observed in Germany. However, since the species is still widespread and inhabits a wide range of habitats, it is considered of least concern. However, locally, suitable habitats rich in open ground can disappear due to afforestation, drainage, eutrophication or succession. Although the Common Groundhopper is usually flightless, it is able to colonise newly created habitats rapidly and, hence, is classified as a pioneer species. It is assumed that it will increasingly suffer from summer droughts as a result of climate change. This is particularly true for populations in dry and warm biogeographical regions.

DANIEL ELIAS

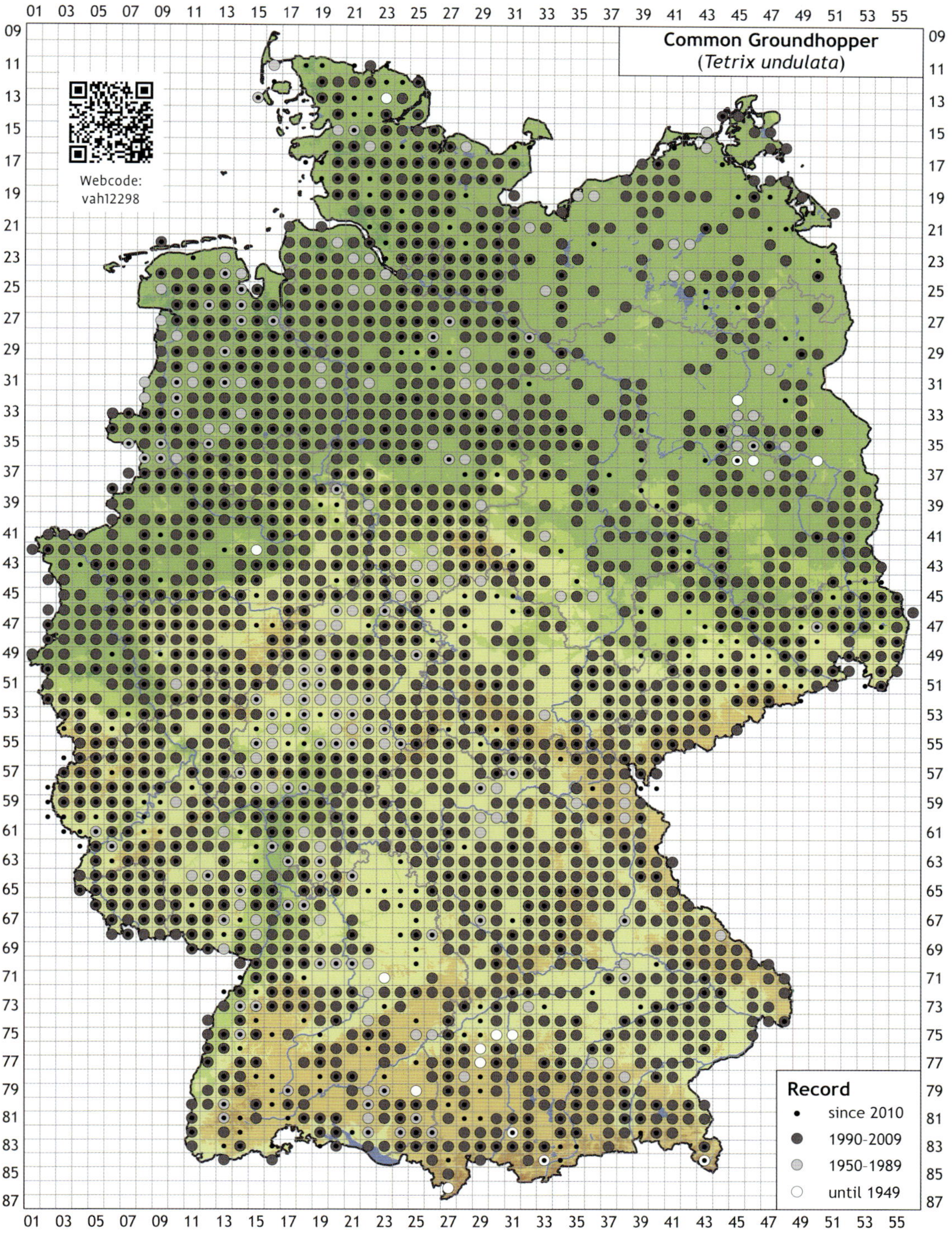
Common Groundhopper
(*Tetrix undulata*)
Webcode:
vah12298
Record
since 2010
1990-2009
1950-1989
until 1949

Sparsely vegetated calcareous grasslands on limestone are a typical habitat of the Two-spotted Groundhopper (North Rhine-Westphalia: Upper Diemel Valley). (Photo credit: Dominik Poniatowski)

Female Two-spotted Groundhopper (*Tetrix bipunctata*). (Photo credit: Thomas Fartmann)

Two-spotted Groundhopper – *Tetrix bipunctata* (Linnaeus, 1758)

Distribution The Two-spotted Groundhopper is among the rare grasshopper species in Germany. Its occurrence is almost completely restricted to hilly and mountainous regions. Most populations are observed along a wide belt from the Diemel Valley and Harz Mountains in the north over to the Franconian Jura and Bavarian Forest in the south and to the Swabian Jura and the Black Forest in the southwest. Further strongholds are located across the alluvial fans of the river Lech and Isar as well as in the Alps up to the alpine zone at 2,000 m a.s.l.
Habitat The main habitats of the Two-spotted Groundhopper are sunny and sparsely vegetated semi-dry and dry grasslands on base-rich, shallow soils. Its habitats are similar to those of the Long-horned Groundhopper, but the Two-spotted Groundhopper prefers a drier and warmer microclimate. This is also reflected by a stronger dependence on shrub and forest edges, which provide shelter against rain and wind. Moreover, the species colonises drained bogs, heathlands, quarries, screes, woodland clearings and light woodlands.
Population trend and threats Like many species of dry grasslands in Germany, the Two-spotted Groundhopper is threatened by the overgrowth of sunny patches of bare ground as a result of abandonment of land use or eutrophication. Due to their low mobility and the strong isolation of the habitat patches, a colonization of newly created habitats rarely occurs. This has been documented by declines across many regions of its German range during recent decades. Hence, the species is considered endangered. A few new records from previously empty grid cells in the recent past can mainly be attributed to populations that have so far been overlooked but hardly to an expansion of the species.

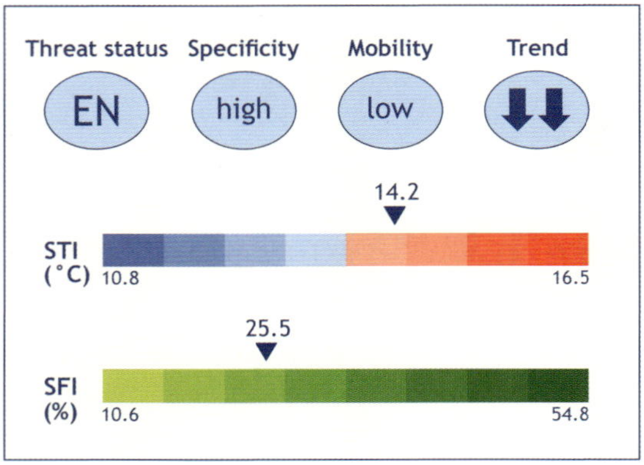

DANIEL ELIAS

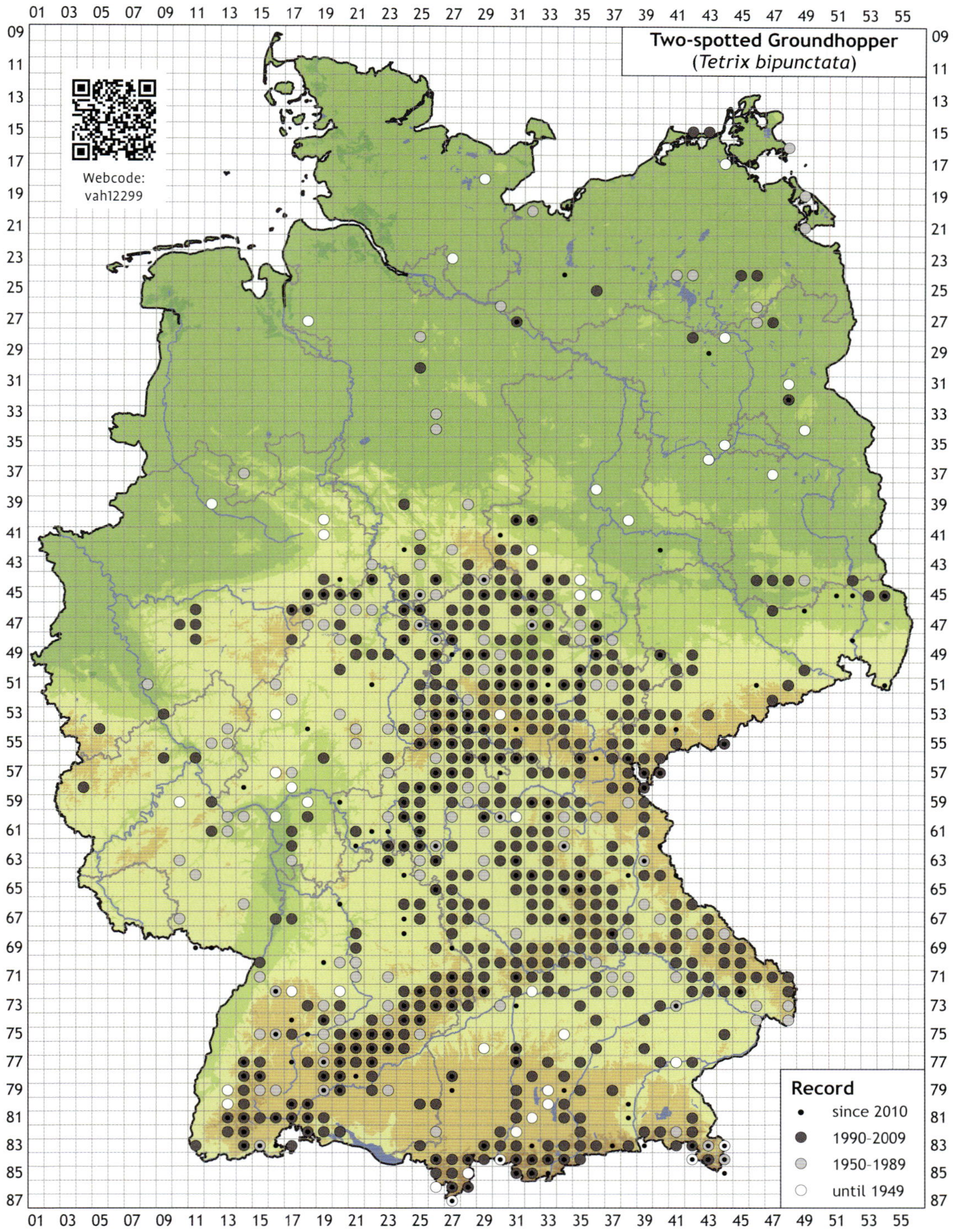
Two-spotted Groundhopper
(*Tetrix bipunctata*)
Webcode:
vah12299
Record
since 2010
1990-2009
1950-1989
until 1949

Female Long-horned Groundhopper (*Tetrix tenuicornis*).
(Photo credit: Dominik Poniatowski)

Sparsely vegetated edges of livestock drifts are among the typical habitats of the Long-horned Groundhopper (Hesse: Lower Diemel Valley).
(Photo credit: Thomas Fartmann).

Long-horned Groundhopper – *Tetrix tenuicornis* (Sahlberg, 1893)

Distribution The Long-horned Groundhopper regularly occurs in Germany on base-rich substrates in the uplands, Alpine Foreland and the Alps up to 1,000 m a.s.l. By contrast, it is very rare in the Northwest and Northeast German Lowland. Here, also base-rich, cohesive soils – for example along the large river valleys or in the young moraine landscape – are pivotal for its occurrence.

Habitat The Long-horned Groundhopper prefers sparsely vegetated dry grasslands on shallow and base-rich, mostly calcareous soils that exhibit a dry and warm microclimate. The dependence on warm microhabitats is particularly pronounced in the cooler northern part of Germany. Moreover, the species occurs in light forests, mesic grasslands and quarries or along embankments or verges. However, within these habitats, the Long-horned Groundhopper is only common where short swards rich in bare ground are present.

Population trend and threats Although the Long-horned Groundhopper colonises a wide range of habitats in Germany, the persistence of open swards within its habitats is threatened in many sites due to abandonment of land use and subsequent succession. Accordingly, a moderate decline has been assumed for recent decades and the species is now considered near threatened. Locally, however, new populations have also been recorded. Currently, it is unclear whether these observations reflect real expansions or solely an increased survey effort. Nevertheless, it is well-known that the Long-horned Groundhopper colonises newly created habitats, for example quarries, much faster than the sedentary Two-spotted Groundhopper.

DANIEL ELIAS

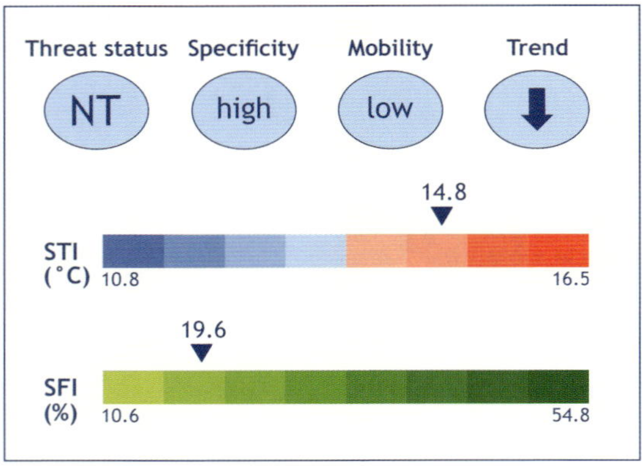

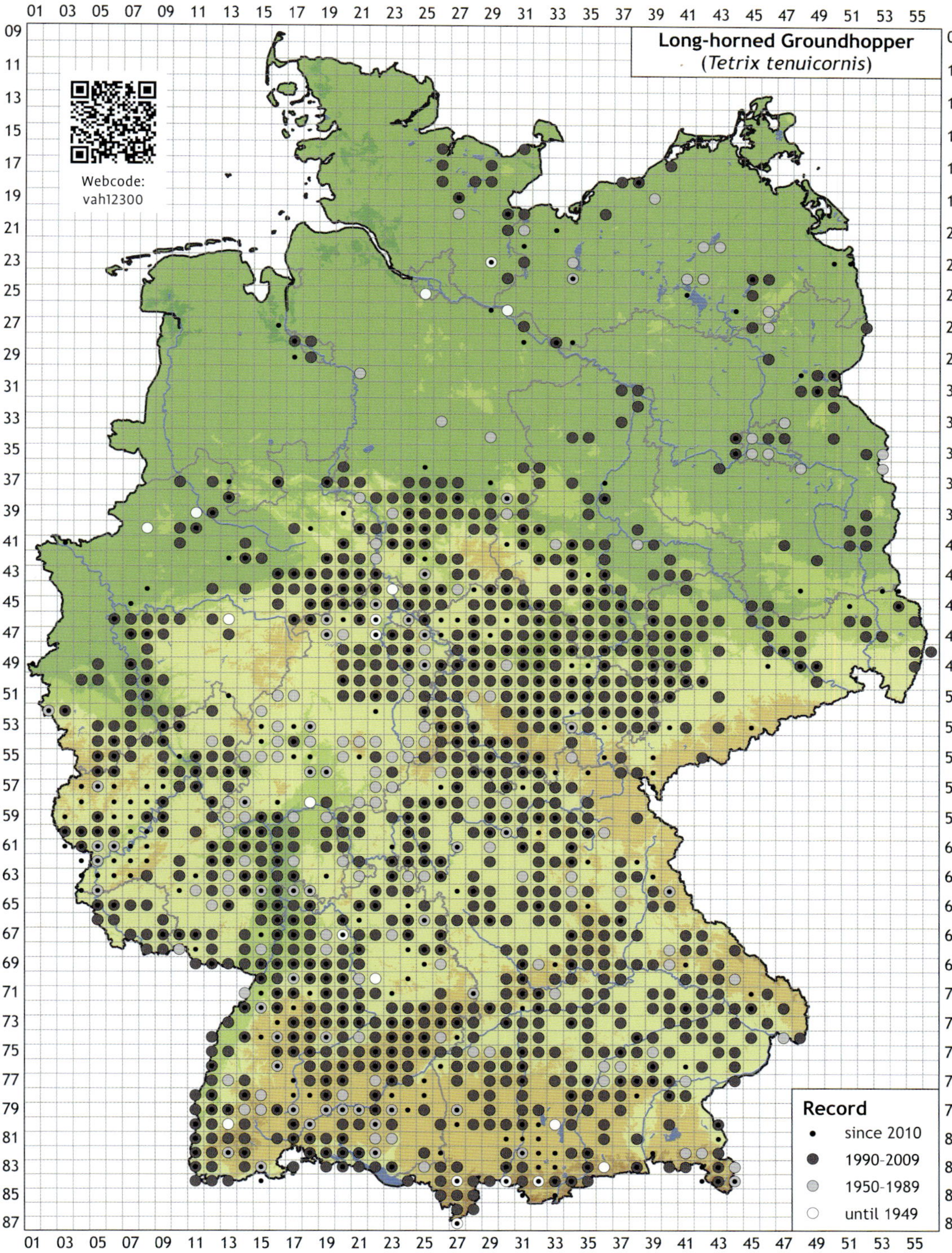
Long-horned Groundhopper
(Tetrix tenuicornis)
Webcode:
vah12300
Record
since 2010
1990-2009
1950-1989
until 1949

Female Italian Locust (*Calliptamus italicus*). (Photo credit: Dominik Poniatowski)

In the Northeast German Lowland, the Italian Locust is nowadays widespread in sandy grasslands rich in bare ground (Brandenburg: Uckermark). (Photo credit: Thomas Fartmann)

Italian Locust – *Calliptamus italicus* (Linnaeus, 1758)

Distribution The Italian Locust is among the most thermophilic and xerophilic grasshopper species in Germany. Accordingly, it occurs in regions with warm summers and low precipitation. These are particularly the valleys of the Upper Rhine, Nahe and Main in SW Germany as well as Brandenburg and adjacent parts of Saxony-Anhalt and Saxony in NE Germany. The STI of 15.5 °C thus ranks among the highest of all indigenous grasshoppers.

Habitat The colonised open habitats of the Italian Locust are diverse but have in common a lack of tall and dense vegetation and a presence of some spots of bare ground. Taller plants serve as shelter against predators, and bare soil is important for oviposition and mating. Characteristic habitats are a wide spectrum of dry grasslands on sandy soils and bedrock as well as fallow arable land and heathlands.

Population trend and threats Until about the year 2000, only a few relict populations of the Italian Locust had remained in those German regions characterised by the warmest summers. However, since then, a strong expansion in SW and NE Germany due to global warming has occurred. This expansion continues today and now even encompasses higher elevations of neighbouring low mountain ranges (e. g. the southern spurs of the Black Forest). Since the species has a high flight capability, it is actually able to colonise remote and isolated habitats. The current German distribution range is even larger than the range documented by historical observations. Consequently, the former endangered species is now considered of least concern.

ANDREAS ZEHM

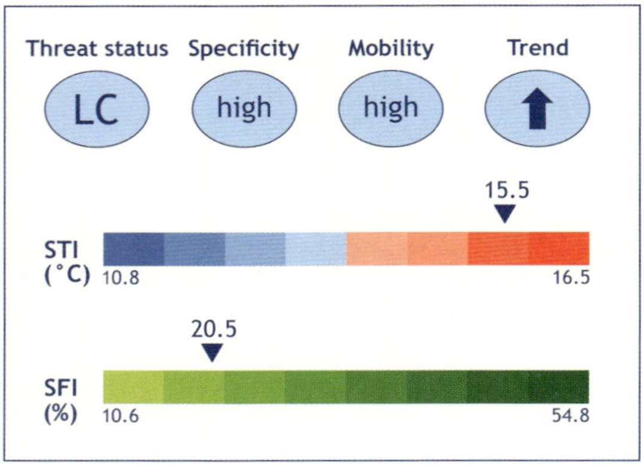

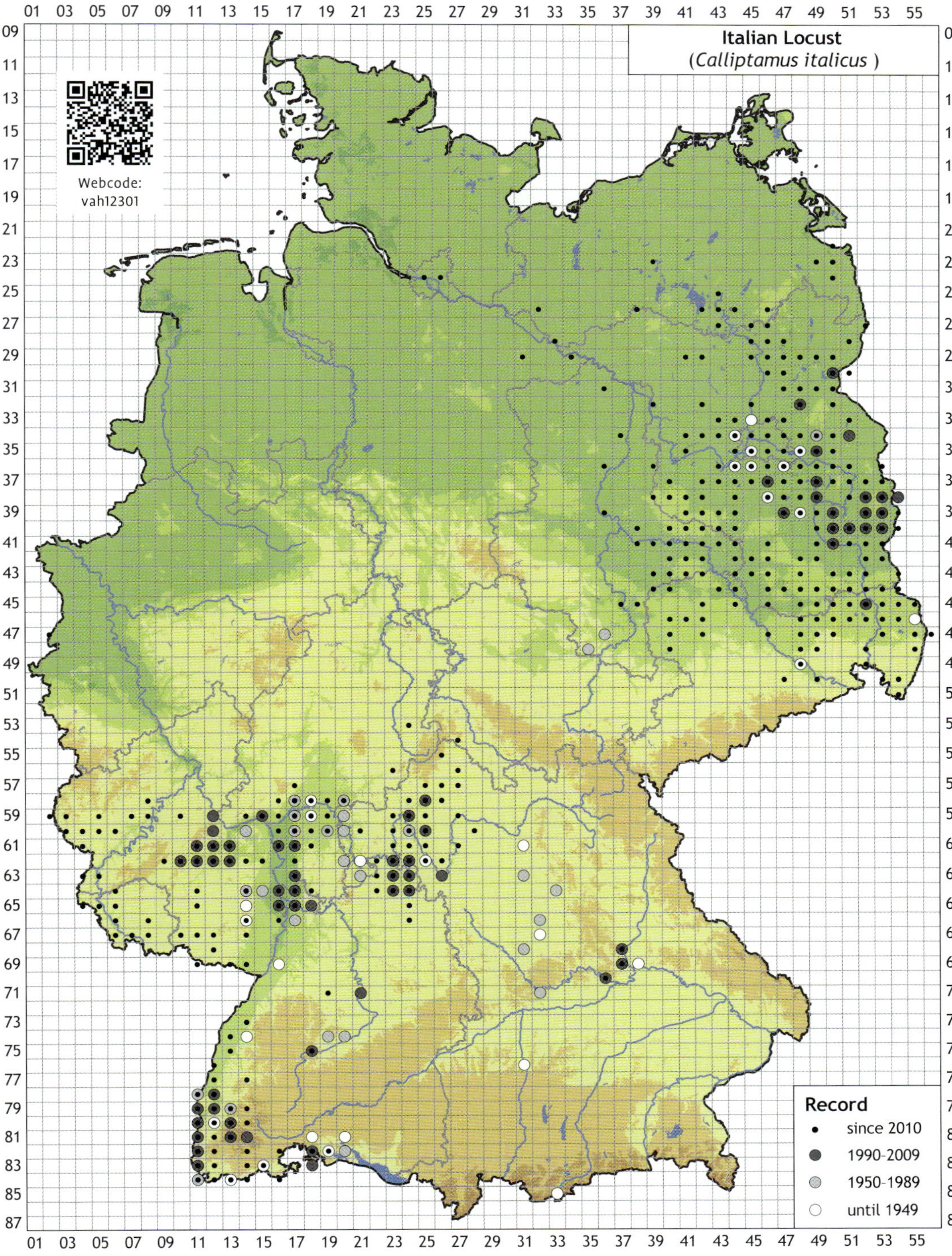
Italian Locust
(Calliptamus italicus)
Webcode:
vah12301
Record
since 2010
1990-2009
1950-1989
until 1949
01 03 05 07 09 11 13 15 17 19 21 23 25 27 29 31 33 35 37 39 41 43 45 47 49 51 53 55
09 11 13 15 17 19 21 23 25 27 29 31 33 35 37 39 41 43 45 47 49 51 53 55 57 59 61 63 65 67 69 71 73 75 77 79 81 83 85 87

Screes and outcrops are the main habitats of the Common Mountain Grasshopper (*Podisma pedestris*) (Baden-Württemberg: Upper Danube Valley).
(Photo credit: Dominik Poniatowski)

Male Common Mountain Grasshopper (*Podisma pedestris*).
(Photo credit: Dominik Poniatowski)

Common Mountain Grasshopper – *Podisma pedestris* (Linnaeus, 1758)

Distribution The Common Mountain Grasshopper is a boreo-montane glacial relict that has its main distribution within Germany in the Bavarian Alps from the Wetterstein Mountains eastwards to the Berchtesgaden Alps. Here, it occurs at elevations of 600 to 1,500 m a.s.l. Consequently, the STI is relatively low at 12.7 °C.

Habitat The Common Mountain Grasshopper is mostly geophilic and inhabits low-growing, sparsely vegetated nutrient-poor grasslands. The soils are permeable. Many German populations are small and isolated, even those within the Alps. An exception are the extensive populations on large-scale, natural screes, such as the Wimbachgries in the Berchtesgaden Alps. Long-winged individuals that are capable of flight occur very rarely. However, they may have contributed to the establishment of temporary populations in the near-natural floodplain of the Upper Isar.

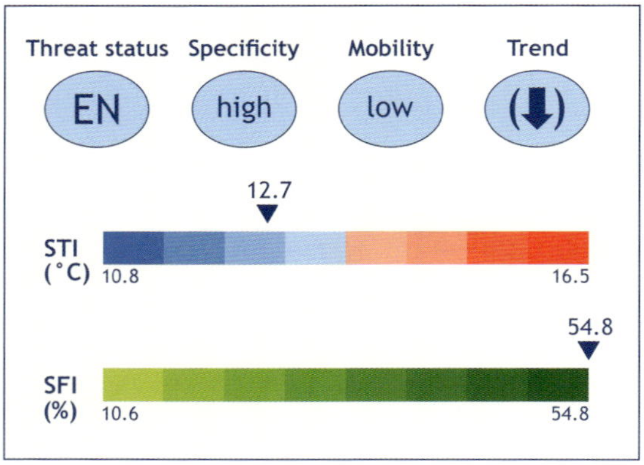

Population trend and threats The relict populations of the Common Mountain Grasshopper outside the Alps have experienced severe losses. To a limited extent, this is also true for some of those in the Alps. Today, in Germany, only eight non-Alpine areas are known to have populations, some of which exhibit several separated populations: seven of them are located in northern Bavaria (mainly Franconian Jura) and one in the Upper Danube Valley in Baden-Württemberg. The last population in Saxony (Königsbrück Heath) has not been confirmed since 2011. The main threat of the species is succession. Germany has a high responsibility for the conservation of the highly isolated outposts of the Common Mountain Grasshopper.

JOHANNES VOITH

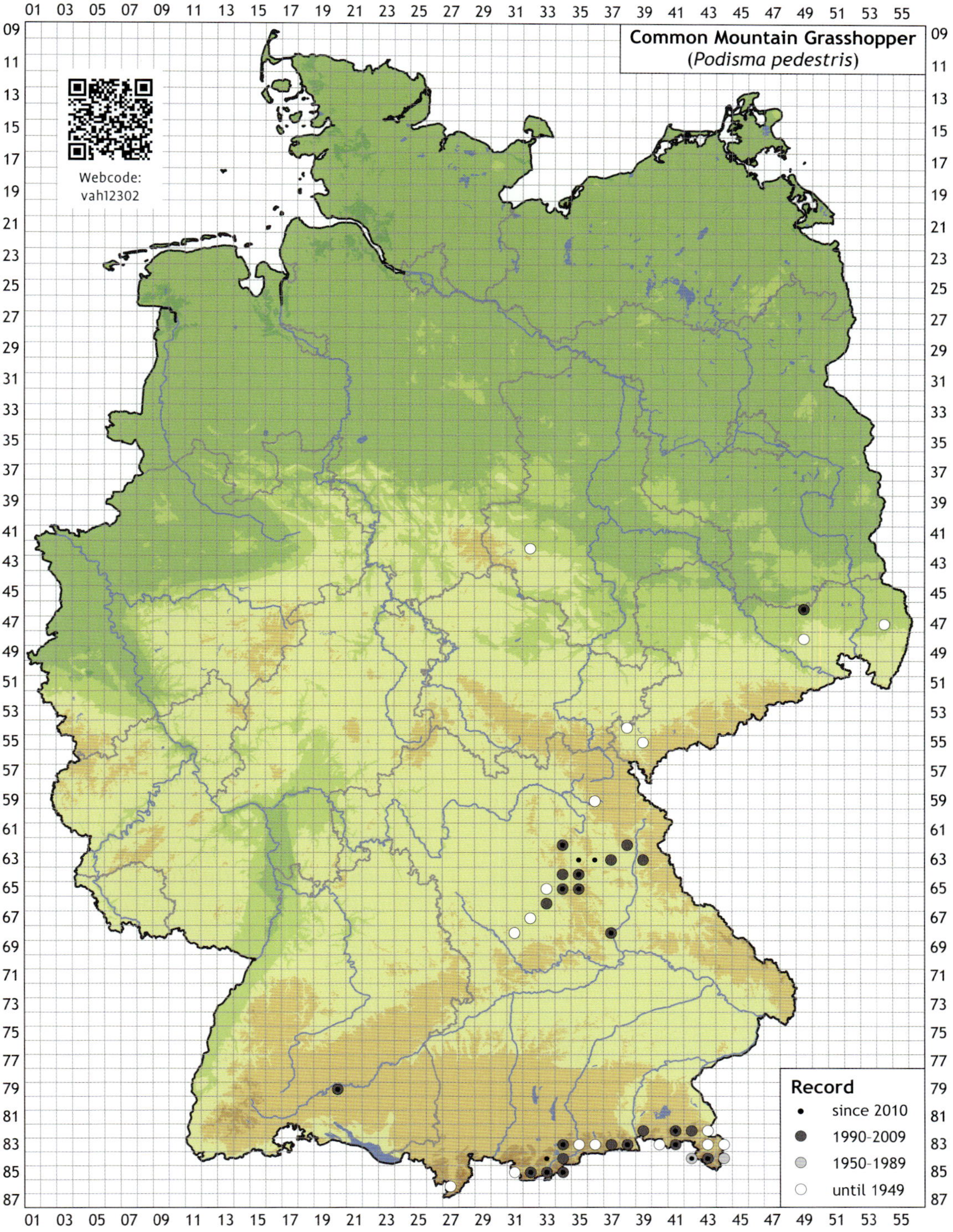
01 03 05 07 09 11 13 15 17 19 21 23 25 27 29 31 33 35 37 39 41 43 45 47 49 51 53 55
09 11 13 15 17 19 21 23 25 27 29 31 33 35 37 39 41 43 45 47 49 51 53 55 57 59 61 63 65 67 69 71 73 75 77 79 81 83 85 87
Common Mountain Grasshopper
(*Podisma pedestris*)
Webcode:
vah12302
Record
since 2010
1990-2009
1950-1989
until 1949

A characteristic habitat of the Green Mountain Grasshopper are subalpine stands of Butterbur (Bavaria: Allgäu Alps). (Photo credit: Dominik Poniatowski)

Male Green Mountain Grasshopper (*Miramella alpina*). (Photo credit: Dominik Poniatowski)

Green Mountain Grasshopper – *Miramella alpina* (Kollar, 1833)

Distribution The Green Mountain Grasshopper is a glacial relict that inhabits the entire Bavarian Alps. Locally, it also occurs in the Alpine Foreland. This is especially true for the strongly montane part that belongs to the Allgäu. Another stronghold is the Black Forest. Correspondingly, the STI is very low at 12.4 °C. The occurrence in the Bavarian Forest has only recently been confirmed. In the neighbouring Czech Republic, a larger distribution hotspot exists. In the Alps the elevational range reaches from 600 to 2,000 m a.s.l. In the Alpine Foreland and the Black Forest the species originally occurred below 500 m a.s.l. Today, the lower elevational distribution limit there is between 600 and 700 m a.s.l.

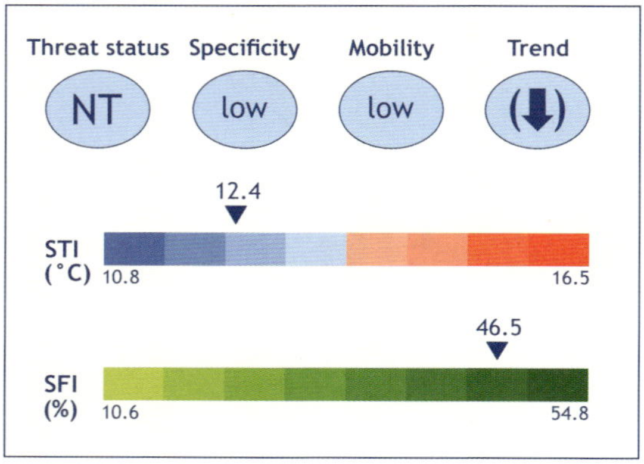

Habitat In heterogeneous open and semi-open habitats of the Alps and Black Forest, the Green Mountain Grasshopper is widespread. Here it inhabits mountain pastures, stands of tall forbs, clearings and the fringes of water bodies and woodlands. A prerequisite for its occurrence is a cool and humid microclimate. The vegetation within the habitats is often dense and rich in herbs; mostly broadleaved plant species (e. g., Butterbur) dominate, on which the grasshoppers reside.

Population trend and threats Due to the dependence on cool and moist habitats, the Green Mountain Grasshopper is assumed to suffer from climate change. Population losses at lower elevations in the Black Forest and eastern Alpine Foreland confirm this assumption. However, changes in land use, such as afforestation, abandonment and intensification of grassland management, have also been responsible for losses. However, within the cool and moist Alpine Foreland of the Allgäu and in the Alps, the status of the species is currently stable.

JOHANNES VOITH

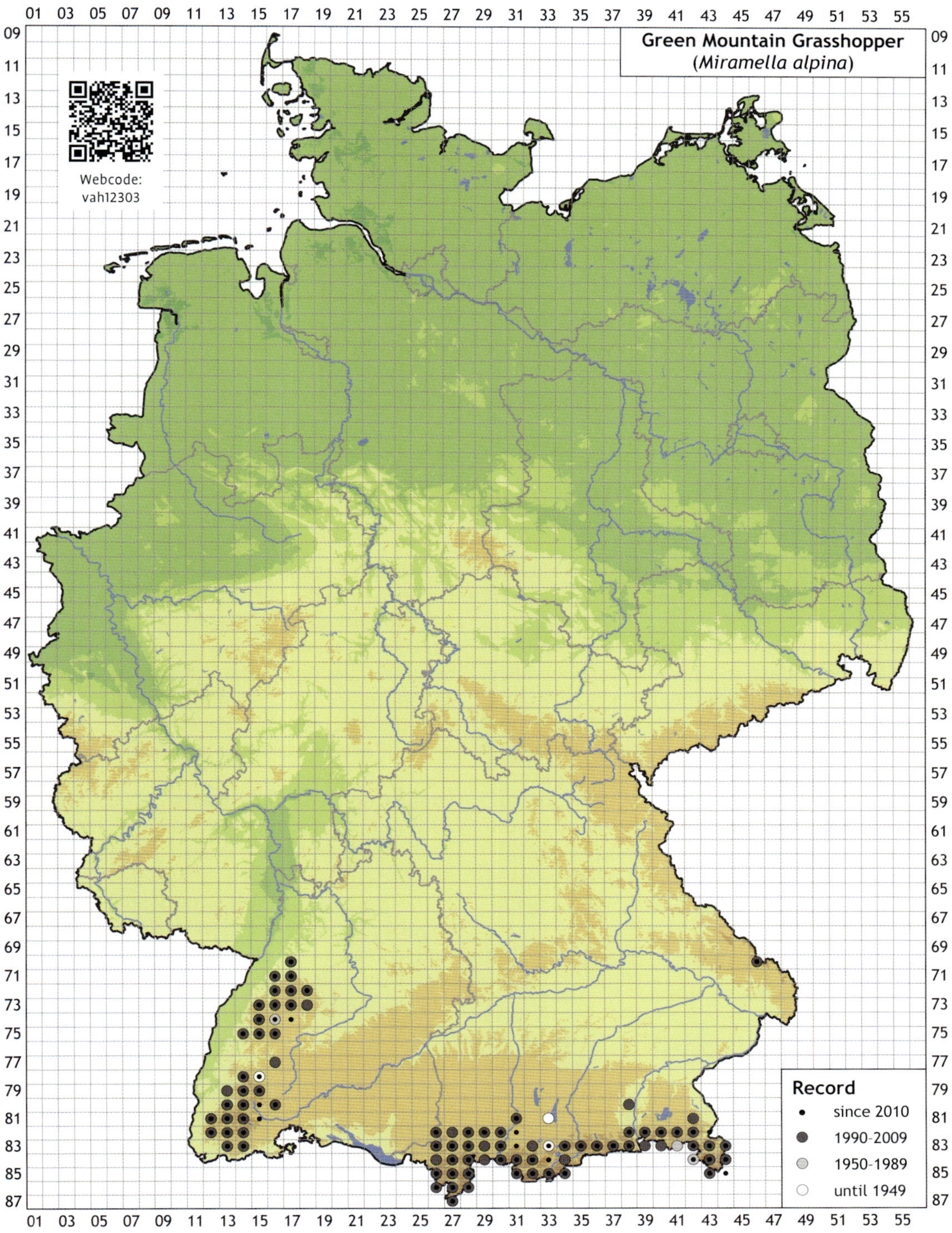
Green Mountain Grasshopper
(*Miramella alpina*)
Webcode:
vah12303
Record
since 2010
1990-2009
1950-1989
until 1949

Valley meadows characterised by low land-use intensity and a warm-moist microclimate were already colonised in the 20th century and are still core habitats of the Leek Grasshopper (Bavaria: Donau Valley). (Photo credit: Jens Sachteleben).

Female Leek Grasshopper (*Mecostethus parapleurus*). (Photo credit: Dominik Poniatowski)

Leek Grasshopper – *Mecostethus parapleurus* (Hagenbach, 1822)

Distribution The Leek Grasshopper reaches its northern range margin in Germany. Here, its distribution is largely restricted to the Southwestern Uplands, Alpine Foreland, Alps and the southernmost part of the Eastern Uplands (Danube Valley and adjacent areas). The strongholds in these biogeographic regions are currently the Upper Rhine Valley along with the adjacent Black Forest, the Lake Constance Basin and the southeast of Bavaria.

Habitat The Leek Grasshopper prefers warm and moist open habitats that are characterised by low land-use intensity, such as wet and periodically wet grasslands. Moreover, on moist soils, stands of tall forbs and ruderal vegetation as well as sparse reed beds or sedge reeds are colonised. However, particularly in the distribution centres in Germany, more intensively used grasslands are also inhabited. In dry grasslands, the Leek Grasshopper only occurs if taller vegetation is present and grasslands on moister soils are adjoining. Originally, the species was only present in the lowlands. Today, however, it has been recorded in the Alps at elevations of 1,600 m a.s.l.

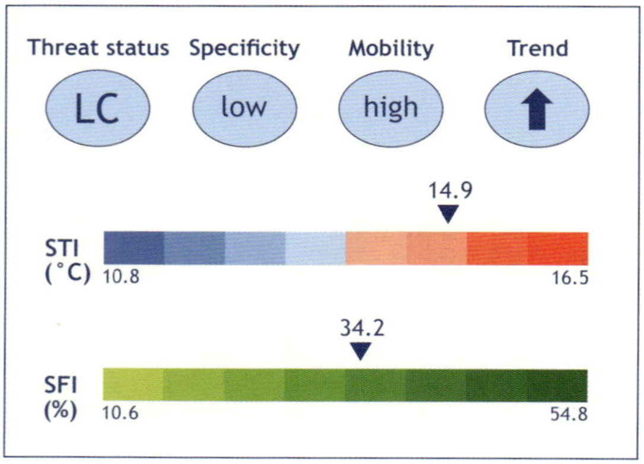

Population trend and threats By the end of the 20th century, the Leek Grasshopper was considered endangered in Germany. The overall status was better only in parts of Baden-Württemberg. At the beginning of the new millennium, a marked population recovery began, leading to a significant increase in the distribution area, especially in southeastern Bavaria but also in the Black Forest and the foothills of the Swabian Jura. Thus, the Leek Grasshopper belongs to those grasshopper species that benefit from global warming and that will probably continue to expand its range.

JENS SACHTELEBEN

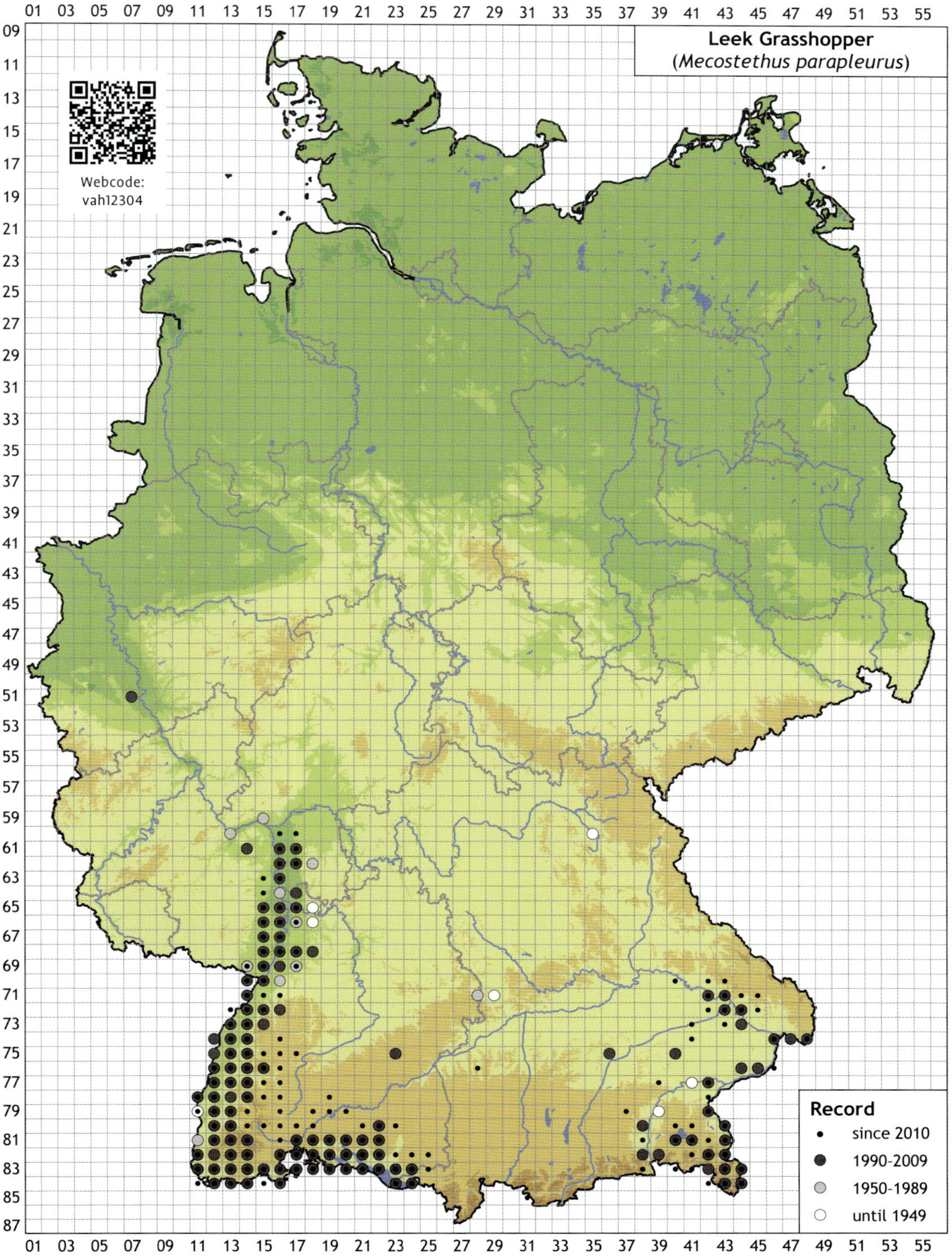
Leek Grasshopper
(*Mecostethus parapleurus*)
Webcode:
vah12304
Record
since 2010
1990-2009
1950-1989
until 1949

Wet grasslands exhibiting a low land-use intensity rank among the most important habitats of the Large Marsh Grasshopper (North Rhine-Westphalia: Münsterland). (Photo credit: Thomas Fartmann)

Female Large Marsh Grasshopper (*Stethophyma grossum*). (Photo credit: Dominik Poniatowski)

Large Marsh Grasshopper – *Stethophyma grossum* (Linnaeus, 1758)

Distribution The Large Marsh Grasshopper occurs in all seven major biogeographical regions in Germany. It mainly inhabits the planar to montane zone and, currently, reaches its upper elevational range limit at about 1,750 m a.s.l. in the Alps. However, the strongholds are located in those regions that are characterised by high precipitation but also not to cool summers. Due to climate change, the slightly thermophilic species has expanded its range during the last two decades in the oceanic northwest of Germany, parts of the uplands and the highest elevations of the Alps.

Habitat The hygrophilic Large Marsh Grasshopper colonises wet to periodically moist open habitats, which often exhibit stands of rushes or sedges. For the embryonic development, it is important that the soils are flooded or saturated with water during winter. Subsequent drying of the sites and warming favour nymphal development later on. The sun-exposed habitats are usually characterised by a heterogeneous vegetation structure. The species prefers particularly wet pastures that exhibit a low land-use intensity but also open mires and riparian zones of water bodies. In improved grassland, the Large Marsh Grasshopper sometimes occurs on a small scale along drainage ditches.

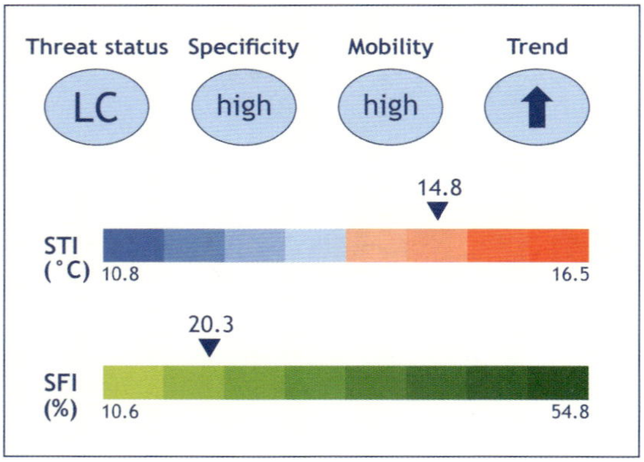

Population trend and threats The Large Marsh Grasshopper is among the common species in Germany. Both the long-term and short-term population trend are clearly positive. Currently, the species benefits from higher winter precipitation followed by warmer springs due to climate change. The protection of wet pastures that are characterised by rough grazing and the stabilization of high water levels are of vital importance for the long-term viability of the species.

ARNE DREWS

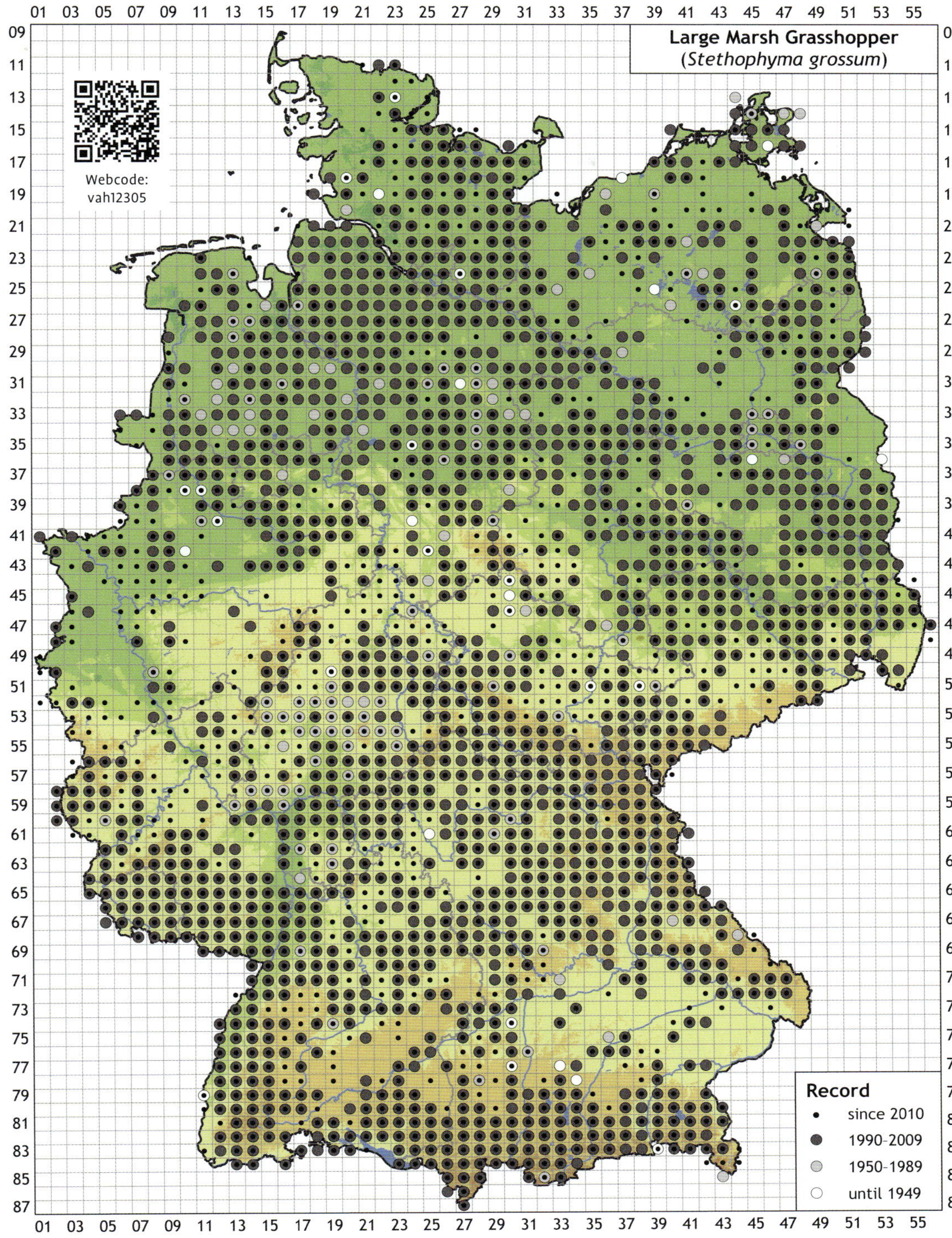
Large Marsh Grasshopper
(Stethophyma grossum)
Webcode:
vah12305
Record
since 2010
1990-2009
1950-1989
until 1949

Overall, heterogeneous calcareous grasslands with at least some patches of bare ground are characteristic habitats of the Rattle Grasshopper (Baden-Württemberg: Upper Donau Valley).
(Photo credit: Dominik Poniatowski)

Male Rattle Grasshopper (*Psophus stridulus*).
(Photo credit: Thomas Fartmann).

Rattle Grasshopper – *Psophus stridulus* (Linnaeus, 1758)

Distribution Today, the occurrence of the Rattle Grasshopper is almost completely restricted to the southern half of Germany and there mainly to the Southwestern and Eastern Uplands as well as the Alps. However, the species occurs frequently only in the Alps and along a belt from the southern Black Forest to the Swabian and Franconian Jura to the Rhön Mountains, then to the northern foothills of the Thuringian Forest and Middle Saale Valley.

Habitat The Rattle Grasshopper inhabits dry, sun-exposed and water-permeable habitats. Most of them are calcareous grasslands. However, acidic grasslands and light pine woodlands are also colonised. Accordingly, the SFI is relatively high with 35.3 %. In former times, heathlands were an additional habitat of the species in the North German Lowland. High-quality habitats exhibit small-scale mosaics of bare soil, short turf and some taller vegetation.

Population trend and threats In large parts of Germany, especially in its northern half, the Rattle Grasshopper had already disappeared by the mid-20th century. By the 1990s, the decline began in its current distribution centres; solely the populations in the Alps still seem to be stable. In former times, habitat destruction was the main cause of decrease. By contrast, today, it is the abandonment of grazing. Even a few years without management deteriorate habitat quality so much that successful reproduction is no longer possible. Since females are flightless, the colonisation of restored habitats is only possible if source populations directly adjoin.

JENS SACHTELEBEN

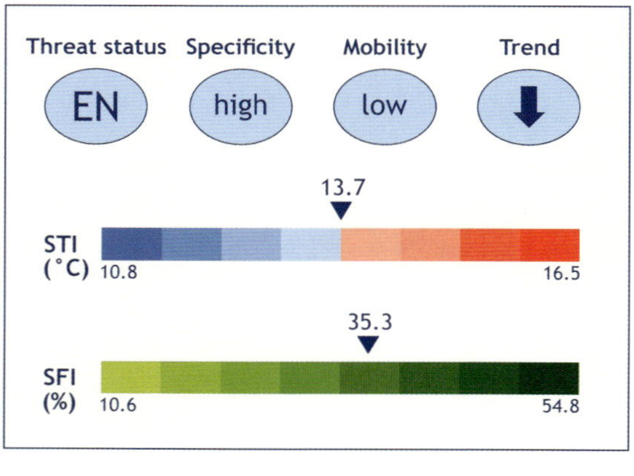

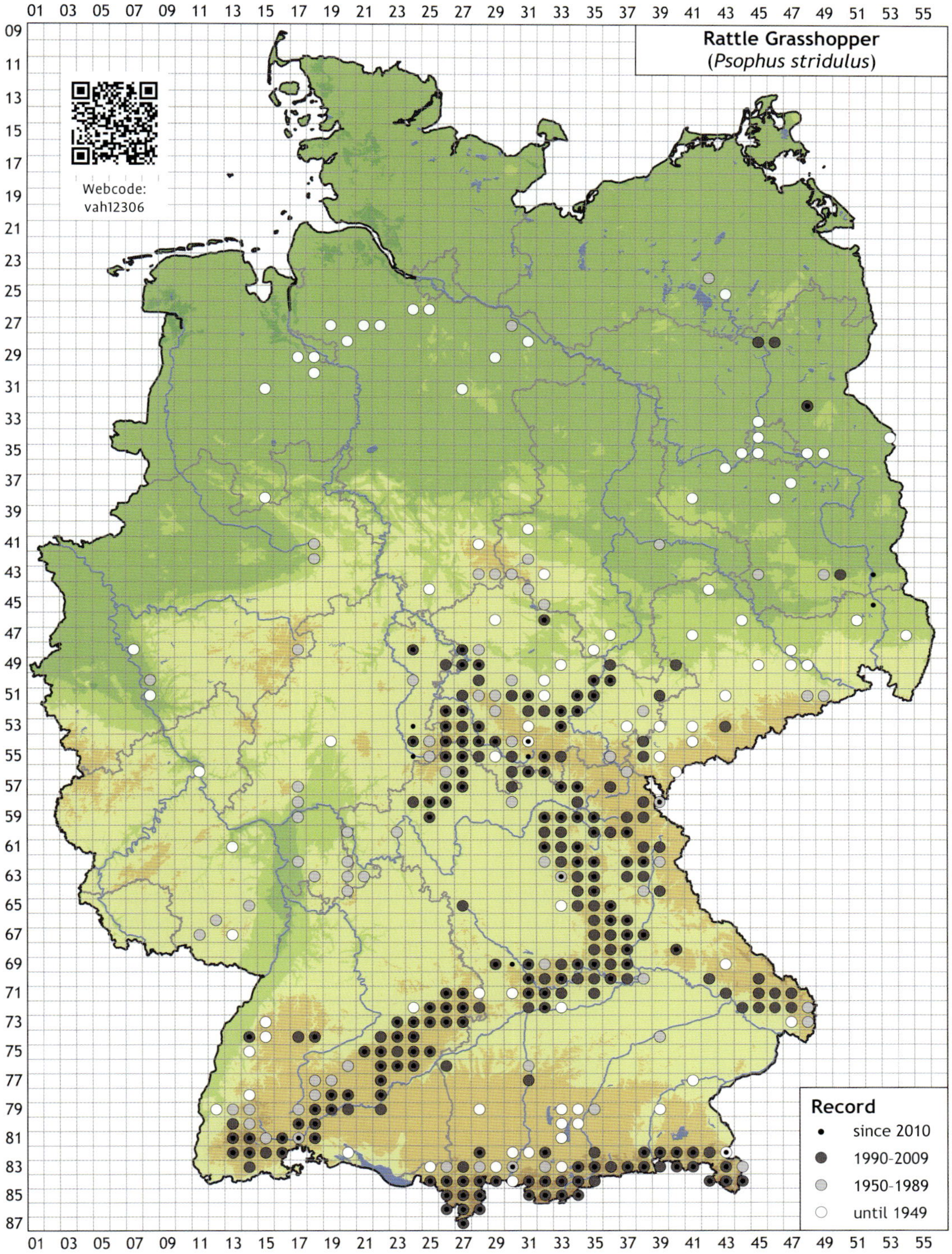
Rattle Grasshopper
(*Psophus stridulus*)
Webcode:
vah12306
Record
since 2010
1990-2009
1950-1989
until 1949

Female Blue-winged Grasshopper (*Oedipoda caerulescens*). (Photo credit: Thomas Fartmann).

Heathlands rich in bare soil are typical habitats of the Blue-winged Grasshopper (Lower Saxony: Lüneburg Heath). (Photo credit: Jonas Brüggeshemke)

Blue-winged Grasshopper – *Oedipoda caerulescens* (Linnaeus, 1758)

Distribution The Blue-winged Grasshopper occurs in all German federal states. However, those regions that are characterised by warm summers are strongly preferred. This is also reflected by a relatively high STI of 15.2 °C. Accordingly, populations above 600 m a.s.l. and in the northwestern and northern parts of Germany, which are characterised by an oceanic climate, are still rare.

Habitat The Blue-winged Grasshopper colonises dry and warm habitats exhibiting sparse vegetation and bare soil. Particularly pronounced is the preference for such early-successional stages in those regions in Germany that are characterised by cool summers. In contrast to the xerophilic adults, the nymphs are more mesophilic and can also occur in slightly denser vegetation. Characteristic habitats are alluvial gravel and sand banks, dry and semi-dry grasslands, former surface mining areas, gravel and sand pits, heaps, heathlands, industrial brownfields, open dunes, quarries, railway systems or vineyards. The area requirements of the species are relatively low; even habitats having a size of a few 100 m^2 can contain populations for longer time periods.

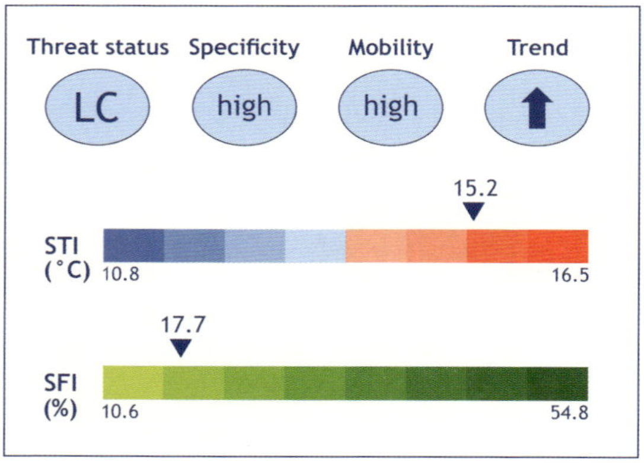

Population trend and threats Due to land-use change in the past, many former habitats of the Blue-winged Grasshopper disappeared. The main drivers of habitat loss were the abandonment of traditional management (burning, grazing or sod-cutting) and afforestation of dry/semi-dry grasslands and heathlands as well as large-scale river regulations and the general eutrophication of the landscape. Despite the high level of habitat fragmentation, the species currently responds to global warming by colonising new habitats within the present range and an expansion in those regions that were formerly characterised by an unsuitable climate.

PETER DETZEL

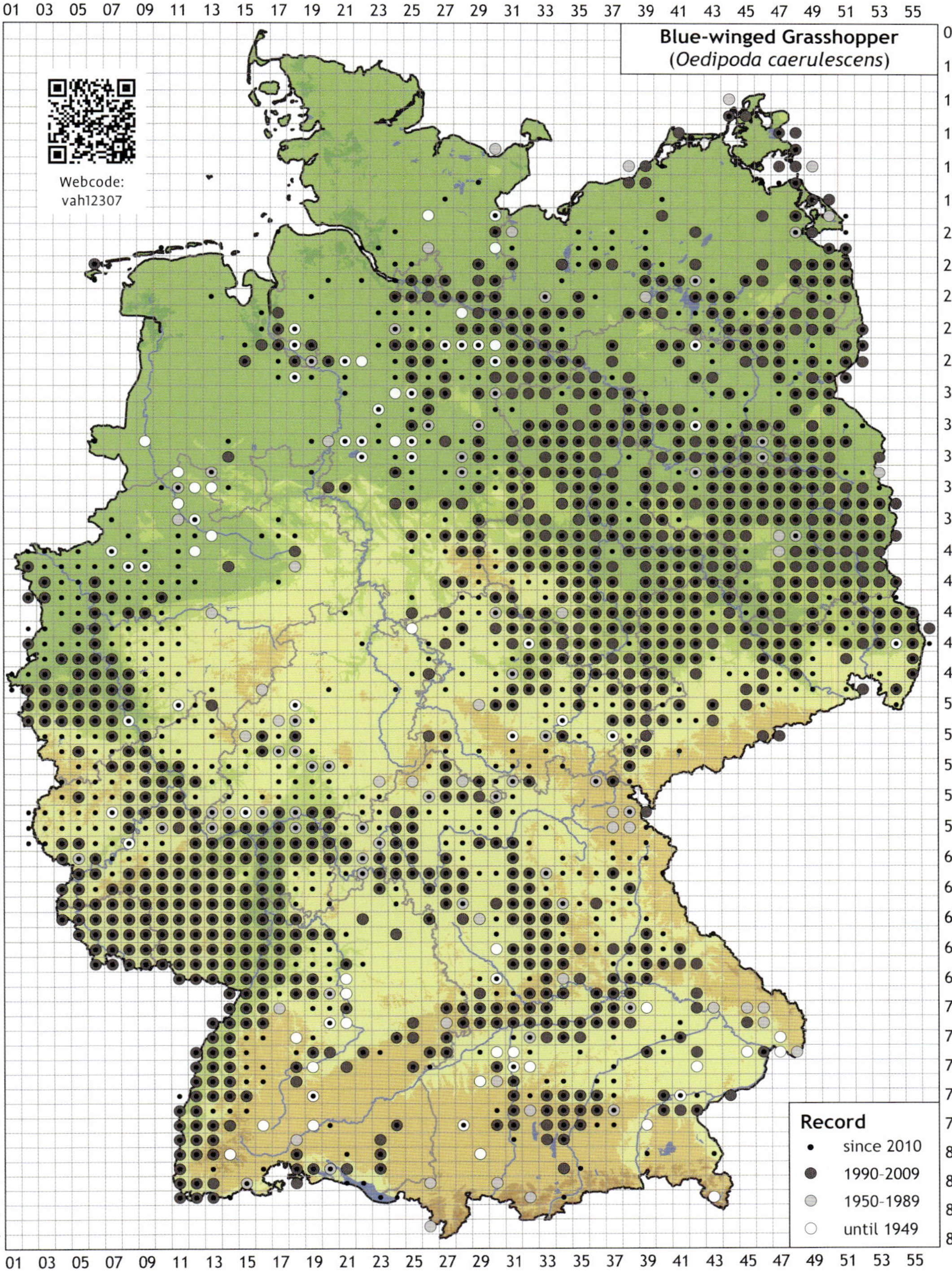
Blue-winged Grasshopper
(Oedipoda caerulescens)
Webcode:
vah12307
Record
since 2010
1990-2009
1950-1989
until 1949
01 03 05 07 09 11 13 15 17 19 21 23 25 27 29 31 33 35 37 39 41 43 45 47 49 51 53 55
09 11 13 15 17 19 21 23 25 27 29 31 33 35 37 39 41 43 45 47 49 51 53 55 57 59 61 63 65 67 69 71 73 75 77 79 81 83 85 87

South-facing screes are characteristic habitats of the Red-winged Grasshopper (Baden-Württemberg: southern Black Forest). (Photo credit: Marcel Kettermann)

Male Red-winged Grasshopper (*Oedipoda germanica*).
(Photo credit: Thomas Fartmann)

Red-winged Grasshopper – *Oedipoda germanica* (Latreille, 1804)

Distribution The occurrence of the sub-Mediterranean Red-winged Grasshopper is restricted to the southern half of Germany and, particularly to those regions that are characterised by the highest temperatures in summer. This is underlined by the relatively high STI of 15.1 °C. The distribution centres are: (i) the Middle Rhine, Nahe and Lower Moselle Valley, (ii) Lower Franconia and the Tauber Valley and (iii) the southern Franconian Jura.

Habitat Characteristic habitats of the Red-winged Grasshopper are calcareous grasslands on skeletal soils and quarries as well as screes or vineyards on shallow soils. Due to their mostly south-facing slopes and very low vegetation cover (< 30 %), during sunny summer days, the habitats exhibit an extremely hot microclimate. All populations in Germany are (i) relicts of larger, continuous populations from the past, (ii) highly isolated and (iii) inhabit mostly very small areas (< 1,000 m^2).

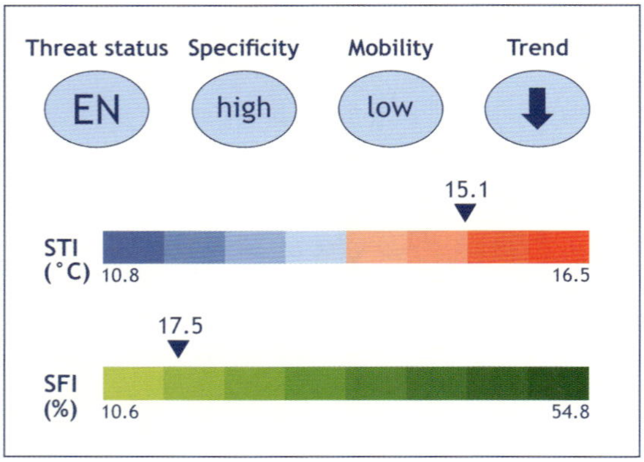

Population trend and threats Due to their very specific habitat requirements, the Red-winged Grasshopper has always had a very scattered distribution in Germany. Nevertheless, today, it is much rarer than in the past. The abandonment of traditional grazing, in particular transhumant sheep grazing, and the subsequent succession drove their strong decline. Many of the remaining populations occur in nature reserves, where the required early-successional stages are actively maintained by shrub removal, creation of bare soil or grazing by goats. Although the successional speed is generally low due to the extreme environmental conditions, habitats that are currently without management and host very small populations have an especially high risk of extinction in future.

PETER DETZEL

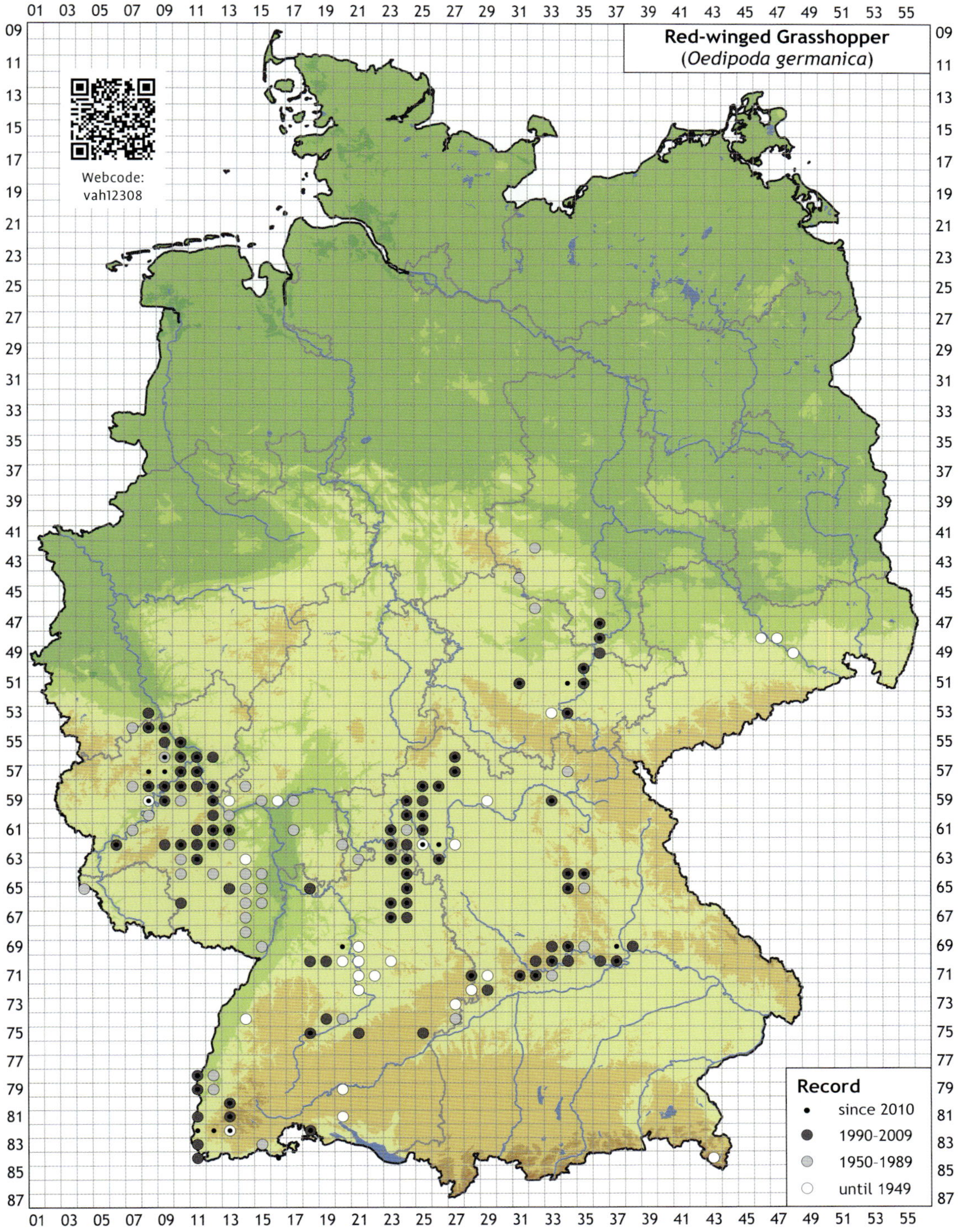
Red-winged Grasshopper
(*Oedipoda germanica*)
Webcode:
vah12308
Record
since 2010
1990-2009
1950-1989
until 1949

Female of Speckled Buzzing Grasshopper (*Bryodemella tuberculata*).
(Photo credit: Jürgen Fischer)

Sparsely vegetated gravel banks of near-natural Alpine rivers are characteristic habitats of the Speckled Buzzing Grasshopper (Bavaria: Upper Isar Valley).
(Photo credit: Michael Reich)

Speckled Buzzing Grasshopper – *Bryodemella tuberculata* (Fabricius, 1775)

Distribution The Speckled Buzzing Grasshopper reaches the western border of its Eurasian range in Germany. Today, it occurs here only in the Bavarian Alps. In the North German Lowland, it had already become extinct more than 50 years ago. Most current records stem from the tributaries of the rivers Ammer, Loisach and Isar in elevations of 700 to 950 m a.s.l.

Habitat The current habitats of the Speckled Buzzing Grasshopper are gravel banks along near-natural sections of rivers and gravel channels with periodic run-off (so-called 'Griese'). Due to fluvial dynamics, they are characterised by pioneer vegetation covering only about 10–50% of the ground.

Population trend and threats The strongest decline of the Speckled Buzzing Grasshopper in Germany already took place before 1950, when it disappeared in the North German Lowland as well as along the tributaries of the river Iller and at the river Lech north of Füssen in southern Bavaria. In the lowlands, the extinction was mainly the result of land-use change in the heathlands. By contrast, in the Alps, losses were particularly driven by hydraulic engineering. The effects of these engineering measures still affect the remaining populations today. Especially reduced discharge dynamics due to reservoirs and water diversion are highly negative. Hence, successional speed increases and habitats become overgrown. However, all embankments that accelerate the discharge are also problematic. They increase the flooding frequency at the gravel banks, and pioneer vegetation, which was stable for long time periods, disappears. As a consequence of both developments, refugia for the colonisation of new or vacant gravel banks are lacking and the remaining local populations become further isolated. The Speckled Buzzing Grasshopper is now considered threatened with extinction in Germany.

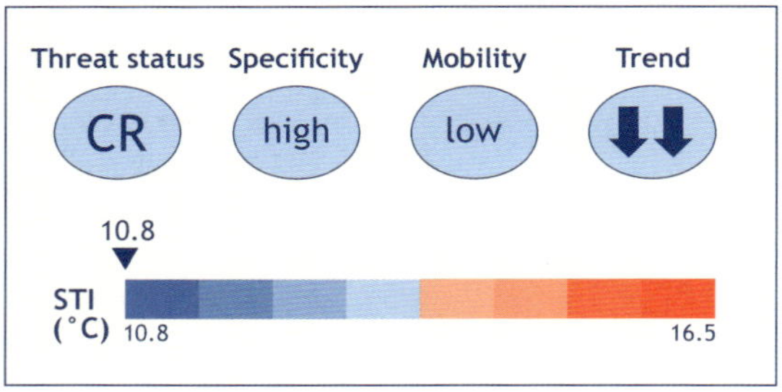

MICHAEL REICH

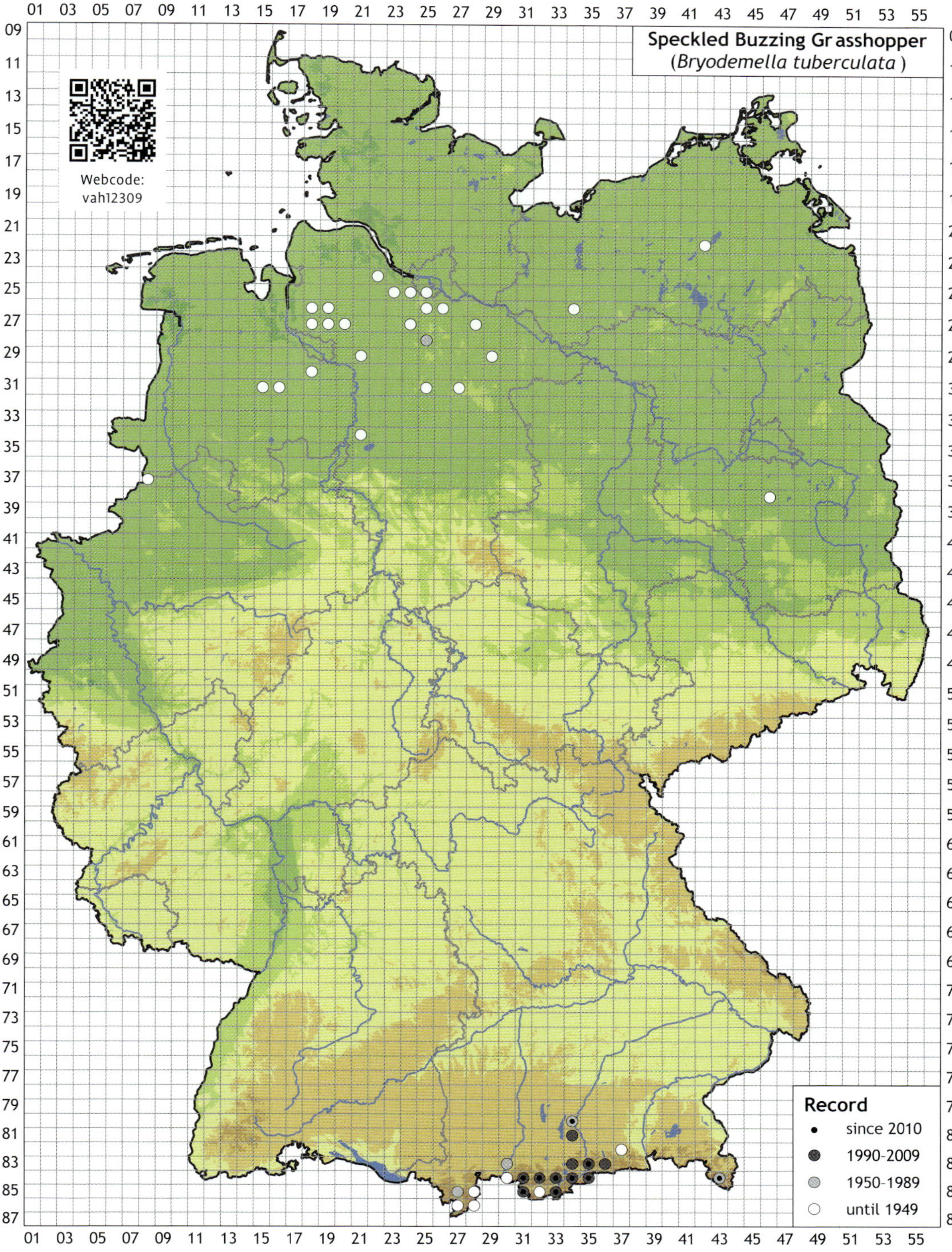
Speckled Buzzing Gr asshopper
(*Bryodemella tuberculata*)
Webcode:
vah12309
Record
since 2010
1990-2009
1950-1989
until 1949

Female Blue-winged Sand Grasshopper (*Sphingonotus caerulans*).
(Photo credit: Thomas Fartmann)

Gravel banks are important natural habitats of the Blue-winged Sand Grasshopper (Baden-Württemberg: southern Upper Rhine Valley).
(Photo credit: Marcel Kettermann)

Blue-winged Sand Grasshopper – *Sphingonotus caerulans* (Linnaeus, 1767)

Distribution An STI of 15.4 °C characterises the Blue-winged Sand Grasshopper as highly thermophilic. In Germany, the species was very rare for a long time and occurred only in a few warm biogeographical regions. The strongholds were the Upper Rhine Valley, the Middle Franconian Basin and the southern part of the Northeast German Lowland. However, due to climate change, the Blue-winged Sand Grasshopper has recently greatly expanded its range. The high capability of flight has even enabled the establishment of new populations in strongly isolated habitat patches. So far, nearly all records are located below 600 m a.s.l.

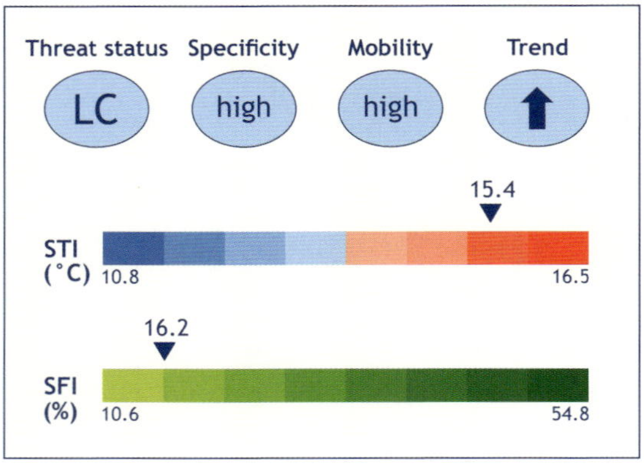

Habitat As a thermophilic pioneer species, the Blue-winged Sand Grasshopper colonises very sparsely vegetated, dry habitats rich in bare ground. Among them are gravel pits, heaps, industrial brownfields, military training areas, post-mining landscapes, quarries and railway systems. Natural habitats include gravel or sand banks along rivers. Overall, the area requirements for viable populations are relatively high in this species.

Population trend and threats In the second half of the 20th century, the Blue-winged Sand Grasshopper suffered severe population losses due to habitat destruction. Hence, in the early 2000s, the species was still considered endangered. However, the high capability of flight has enabled the species to expand its range rapidly due to climate change. Even a passive dispersal of eggs through gravel or sand transport seems to be possible. Currently, the Blue-winged Sand Grasshopper is considered of least concern, even though habitats are still destroyed from time to time at the local level.

ANDREAS KRONSHAGE

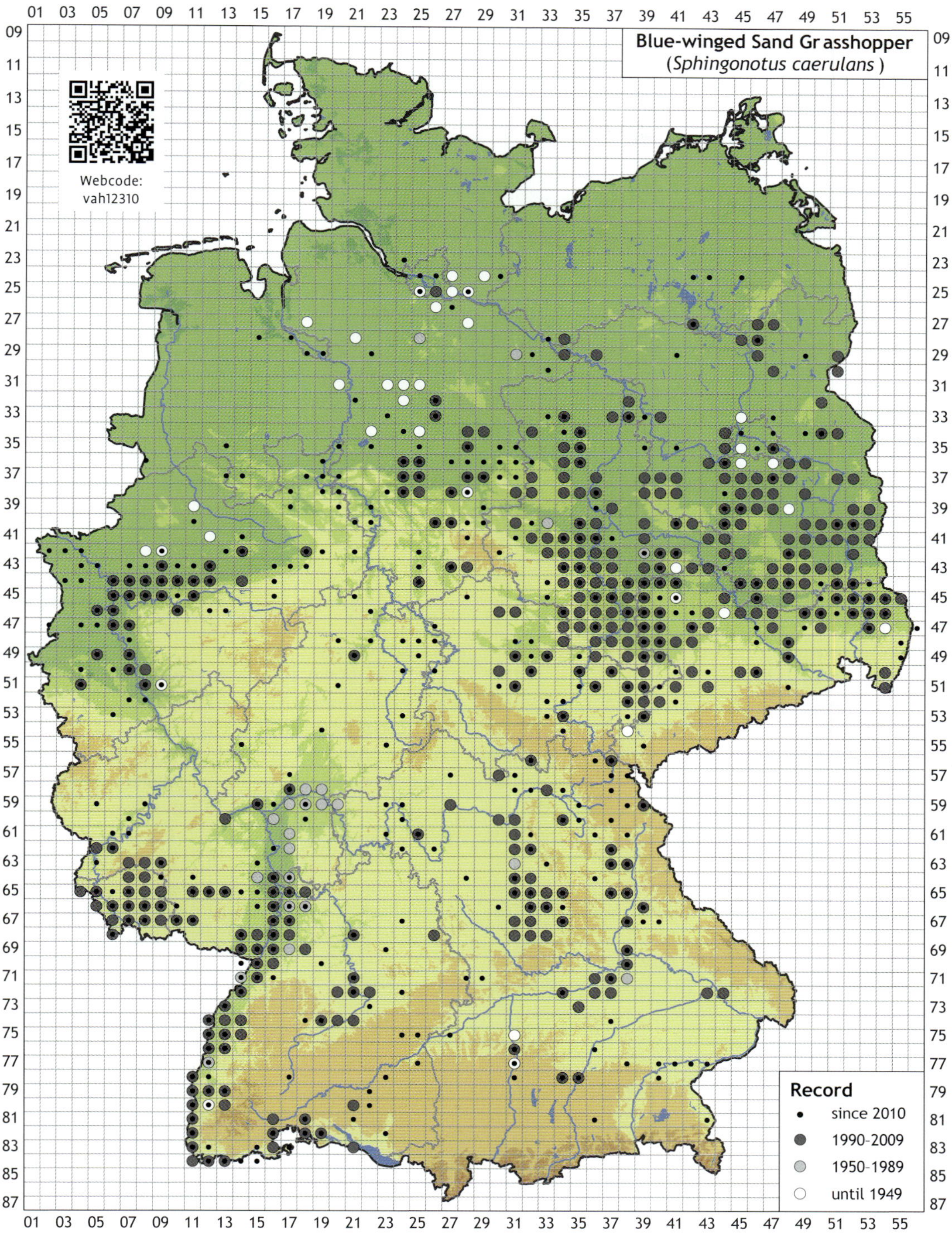
Blue-winged Sand Gr asshopper
(*Sphingonotus caerulans*)
Webcode:
vah12310
Record
since 2010
1990-2009
1950-1989
until 1949
01 03 05 07 09 11 13 15 17 19 21 23 25 27 29 31 33 35 37 39 41 43 45 47 49 51 53 55
09 11 13 15 17 19 21 23 25 27 29 31 33 35 37 39 41 43 45 47 49 51 53 55 57 59 61 63 65 67 69 71 73 75 77 79 81 83 85 87

Habitat of the Migratory Locust in the floodplain of the river Durance (France). (Photo credit: Marcel Kettermann)

Female Migratory Locust (*Locusta migratoria*; solitary phase). (Photo credit: Thomas Fartmann)

Migratory Locust – *Locusta migratoria* (Linnaeus, 1758)

Distribution In Germany, large swarms of the Migratory Locust from the Pontic-Pannonian region were regularly observed between the 6th and 19th century. The migrating animals of the so-called migratory phase probably founded small populations, some of which existed for several years. For example, the occurrence of sedentary individuals in the Coerheide near the city of Münster (North Rhine-Westphalia) from 1875 to 1899 is documented. A last swarm was spotted in Saxony in 1926. A short time later, sedentary animals were no longer detected in Germany. The last observation of a population was from Westphalia: in 1932, three individuals of the solitary phase were found in a typical habitat near the city of Borken. The few more recent records all refer to the migratory phase. It is very likely that these are escaped animals from terrarium breeding.

Habitat The Migratory Locust is both thermophilic and hygrophilic: Typical breeding habitats include riparian zones of rivers and lakes with sparse vegetation as well as floodplains and wetlands in warmer regions. For egg-laying, bare ground, which is characterised by high soil moisture, is required. Due to their high mobility, adults can also be found in drier habitats.

Population trend and threats The Migratory Locust is considered to be extinct in Germany as of 1932. Two developments are probably responsible for this: in Germany, the large-scale drainage of wetlands and the systematic regulation of rivers at the end of the 19th and beginning of the 20th century can be seen as the causes of the decline. As a result of these measures, habitats had gradually disappeared. Another driver could have been the considerable drainage of breeding habitats in the Danube Delta in the first half of the 20th century. This prevented mass reproduction, which is necessary for the formation of swarms. Accordingly, any immigration from southeastern Europe that may have supported the locust populations in Germany no longer occurred.

DOMINIK PONIATOWSKI

Threat status	Specificity	Mobility	Trend
EX	low	high	† 1932

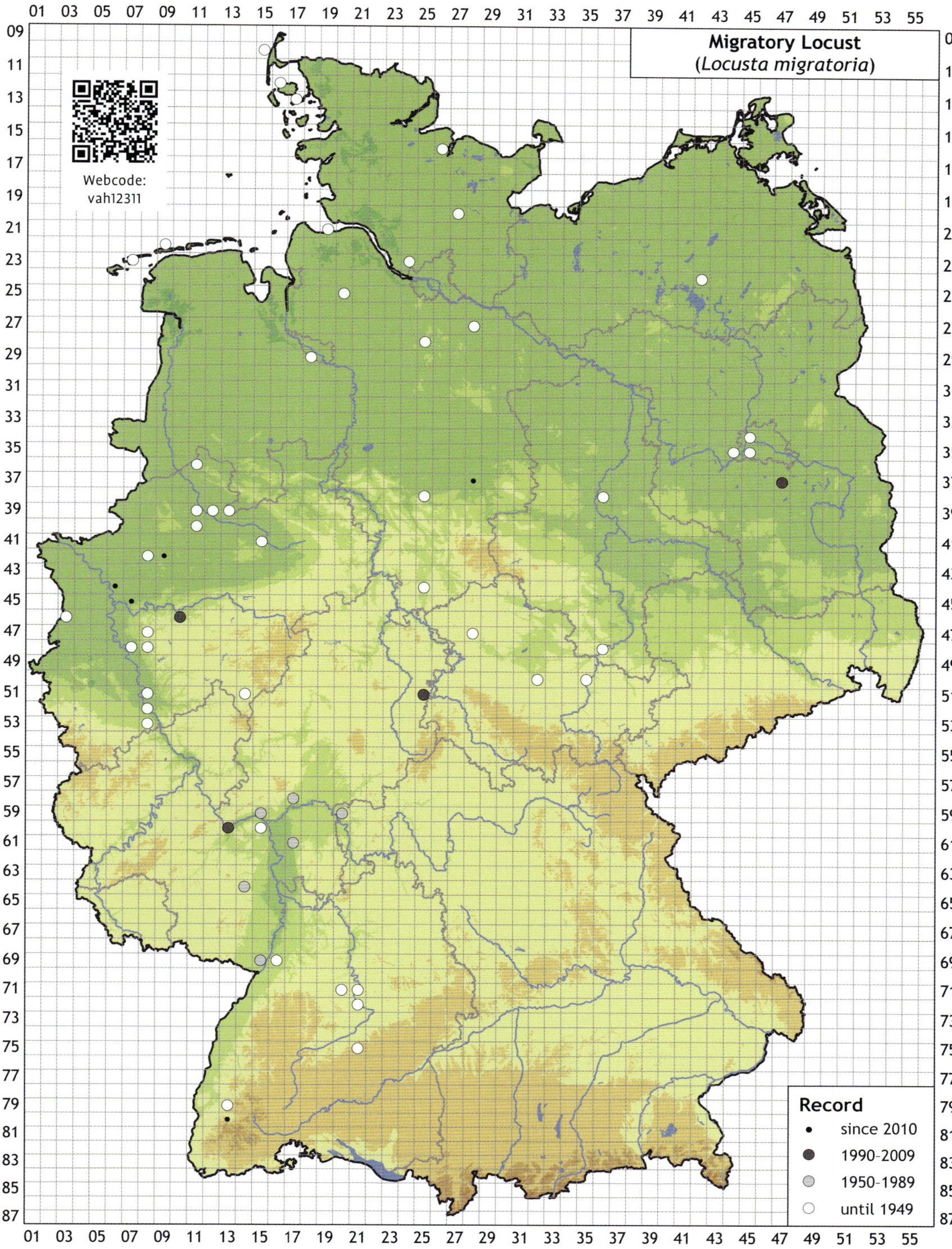
Migratory Locust
(Locusta migratoria)
Webcode:
vah12311
Record
since 2010
1990-2009
1950-1989
until 1949

Female Slender Green-winged Grasshopper (*Aiolopus thalassinus*). (Photo credit: Marcel Kettermann)

Pioneer riparian vegetation in gravel pits is a characteristic habitat of the Slender Green-winged Grasshopper (Baden-Württemberg: Upper Rhine Valley). (Photo credit: Marcel Kettermann)

Slender Green-winged Grasshopper – *Aiolopus thalassinus* (Fabricius, 1781)

Distribution The occurrences of the Slender Green-winged Grasshopper were confined to the very warm Upper Rhine Valley, particularly its northern part, for a long time. In recent years, however, the species has expanded its range to neighbouring warm regions due to climate change. The strong dependence of the Slender Green-winged Grasshopper on very warm conditions during summer is reflected by an STI of 16.5 °C, which is the highest value of all native grasshopper species.

Habitat The Slender Green-winged Grasshopper prefers open, sunlit habitats that are periodically damp to moist and rich in bare ground. For example, it inhabits sparsely vegetated alluvial grasslands, riparian pioneer vegetation and open sedge swamps. Moreover, the species is found in extraction areas such as gravel pits and sand quarries as well as inland dunes. Due to the hygrophilic nature of the early developmental stages, these habitats are often also located near water bodies or they contain periodically flooded or damp areas. The adults, by contrast, are less demanding with regard to soil moisture and can also be found in very dry habitats.

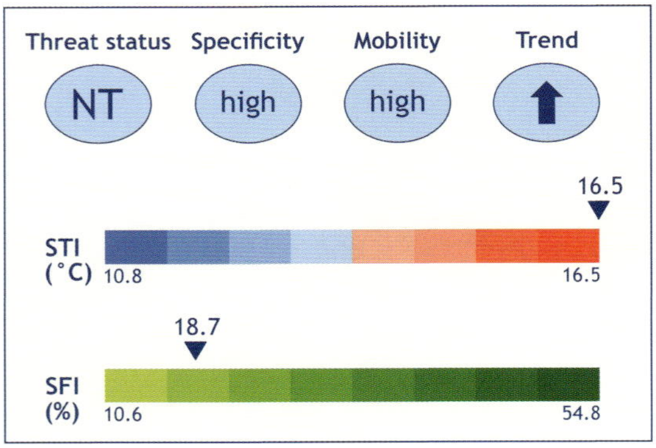

Population trend and threats In Germany, the bare-ground-rich habitats of the Slender Green-winged Grasshopper are frequently lost through construction measures, recultivation or succession. Additionally, many habitats are subject to substantial pressure from recreational activities. This regularly leads to local population declines or extinctions. While the species is suffering a moderate decline in the long term, it has shown a positive population trend in recent years induced by global warming. Though the Slender Green-winged Grasshopper has a very high mobility, the limited availability of suitable habitats may limit a further expansion of the species. According to the most recent German red list assessment, the species is currently considered near threatened.

FELIX HELBING & FRANZ LÖFFLER

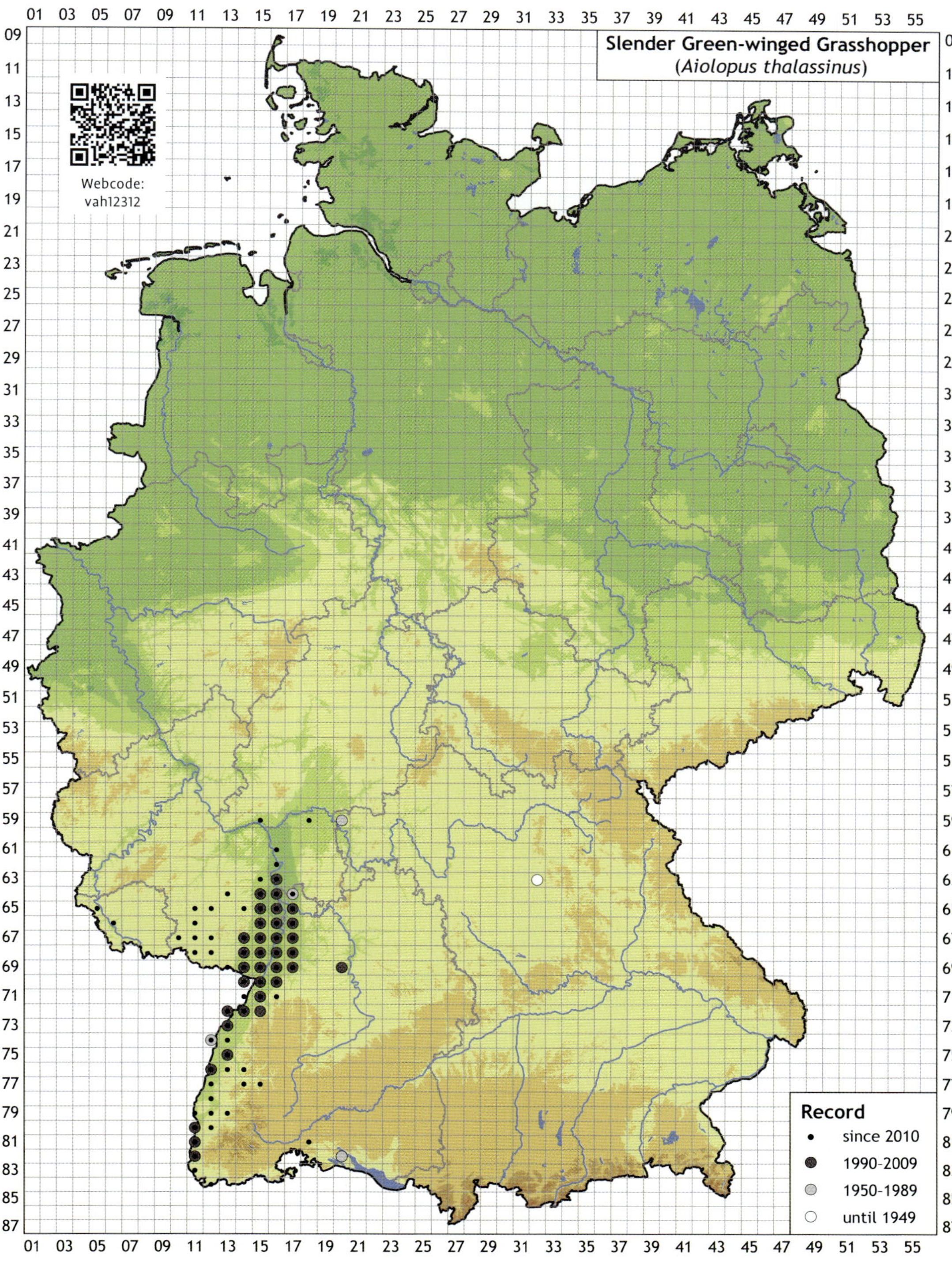
Slender Green-winged Grasshopper
(Aiolopus thalassinus)
Webcode:
vah12312
Record
since 2010
1990-2009
1950-1989
until 1949

Wet mud banks at a tributary of the Rhone river are the habitat of the River Blue-legged Grasshopper (Switzerland: Rhone Valley). (Photo credit: Christian Roesti)

Male River Blue-legged Grasshopper (*Epacromius tergestinus*). (Photo credit: Christian Roesti)

River Blue-legged Grasshopper – *Epacromius tergestinus* (Megerle von Mühlfeld, 1825)

Distribution The few records of the River Blue-legged Grasshopper in the Alpine region are strongly isolated relict populations of the subspecies *Epacromius tergestinus ponticus*. The species' core range is located in Central and East Asia. In Germany, the River Blue-legged Grasshopper was formerly recorded only in the floodplain of the river Lech south of Augsburg in the Bavarian Alpine Foreland. There, it was observed for the last time in 1941.

Habitat Along the Bavarian river Lech, the River Blue-legged Grasshopper once colonised gravel banks, where it was even the dominant grasshopper species. Within the Alps, it inhabits muddy-sandy gravel banks within dynamic floodplains of Alpine rivers. The River Blue-legged Grasshopper is considered a highly specialised pioneer species. The large-scale presence of open alluvial gravel banks characterised by natural dynamics and moist sandy areas is critical to the long-term viability of the species. For successful reproduction, a constant soil moisture and a low vegetation cover are essential.

Population trend and threats In Germany, the River Blue-legged Grasshopper became extinct as a result of extensive river regulation on the river Lech between 1933 and 1940. The species has almost vanished even from the Alpine region because of the large-scale loss of natural floodplains. In particular, hydraulic engineering and gravel extraction have led to a severe loss of natural dynamics and a lack of suitable habitats along the Alpine rivers. Remnant populations only exist in the French Alps. In Switzerland, the species was reintroduced following river restoration at the Rhone River in the Pfyn Forest region. Due to the lack of suitable habitats and geographical restriction, the River Blue-legged Grasshopper is one of the most endangered animal species in the EU.

FRANZ LÖFFLER

Threat status	Specificity	Mobility	Trend
EX	high	high	† 1941

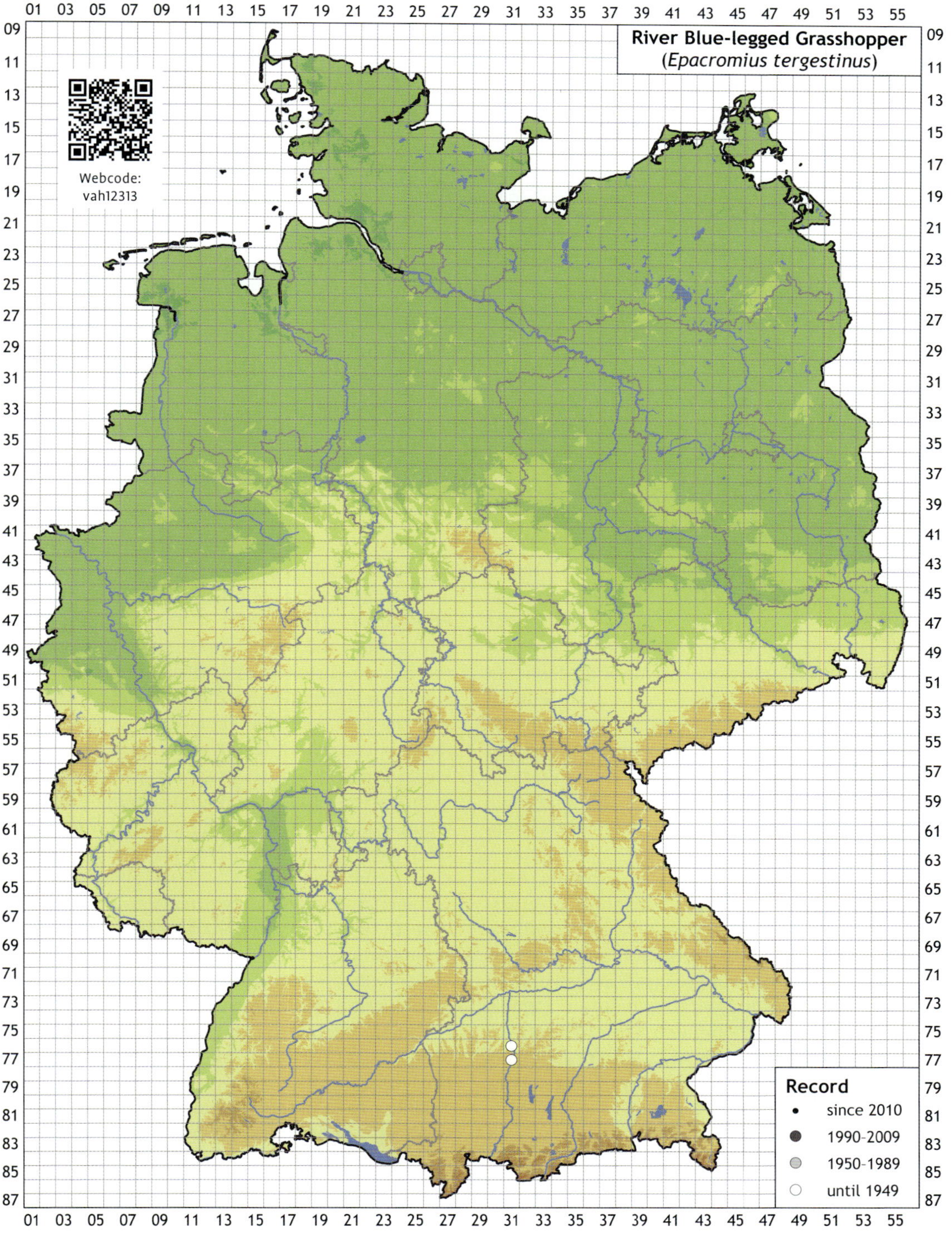
River Blue-legged Grasshopper
(Epacromius tergestinus)
Webcode:
vah12313
Record
since 2010
1990-2009
1950-1989
until 1949

Rough pasture that is colonised by the last German population of the Large Banded Grasshopper (Baden-Württemberg: Swabian Jura).
(Photo credit: Carsten Wagner)

Male Large Banded Grasshopper (*Arcyptera fusca*).
(Photo credit: Thomas Fartmann)

Large Banded Grasshopper – *Arcyptera fusca* (Pallas, 1773)

Distribution The Large Banded Grasshopper originally occurred in the German federal states of Baden-Württemberg and Bavaria. The populations were located in the northern Black Forest (Göttelfingen), at the highest elevations of the Swabian Jura (Irndorf Hardt Nature Reserve and Heuberg Military Training Area) and the pre-Alpine alluvial grasslands of the river Lech (Augsburg Municipal Forest Nature Reserve) and Isar (Munich Gravel Plain: Garching Heath). Today, only the population on the Heuberg Military Training Area in the Swabian Jura still exists.

Habitat The montane species inhabits both grazed and mown but always nutrient-poor grasslands that are characterised by a low land-use intensity. A heterogeneous habitat structure exhibiting larger patches of low-growing vegetation and some bare ground is vital to maintain a viable population. The microclimate of the habitats exhibits continental characteristics with pronounced solar radiation and, hence, warming during the day but strong cooling at night; even in summer, frost occasionally occurs at night.

Population trend and threats The former populations mostly disappeared long ago; solely the aforementioned one in the Augsburg Municipal Forest remained until 2001. Due to its rarity and the low dispersal capability – macropterous females capable of flight occur only rarely – the Large Banded Grasshopper is highly threatened by any kind of environmental change. Land-use change and the subsequent succession were very likely responsible for the extinction of the original populations: especially (i) the abandonment of wood pasture and litter gathering (northern Black Forest), (ii) the conversion of low-intensity grazing to mowing late in the season (Irndorf Hardt) and (iii) the cessation of military use associated with an introduction of rotational mowing late in the season at an interval of two to three years (Augsburg Municipal Forest) were the drivers of the population collapses. By contrast, at the Heuberg Military Training Area in the Swabian Jura, a combination of sheep grazing and military shooting/blasting maintains the preferred habitat structures.

PETER DETZEL

Threat status	Specificity	Mobility	Trend
CR	high	low	⬇

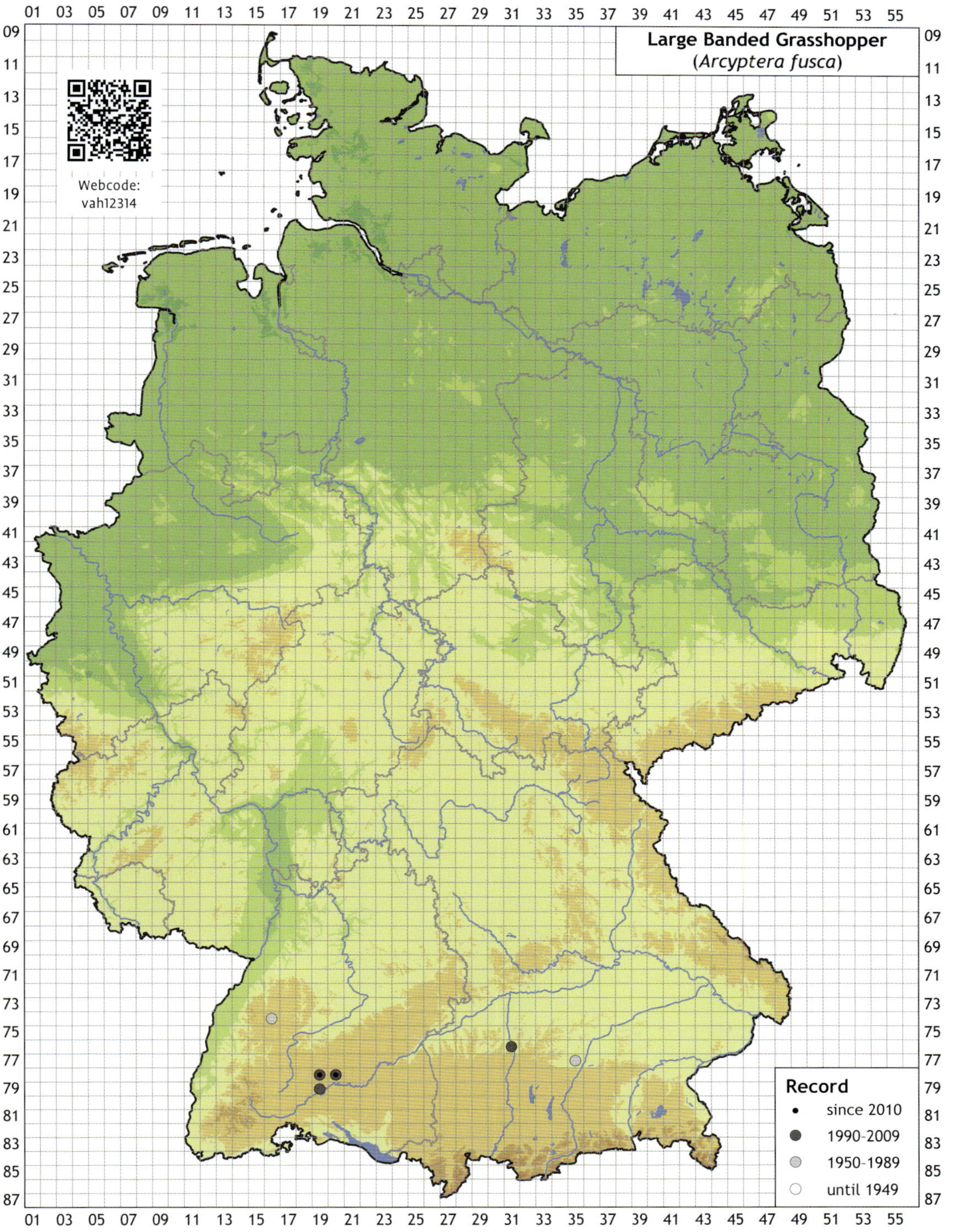
Large Banded Grasshopper
(*Arcyptera fusca*)
Webcode:
vah12314
Record
since 2010
1990-2009
1950-1989
until 1949

Large areas of open steppe grasslands in the Pannonian Basin are among the last strongholds of the Eastern Banded Grasshopper in Central Europe (Hungary: Eastern Bakony Mountains). (Photo credit: Zoltán Kenyeres)

Male Eastern Banded Grasshopper (*Arcyptera microptera*). (Photo credit: Petr Kočárek)

Eastern Banded Grasshopper – *Arcyptera microptera* (Fischer von Waldheim, 1833)

Distribution The Eastern Banded Grasshopper is a Euro-Siberian species that is common across the continental steppe and forest steppe zone of Central and Eastern Asia. By contrast, in Europe, the species is usually very rare. Only from the Pannonian Basin in Hungary to the Balkans does the species occur more frequently. In Germany, the Eastern Banded Grasshopper was historically only found in more continental biogeographical regions that are characterised by dry conditions and warm summers. The records came from the northern Upper Rhine Valley (Mainzer Sand, Griesheimer Sand), the centre of the Northeast German Lowland and the northern Franconian Jura. The species was last observed in 1957 in the Griesheimer Sand near Darmstadt.

Habitat The Eastern Banded Grasshopper is strongly xero-thermophilic. It prefers open, sunlit habitats such as inland dunes as well as sandy and rocky dry grasslands with spare vegetation. A high proportion of bare ground is essential for oviposition. This ensures a strong warming of the habitats on sunny days, which favours a successful development of the eggs and later the nymphs. Within the Asian core distribution range, the species inhabits mainly steppes. There, under favourable climatic conditions, eruptive population outbreaks usually resulting in a rapid spread of the Eastern Banded Grasshopper can occur, with the risk of high losses of the grain harvest.

Population trend and threats As a result of the large-scale loss of extensive, sparsely vegetated dry habitats during the 20th century, the populations of the Eastern Banded Grasshopper have declined sharply throughout Europe. The last Central European populations are now located in Hungary.

FRANZ LÖFFLER

Threat status	Specificity	Mobility	Trend
EX	high	high	† 1957

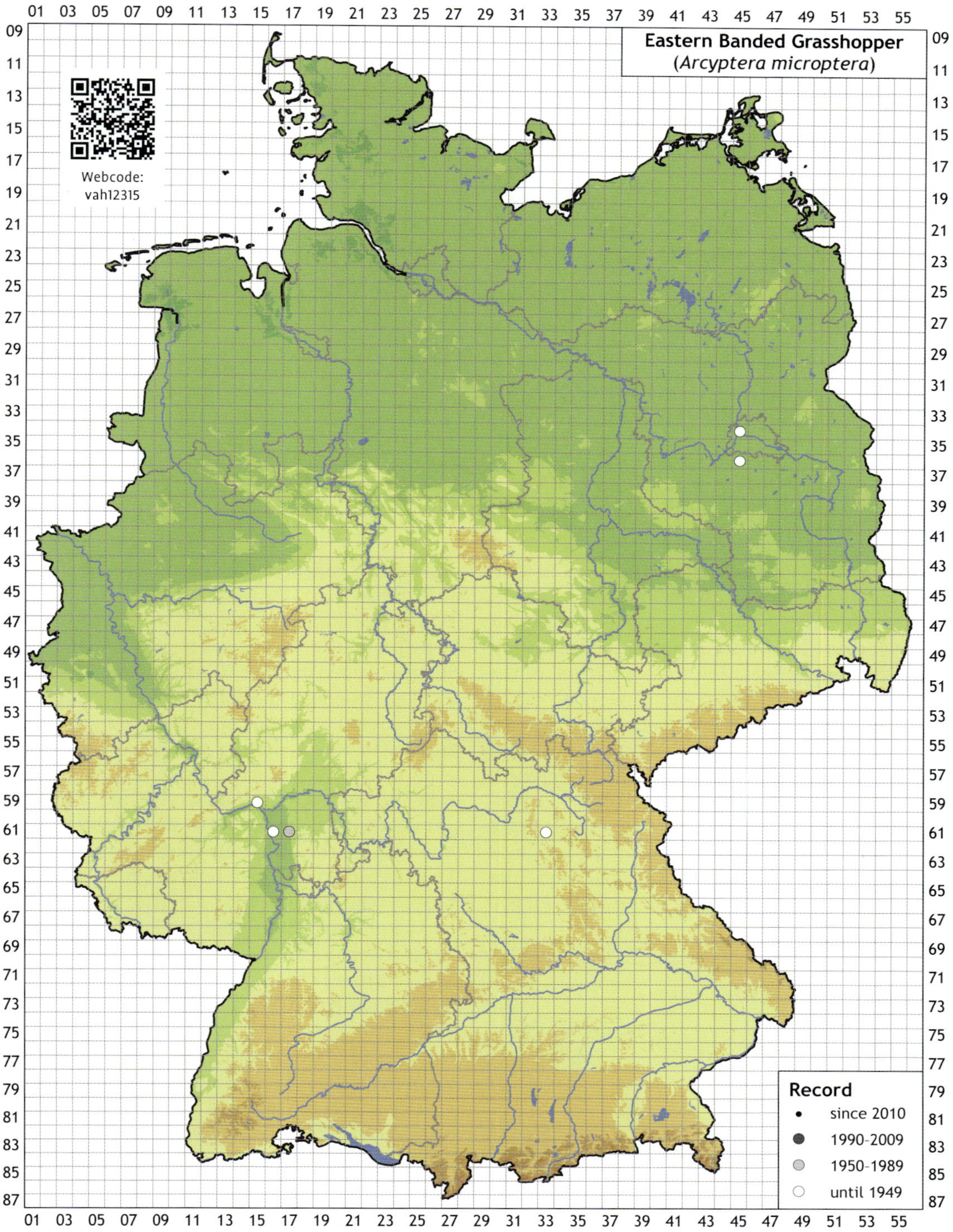
Eastern Banded Grasshopper
(*Arcyptera microptera*)
Webcode:
vah12315
Record
since 2010
1990-2009
1950-1989
until 1949

Long-winged female Large Gold Grasshopper (*Chrysochraon dispar*).
(Photo credit: Dominik Poniatowski)

Woodland clearings have recently become an important habitat for the Large Gold Grasshopper in the low mountain ranges (North Rhine-Westphalia: Eifel Mountains).
(Photo credit: Marcel Kettermann)

Large Gold Grasshopper – *Chrysochraon dispar* (Germar, [1834])

Distribution During the last 30 years, the thermophilic Large Gold Grasshopper has expanded its range in Germany northwestwards and to higher elevations due to climate change. The important role of warm summers for the occurrence of the species is reflected by an above average STI of 14.9 °C. Currently, larger distribution gaps only occur in cooler, higher elevations of some low mountain ranges and the Alps as well as in the most oceanic northwest and north of the country. In contrast to many other grasshopper species, this range expansion did not only take place in the lowlands and along valley bottoms, but also on a broad front across mountain ridges. In Particular, long-winged individuals capable of flight were responsible for this spread.

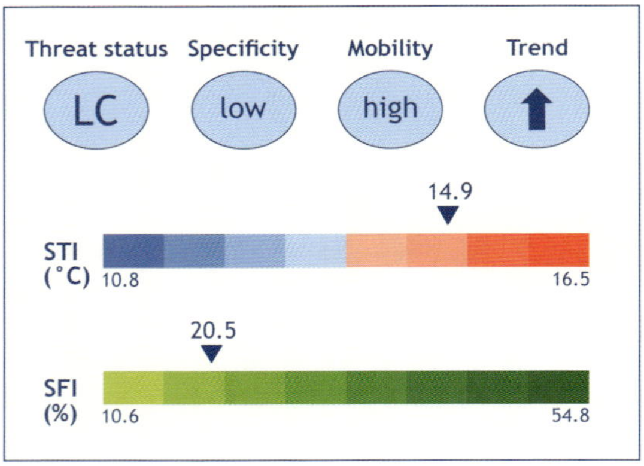

Habitat The Large Gold Grasshopper is slightly hygrophilic and inhabits mainly tall and dense swards of open habitats. Typical habitats are ditches, fens, stands of tall forbs or wet grasslands. In rainy regions, the species can also be found on fresh sites, for example in woodland clearings, or even in semi-dry grasslands. In contrast to most other grasshoppers, the Large Gold Grasshopper deposits its eggs in plant stems or dead wood and not into the soil. Consequently, it depends on low-intensity land use or land where management has become abandoned. Even litter accumulation on the ground is tolerated; on fresh or even dry sites, it results in a more humid microclimate, which favours the colonisation by the species.

Population trend and threats In the long- and short-term, the population trend of the Large Gold Grasshopper in Germany is positive due to global warming. With ongoing climate change, the current distribution gaps will very likely also disappear. As a result, the species is considered of least concern.

CARSTEN RENKER

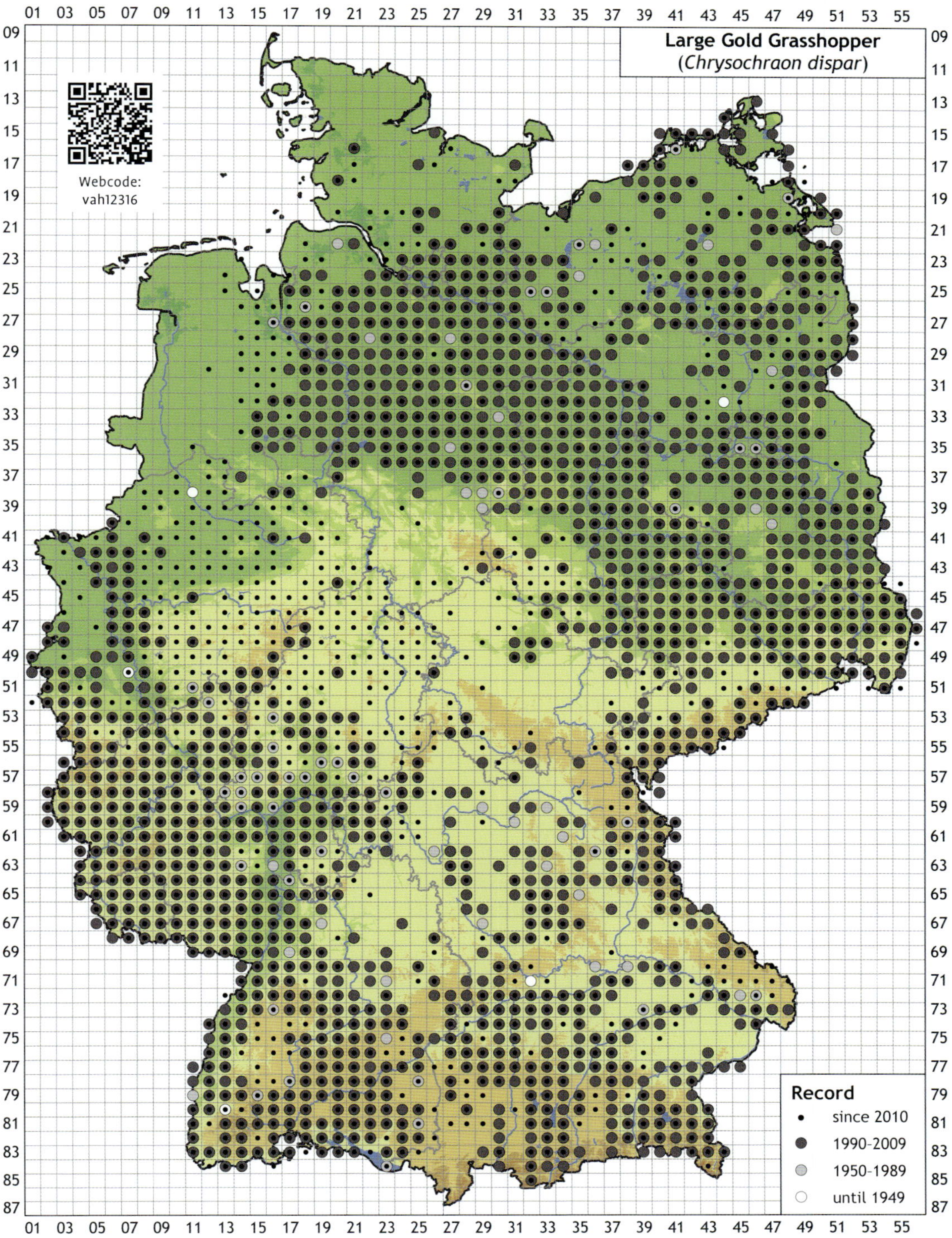
Large Gold Grasshopper
(*Chrysochraon dispar*)
Webcode:
vah12316
Record
since 2010
1990-2009
1950-1989
until 1949
01 03 05 07 09 11 13 15 17 19 21 23 25 27 29 31 33 35 37 39 41 43 45 47 49 51 53 55
09 11 13 15 17 19 21 23 25 27 29 31 33 35 37 39 41 43 45 47 49 51 53 55 57 59 61 63 65 67 69 71 73 75 77 79 81 83 85 87

In calcareous grasslands of the Alps that are characterised by low land-use intensity, the Small Gold Grasshopper is common (Bavaria: Niederwerdenfelser Land). (Photo credit: Gwydion Scherer)

Male Small Gold Grasshopper (*Euthystira brachyptera*). (Photo credit: Dominik Poniatowski)

Small Gold Grasshopper – *Euthystira brachyptera* (Ocskay, 1826)

Distribution The Small Gold Grasshopper is restricted to the southeastern part of Germany. The species is widespread in the Alps and Alpine Foreland as well as in the south and east of the Southwestern and Eastern Uplands. Moreover, there are locally many records in the south of the Northeast German Lowland.

Habitat The Small Gold Grasshopper inhabits a wide variety of tall-growing open habitats. On the one hand, the species thrives in dry habitats such as calcareous grasslands. Here, it prefers especially young fallows. On the other hand, the Small Gold Grasshopper occurs in wet habitats such as litter meadows and mires. In grasslands, an SFI of 26.3 % characterises the species as an indicator for low land-use intensity. Further typical habitats are clear-cuts, forest edges, glades and windthrows.

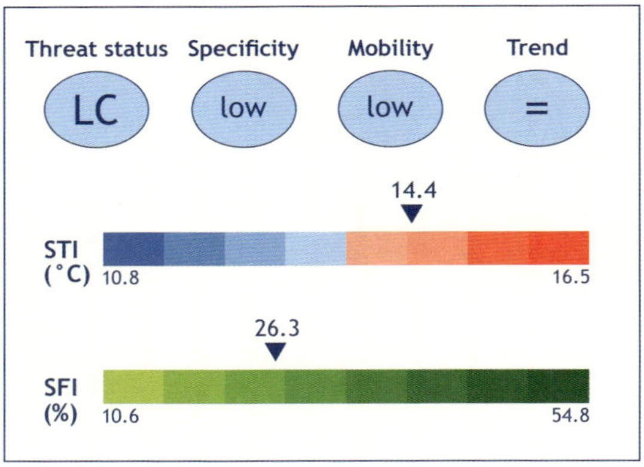

Population trend and threats In recent decades, there have been only minor changes in the populations of the Small Gold Grasshopper in Germany. However, there are strong regional differences. Windthrows and bark beetle calamities have recently created new habitats that have fostered the expansion of the species. By contrast, population declines have been observed in those regions where abandonment of grassland management has led to large-scale shrub encroachment within the habitats. Afforestation of open habitats and intensive grassland management are considered further threats. At the northern and western range margin, habitat fragmentation hampers an expansion of the sedentary species. Due to the regionally very different population states and development forecasts, the Small Gold Grasshopper should be closely monitored in the following years.

GWYDION SCHERER

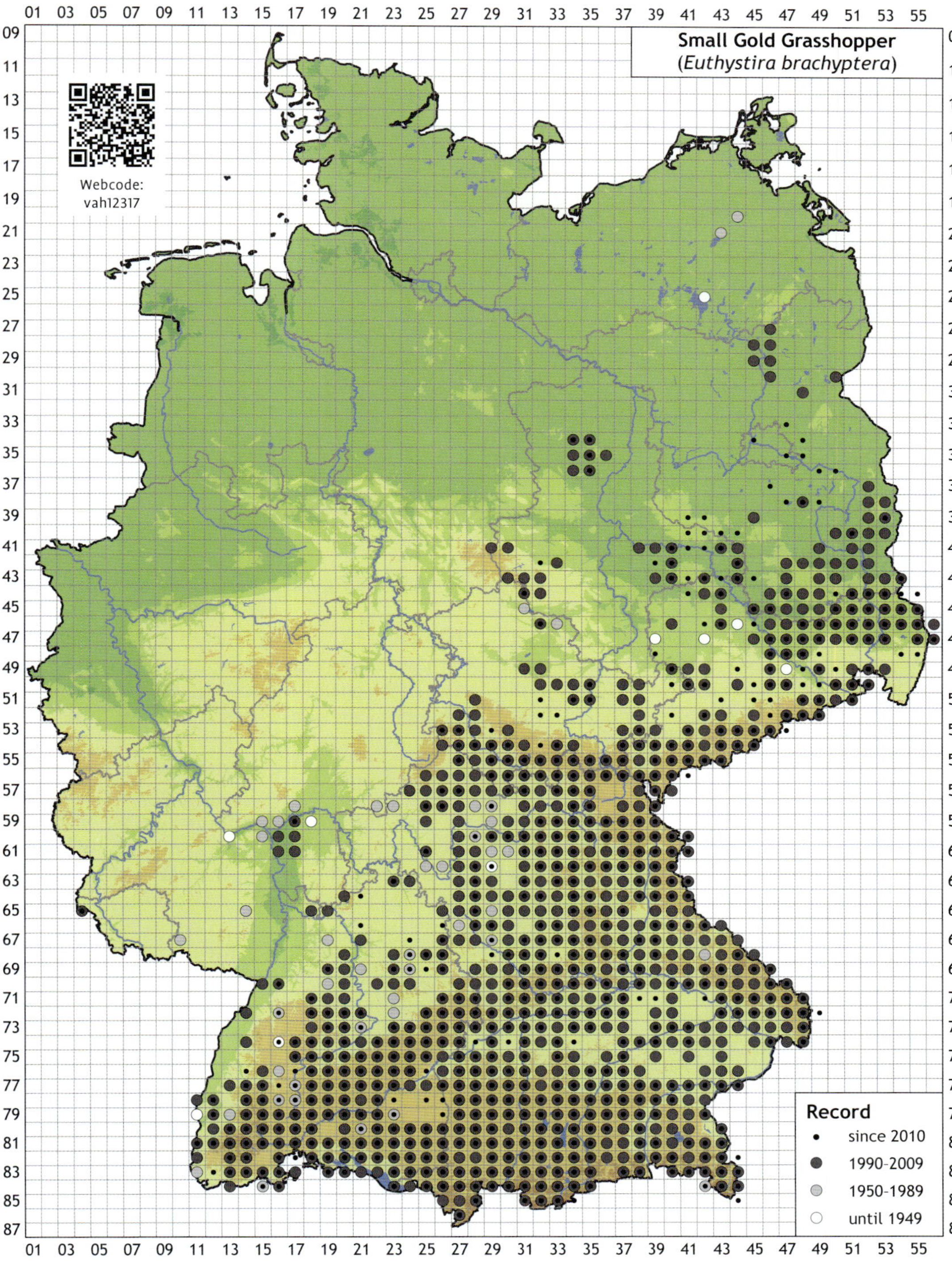
Small Gold Grasshopper
(*Euthystira brachyptera*)
Webcode:
vah12317
Record
since 2010
1990-2009
1950-1989
until 1949
01 03 05 07 09 11 13 15 17 19 21 23 25 27 29 31 33 35 37 39 41 43 45 47 49 51 53 55
09 11 13 15 17 19 21 23 25 27 29 31 33 35 37 39 41 43 45 47 49 51 53 55 57 59 61 63 65 67 69 71 73 75 77 79 81 83 85 87

Abandoned grasslands and grassy heathlands are a part of the wide spectrum of nutrient-poor habitats of the Common Green Grasshopper (Schleswig-Holstein: Anglia).
(Photo credit: Christian Winkler)

Male Common Green Grasshopper (*Omocestus viridulus*).
(Photo credit: Jürgen Fischer)

Common Green Grasshopper – *Omocestus viridulus* (Linnaeus, 1758)

Distribution The Common Green Grasshopper is widespread in Germany. It occurs in all federal states and across the complete elevation gradient from the planar to the alpine zone (highest record: 2,070 m a.s.l. in the Bavarian Alps). However, the distribution hotspots are those regions that are characterised by high precipitation, particularly higher elevations of the low mountain ranges and Alps. By contrast, the species is missing across large parts of the dry Northeast German Lowland and northern Upper Rhine Valley as well as regions that are characterised by moderate precipitation but very warm summers (e. g. Cologne Bay, northern Baden-Württemberg or southern Upper Rhine Valley)

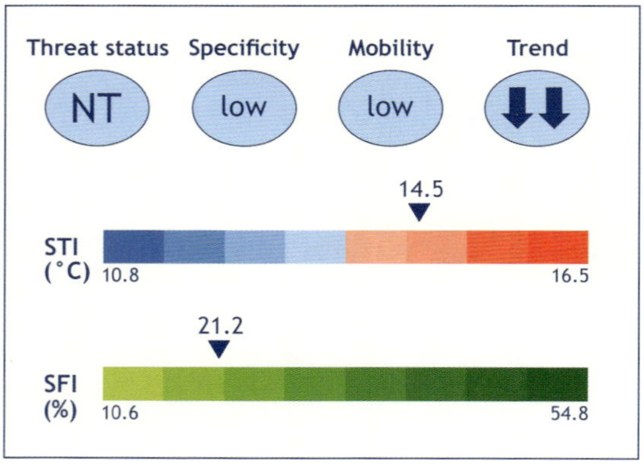

Habitat The Common Green Grasshopper inhabits a wide range of fresh to moist – in rainy regions also dry – open habitats. The sites are mostly nutrient-poor and are characterised by a dense but not too tall herbaceous layer. The species is very sensitive to intensive land use. Hence, the Common Green Grasshopper especially occurs in habitats characterised by low-intensity land use or short periods of abandonment such as nutrient-poor grasslands, grassy heathland, moorland and fringes. In the rainy montane zone, it also colonises semi-dry grasslands.

Population trend and threats In Germany, the Common Green Grasshopper is still considered common, although the short-term population trend is strongly negative. The main drivers of decline are the conversion of grasslands to arable fields, melioration measures and the intensification of land use in grasslands and road verges. In addition, the status of the populations is worsening due to ongoing climate change. Egg-laying occurs relatively early in the growing season at the soil surface. Accordingly, the eggs are poorly protected against desiccation and may suffer from summer drought over long time periods.

CHRISTIAN WINKLER

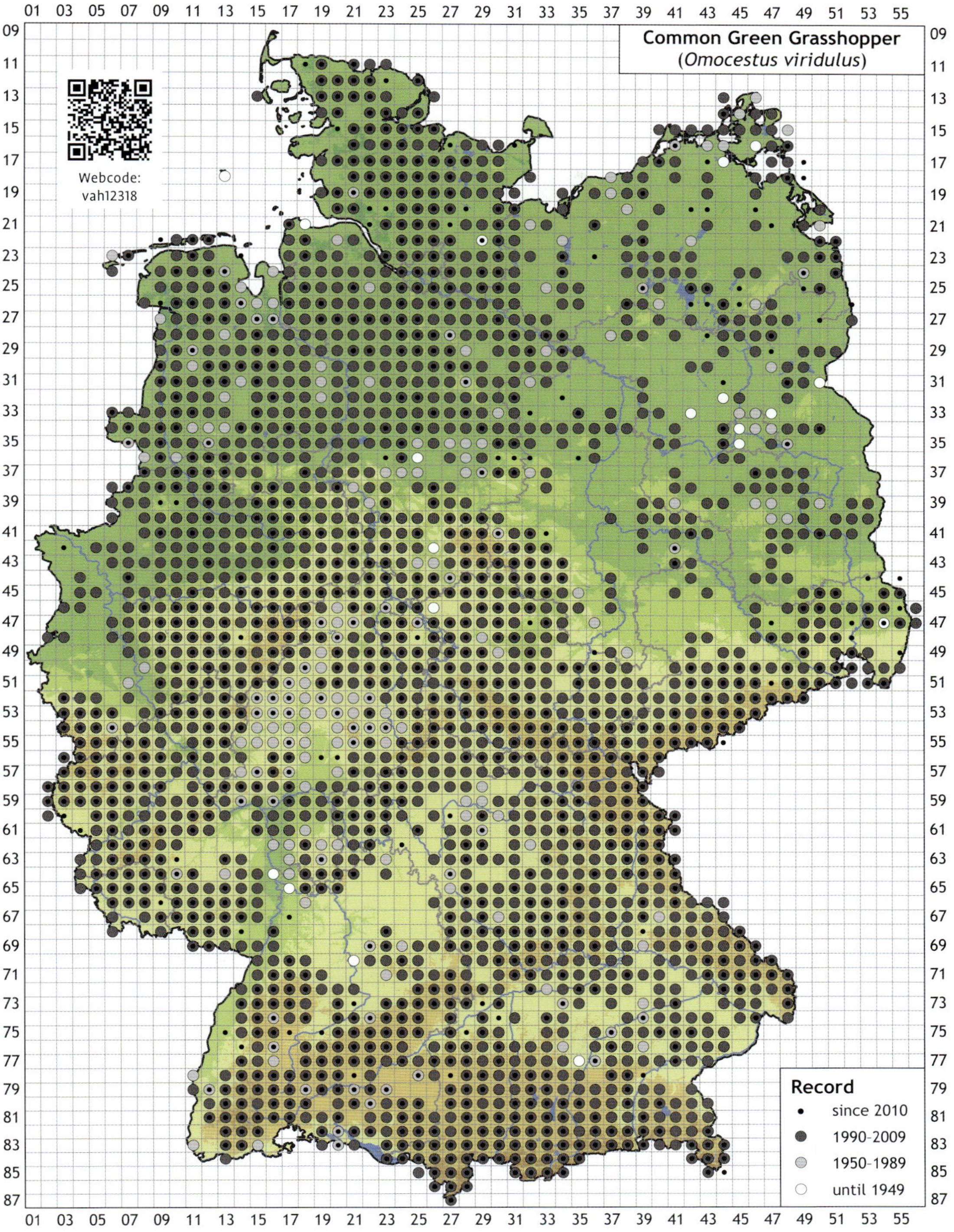
Common Green Grasshopper
(*Omocestus viridulus*)
Webcode:
vah12318
Record
since 2010
1990-2009
1950-1989
until 1949

Former peat extraction sites exhibiting mosaics of Heather (*Calluna vulgaris*) and open peat are a preferred habitat of the Woodland Grasshopper (Schleswig-Holstein: Holsteiner Vorgeest).
(Photo credit: Christian Winkler)

Male Woodland Grasshopper (*Omocestus rufipes*).
(Photo credit: Dominik Poniatowski)

Woodland Grasshopper – *Omocestus rufipes* (Zetterstedt, 1821)

Distribution The Woodland Grasshopper has a strongly scattered distribution in Germany. The strongholds are located in the Northwest German Lowland, Western and Southwestern Uplands, Alpine Foreland and Alps. Here, it occurs from the planar to the upper montane zone; the highest record is from the Feldberg in the Black Forest at 1,395 m a.s.l.

Habitat The Woodland Grasshopper inhabits dry, fresh and periodically wet habitats such as calcareous and acidic grasslands or heathlands. Consequently, the SFI of 27.7 % is relatively high. Strong sun exposure and mosaics of short to medium-high vegetation with large patches of bare soil are typical of the habitats. In northern Germany and in the Alpine Foreland, the species also occurs on drained mires, mostly raised bogs and transition mires. Here, the Woodland Grasshopper colonises sparse moorland and stands of Purple Moor Grass (*Molinia caerulea*). Within the mires, the species benefits from regular burning since it creates and maintains, respectively, the preferred habitat structures (early-successional stages).

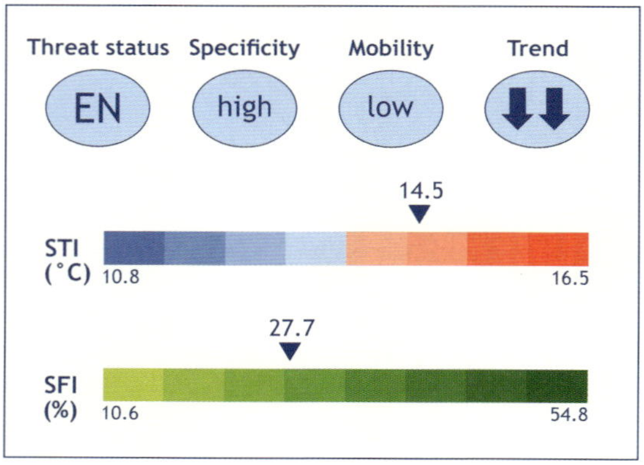

Population trend and threats Since the mid-20th century, the Woodland Grasshopper has suffered severe losses in Germany. Responsible for this decline was the change in land use, which has resulted in a large-scale disappearance of its preferred habitats. Currently, the species is classified as rare and the populations are still declining rapidly. Important causes of the recent decrease are the abandonment of low-intensity land use, the lack of appropriate conservation management and the loss of natural disturbance such as floods (alluvial calcareous grasslands) and fires. A further threat is large-scale rewetting of peat extraction sites.

CHRISTIAN WINKLER

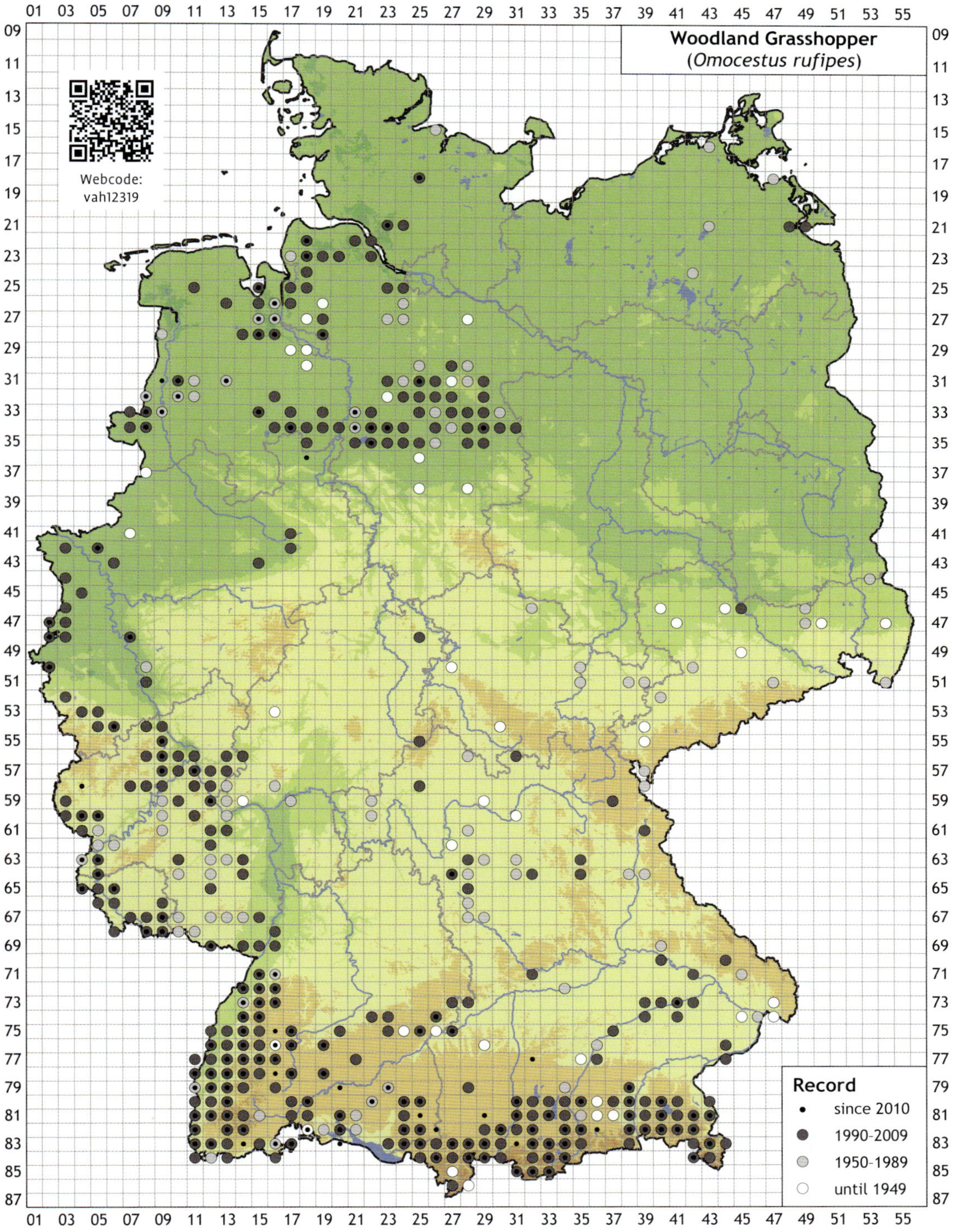
Woodland Grasshopper
(*Omocestus rufipes*)
Webcode:
vah12319
Record
since 2010
1990-2009
1950-1989
until 1949

Sunlit, nutrient-poor grasslands that are rich in tussock grasses are characteristic habitats of the Orange-tipped Grasshopper (Brandenburg: Uckermark). (Photo credit: Thomas Fartmann)

Male Orange-tipped Grasshopper (*Omocestus haemorrhoidalis*).
(Photo credit: Thomas Fartmann)

Orange-tipped Grasshopper – *Omocestus haemorrhoidalis* (Charpentier, 1825)

Distribution Within Germany, the xero-thermophilic Orange-tipped Grasshopper occurs mainly in more continental regions that are characterised by low precipitation and high summer temperatures. This is reflected by an above average STI of 15.1 °C. The majority of the populations are found in the most continental parts of the Northeast German Lowland as well as the Northwest German Lowland (Lüneburg Heath) and the Southwestern Uplands (eastern part of the uplands and Upper Rhine Valley).

Habitat The most important habitat of the Orange-tipped Grasshopper are sunlit, nutrient-poor grasslands that exhibit some bare ground and are often rich in tussock grasses. Many populations are found in semi-dry grasslands on sandy soils. However, acidic grasslands on bedrock, calcareous grasslands and heathlands are also colonised as long as they provide the required habitat structure. Rough grazing by sheep but also cattle grazing were and are still essential for the long-term survival of the populations. By contrast, litter accumulation due to abandonment rapidly causes population losses.

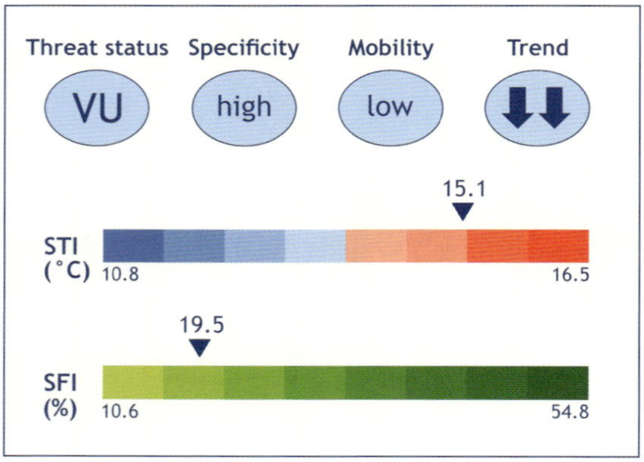

Population trend and threats The German populations of the Orange-tipped Grasshopper have greatly declined in the long- and short-term. A major driver of this was the large-scale cessation of low-intensity grazing systems and the associated dramatic loss of nutrient-poor grasslands due to succession and afforestation. Grazing animals did not only maintain the habitat but also fostered dispersal of the sedentary grasshopper. Along the drift, with its short swards, they enhanced connectivity of the habitat patches and additionally served as vectors (e.g. within the fur of sheep) for the individuals. Accordingly, nearly all current records of the Orange-tipped Grasshopper are from relict populations.

THOMAS FARTMANN

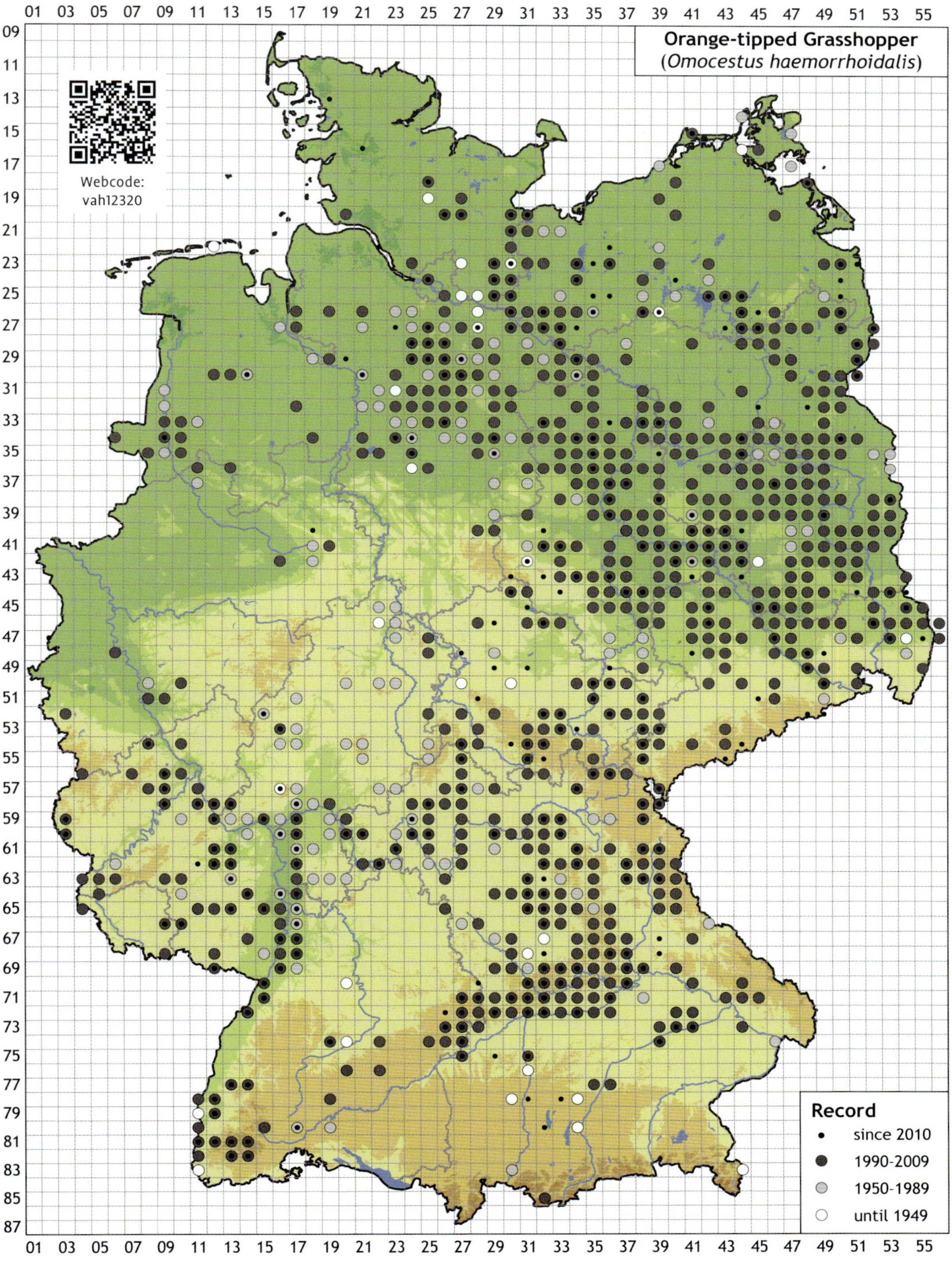
Orange-tipped Grasshopper
(Omocestus haemorrhoidalis)
Webcode: vah12320
01 03 05 07 09 11 13 15 17 19 21 23 25 27 29 31 33 35 37 39 41 43 45 47 49 51 53 55
09 11 13 15 17 19 21 23 25 27 29 31 33 35 37 39 41 43 45 47 49 51 53 55 57 59 61 63 65 67 69 71 73 75 77 79 81 83 85 87
Record
since 2010
1990-2009
1950-1989
until 1949

Calcareous grasslands exhibiting relatively closed, but not too tall swards are typical habitats of the Stripe-winged Grasshopper (Hesse: Diemel Valley).
(Photo credit: Thomas Fartmann)

Male Stripe-winged Grasshopper (*Stenobothrus lineatus*).
(Photo credit: Thomas Fartmann).

Stripe-winged Grasshopper – *Stenobothrus lineatus* (Panzer, [1796])

Distribution Among the four native *Stenobothrus* species in Germany, the slightly thermophilic Stripe-winged Grasshopper is the most widespread and common one. It regularly occurs in all seven major biogeographical regions. Larger distribution gaps are in those regions that are characterised by cool summers, such as the north of the Northwest and Northeast German Lowland or higher elevations of some low mountain ranges, as well as in landscapes where intensive agriculture dominates.

Habitat The Stripe-winged Grasshopper inhabits a wide range of nutrient-poor grasslands and heathlands on dry and warm sites. However, compared with the two related *Stenobothrus* species (*S. nigromaculatus* and *S. stigmaticus*), it prefers taller and nearly closed vegetation, which often consists of rosette and cushion plants or tussock grasses. The xerophilic eggs are deposited below the plants on the ground or into the soil. In general, some heterogeneity in sward height and density is beneficial; this best fulfils the requirements of all developmental stages concerning the vegetation structure and microclimate. In young fallows, the abundance can be high, however, with ongoing succession, the habitats become unsuitable for the species.

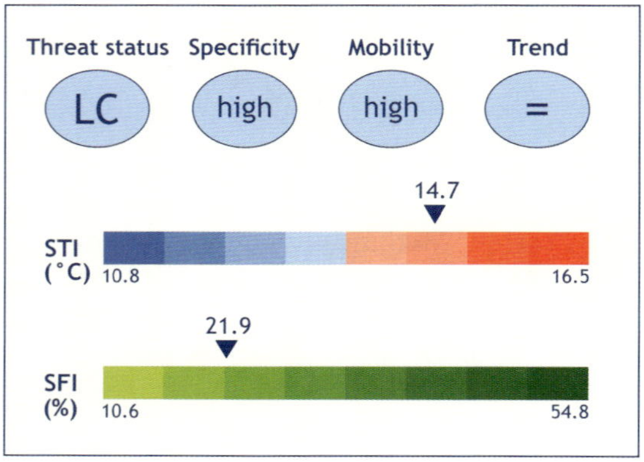

Population trend and threats During the past 200 years, the habitats of the Stripe-winged Grasshopper in Germany, nutrient-poor grasslands and heathlands, have experienced dramatic losses due to abandonment, afforestation or melioration. Currently, the remaining populations are especially threatened by a decreasing habitat quality because of, for example, unsuitable management or abandonment of land use. By contrast, climate change has recently fostered regional spreadings of the species.

THOMAS FARTMANN

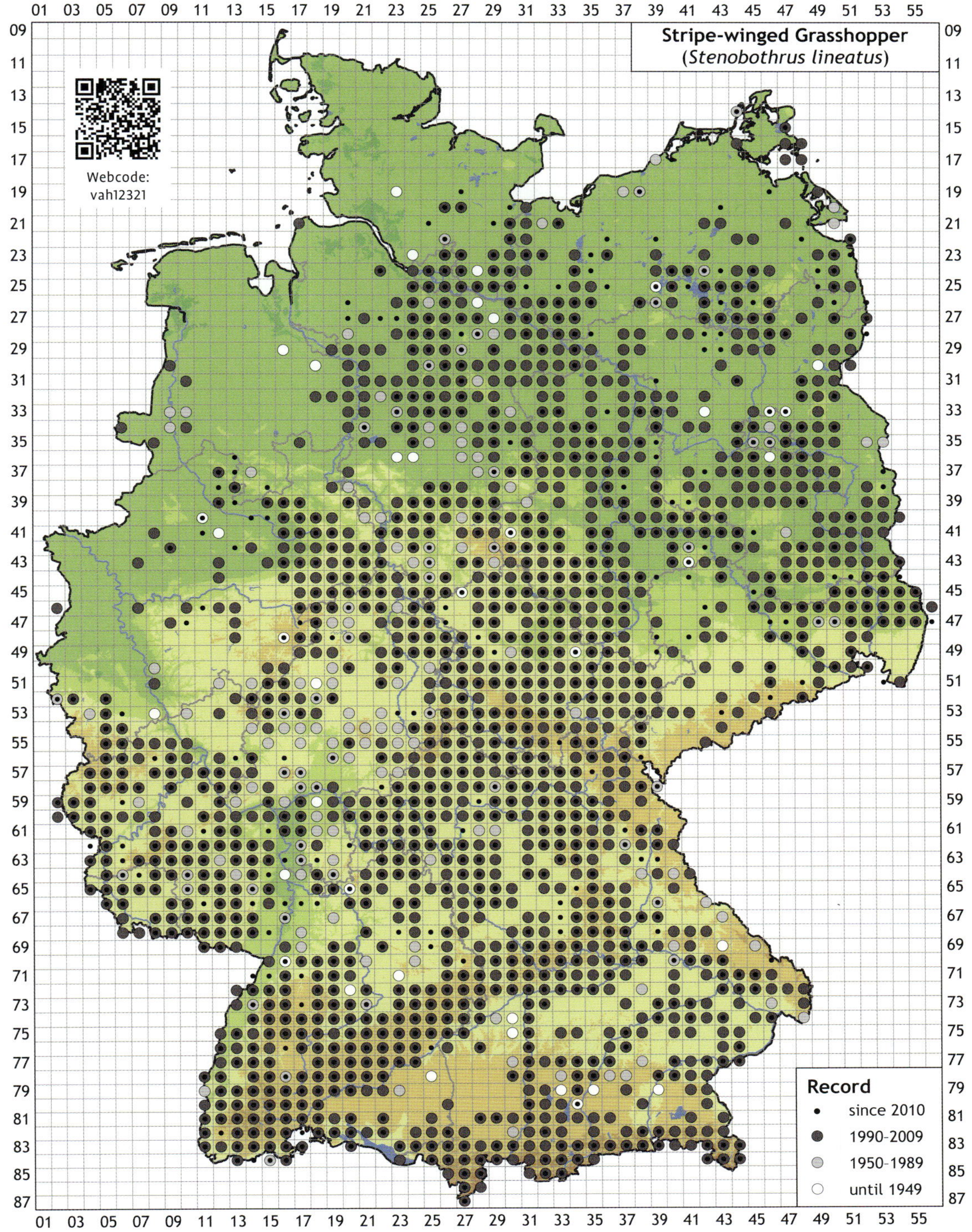
Stripe-winged Grasshopper
(*Stenobothrus lineatus*)
Webcode:
vah12321
Record
since 2010
1990-2009
1950-1989
until 1949

Low-growing steppe grasslands rich in tussock grasses and bare soil are among the important habitats of the Black-spotted Toothed Grasshopper (Thuringia: Kyffhäuser).
(Photo credit: Dominik Poniatowski)

Male Black-spotted Toothed Grasshopper (*Stenobothrus nigromaculatus*).
(Photo credit: Dominik Poniatowski)

Black-spotted Toothed Grasshopper – *Stenobothrus nigromaculatus* (Herrich-Schäffer, 1840)

Distribution The Black-spotted Toothed Grasshopper is a rare species that exhibits a strong dependence on regions with a lot of sunshine. Consequently, it is absent in the north/northwest of Germany, occurs in the Western and Eastern Uplands only in the rain shadow of the mountain ranges and also inhabits the generally sunny regions further south and east.

Habitat The habitat requirements of the Black-spotted Toothed Grasshopper are also very specific. Low-growing, nutrient-poor grasslands as well as heathlands rich in bare soil of sufficient size are preferred. All populations are relicts that almost always owe their occurrence to the continuity of traditional grazing (rough grazing by sheep but also cattle grazing). As a result, the SFI is very high with 30.3 %.

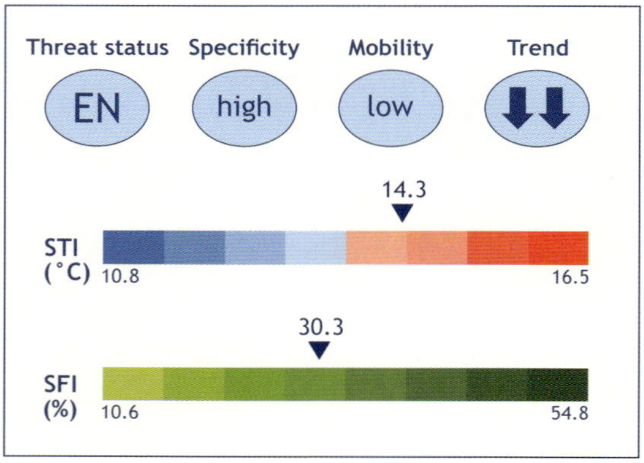

Population trend and threats The populations of the Black-spotted Toothed Grasshopper have strongly declined in the past and still continue to decrease. Most of the populations comprise fewer than 100 individuals and are highly isolated. The abandonment of traditional grazing has been mainly responsible for this; many former habitats are now encroached by shrubs or afforested. Due to the cessation of transhumant sheep grazing, dispersal between populations has been halted. Females of the Black-spotted Toothed Grasshopper are flightless. Hence, they rely on low-growing vegetation (e. g. along the drift) or vectors (e. g. within the fur of sheep) for dispersal. Besides habitat fragmentation, the remaining populations are threatened by undergrazing and mowing as a conservation measure. Both lead to moss encroachment and the loss of the preferred habitat structures.

PETER DETZEL

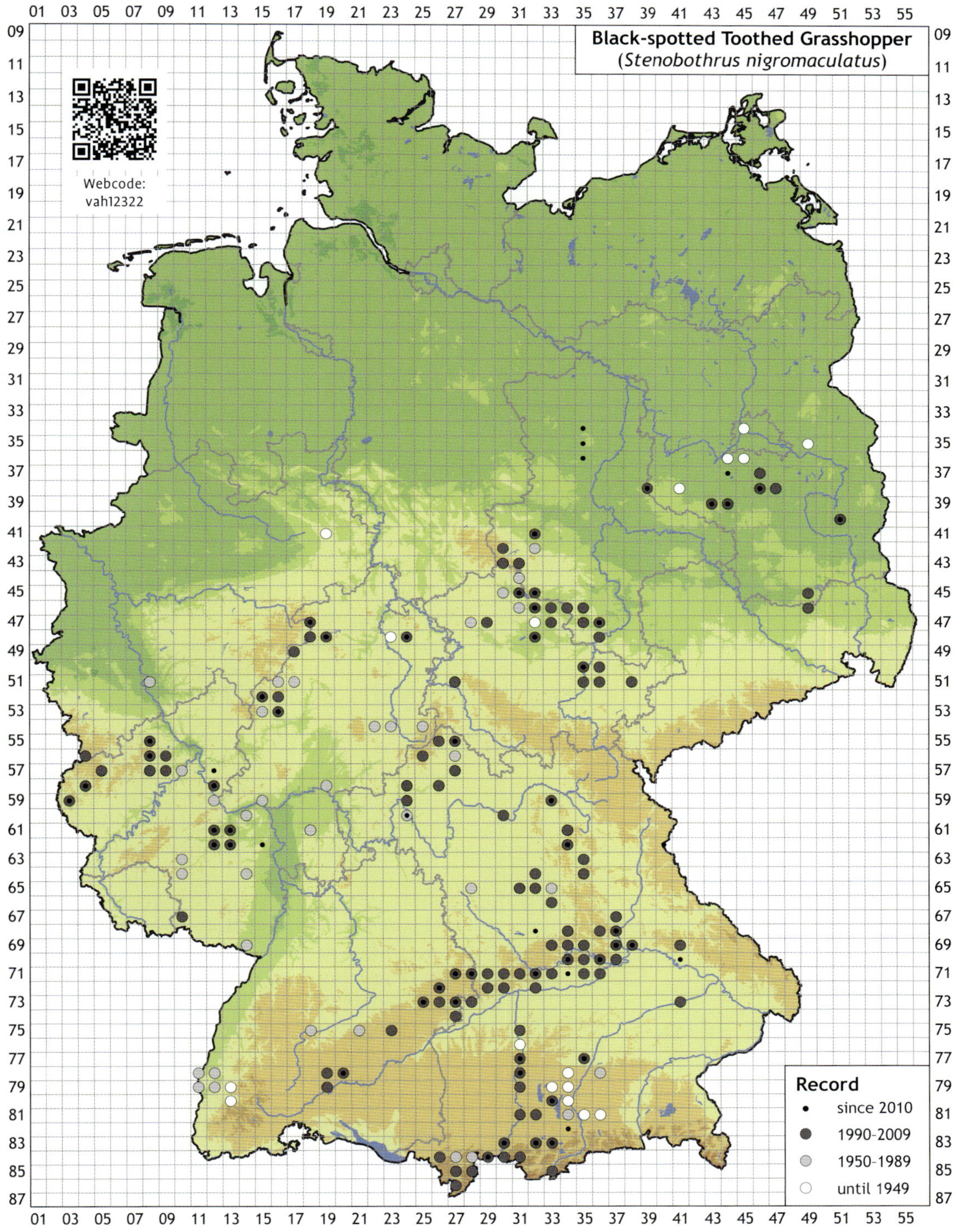
Black-spotted Toothed Grasshopper
(*Stenobothrus nigromaculatus*)
Webcode:
vah12322
Record
since 2010
1990-2009
1950-1989
until 1949

Nutrient-poor grasslands on sandy soils exhibiting very short swards are among the typical habitats of the Lesser Mottled Grasshopper (Lower Saxony: Emsland). (Photo credit: Thomas Fartmann)

Male Lesser Mottled Grasshopper (*Stenobothrus stigmaticus*). (Photo credit: Thomas Fartmann)

Lesser Mottled Grasshopper – *Stenobothrus stigmaticus* (Rambur, 1838)

Distribution Except in the Alps, the Lesser Mottled Grasshopper has a scattered distribution in all major biogeographical regions in Germany. Concentrations of records occur where traditional grazing systems have remained on a larger scale. Examples are the southern Black Forest, Swabian and Franconian Jura, eastern part of the Rothaar Mountains, Emsland or the Lüneburg Heath. By contrast, in very continental, dry regions – for example in large parts of the Northeast German Lowland – the species is very rare.

Habitat The Lesser Mottled Grasshopper typically inhabits sunlit, nutrient-poor grasslands or heathlands exhibiting very short swards with some bare ground and that are often rich in tussock grasses. Eggs are deposited into the soil or underneath plant material on the ground. Grazing, sometimes even more intensive, plays a vital role for the long-term viability of the populations. The relatively small species is even able to survive on tiny patches of 50 m^2 or more as long as the habitat quality is favourable.

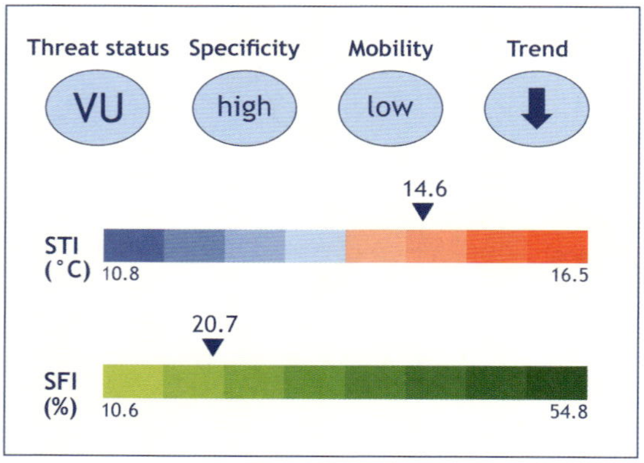

Population trend and threats Due to the strong dependence on low-intensity grazing, the Lesser Mottled Grasshopper has greatly suffered from the collapse of traditional grazing systems since the 19th century in Germany. Many habitats of former pasture landscapes are now covered by woodland, and in the case of heathlands, they have also been converted to arable fields or improved grassland. The remaining relict populations are often very small and frequently suffer from an extinction debt due to nitrogen deposition and no or unsuitable land use. Since the females of the Lesser Mottled Grasshopper are flightless, the colonisation of new habitats depends on suitable corridors or passive dispersal (e. g. in the fur of sheep).

THOMAS FARTMANN

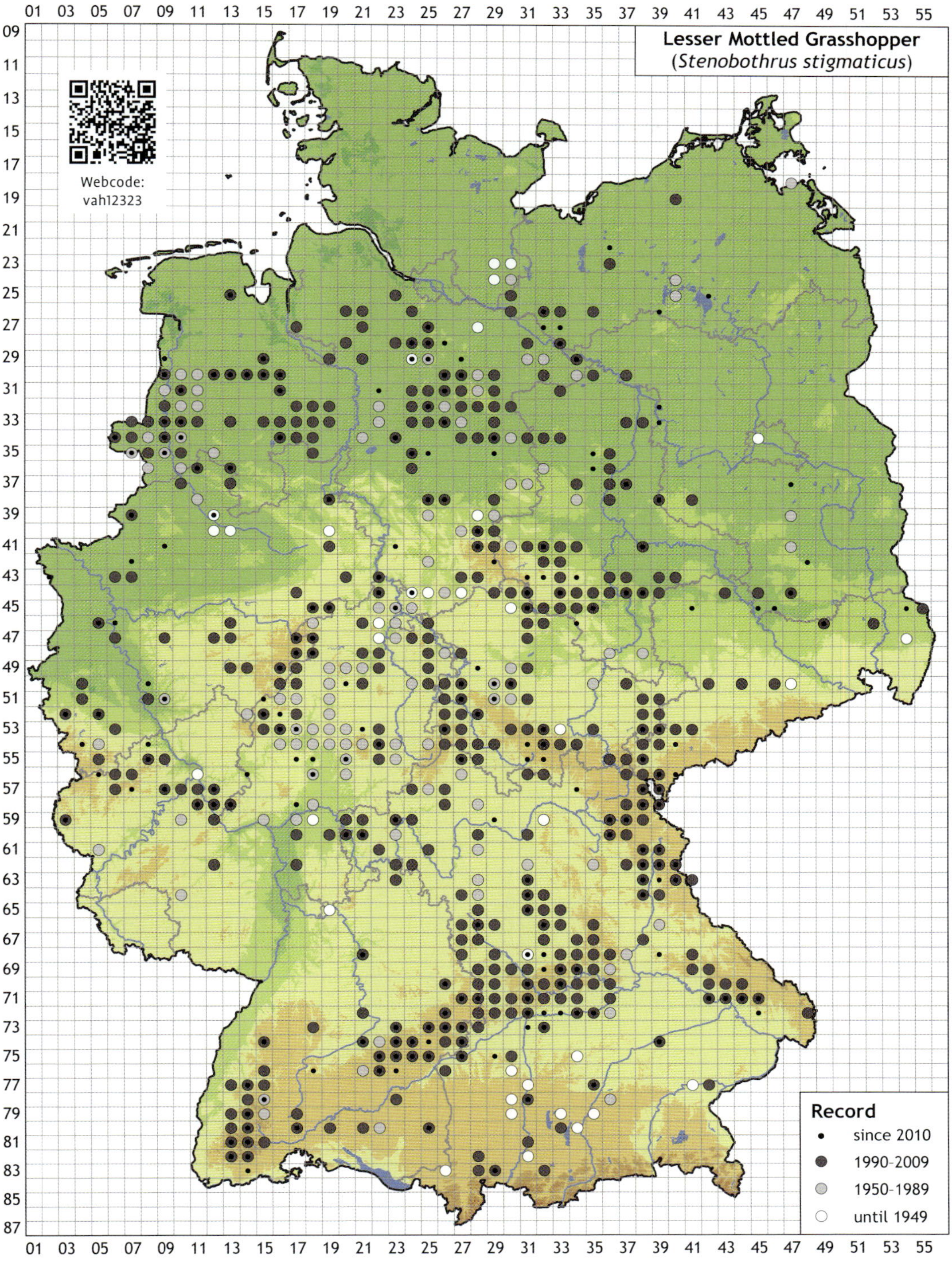
Lesser Mottled Grasshopper
(Stenobothrus stigmaticus)
Webcode:
vah12323
Record
since 2010
1990-2009
1950-1989
until 1949
01 03 05 07 09 11 13 15 17 19 21 23 25 27 29 31 33 35 37 39 41 43 45 47 49 51 53 55
09 11 13 15 17 19 21 23 25 27 29 31 33 35 37 39 41 43 45 47 49 51 53 55 57 59 61 63 65 67 69 71 73 75 77 79 81 83 85 87

Among the characteristic habitats of the Pygmy Toothed Grasshopper are calcareous grasslands (Saxony-Anhalt: northern Harz Foreland).
(Photo credit: Martin Schädler)

Female Pygmy Toothed Grasshopper (*Stenobothrus crassipes*).
(Photo credit: Dominik Poniatowski)

Pygmy Toothed Grasshopper – *Stenobothrus crassipes* (Charpentier, 1825)

Distribution The Pygmy Toothed Grasshopper has a Pannonian main distribution range. In Germany, due to postglacial disjunction, it only occurs in a few exclaves surrounding the Harz Mountains. There, it is only documented in the colline zone (160–320 m a.s.l.) of the northern foothills of the Harz Mountains (eastern Huy Hills) and the northern Thuringian Basin (western Kyffhäuser Hills and eastern spurs of the Hainleite Hills). Both distribution areas are characterised by a subcontinental climate with low precipitation, pronounced annual temperature amplitude and warm summers. The Pygmy Toothed Grasshopper is usually short-winged and flightless; hence, it is considered a philopatric species.

Habitat The Pygmy Toothed Grasshopper inhabits dry and warm habitats on dolomite/gypsum (Kyffhäuser) and shell limestone (Hainleite, Huy) such as nutrient-poor grasslands with short and sparse swards. However, occasionally, it also occurs in abandoned dry and mesic grasslands exhibiting taller swards. Originally and partly still today, the habitats were used as sheep pastures. Additionally, parts of the Huy were a former military training area.

Population trend and threats All German populations of the Pygmy Toothed Grasshopper are likely very old and were overlooked for long time periods. The extremely rare species occurs only in six grid cells and in a few dozen, mostly small populations that are equally distributed across the two distribution areas Kyffhäuser/Hainleite and Huy. The short-term trend of the populations is considered stable. However, in the long term, a moderate decline is assumed since many nutrient-poor grasslands (particularly at the Huy) are now encroached by woody plants or have been afforested. Knowledge on the status of the remaining populations is scarce. Since Germany is particularly responsible for the conservation of the highly isolated exclaves of the species, continuous population monitoring is greatly needed.

GÜNTER KÖHLER

Threat status	Specificity	Mobility	Trend
EN	high	low	=

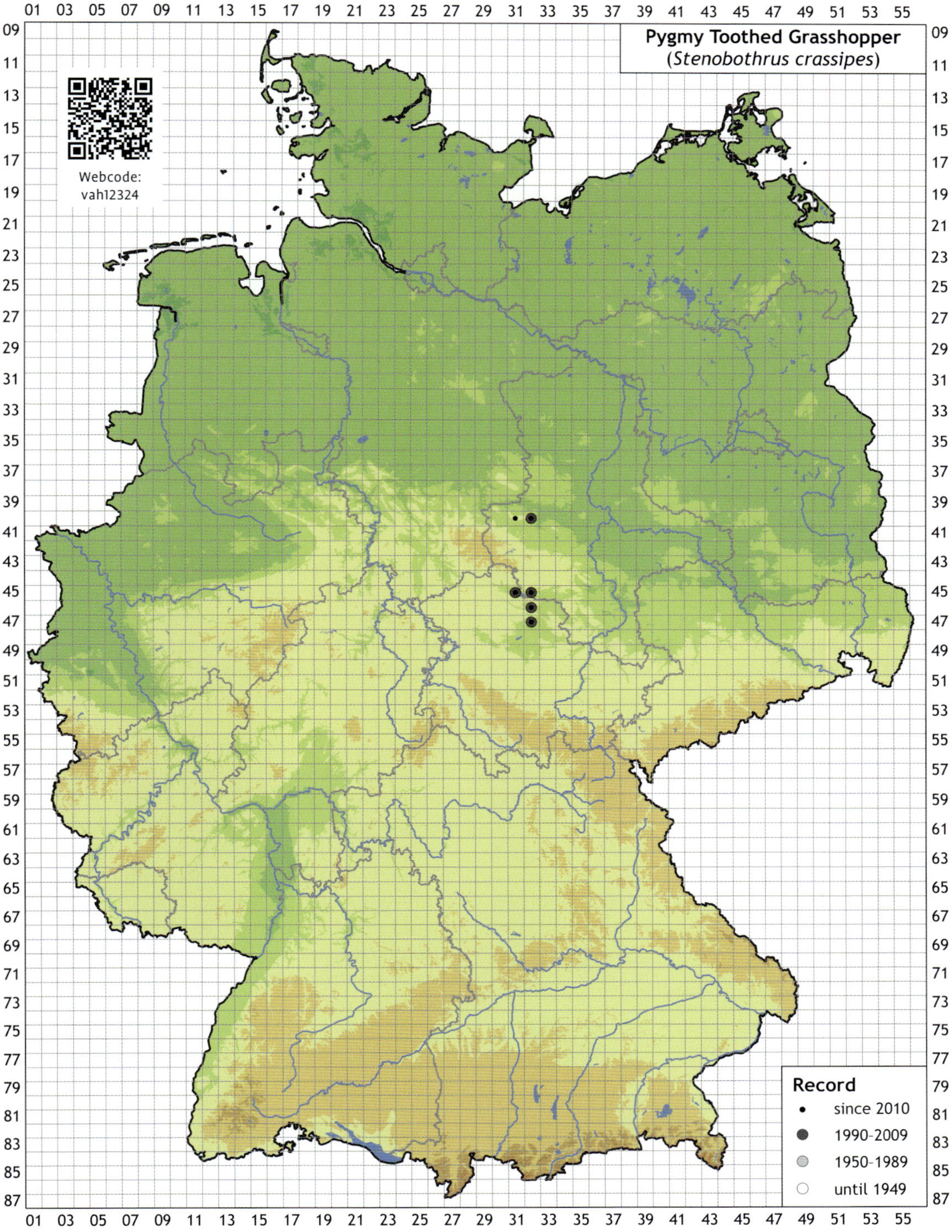
Pygmy Toothed Grasshopper
(Stenobothrus crassipes)
Webcode:
vah12324
Record
since 2010
1990-2009
1950-1989
until 1949

Sunny and sparse Alpine grasslands are preferred habitats of the Club-legged Grasshopper (Bavaria: Allgäu Alps). (Photo credit: Johannes Voith)

Male Club-legged Grasshopper (*Gomphocerus sibiricus*). (Photo credit: Thomas Fartmann)

Club-legged Grasshopper – *Gomphocerus sibiricus* (Linnaeus, 1767)

Distribution In Germany, the occurrence of the Club-legged Grasshopper is restricted to the Bavarian Alps. The STI is correspondingly low at 11.2 °C. The Allgäu Alps exhibit the highest record density. Until a few years ago, the Berchtesgaden Alps represented a striking postglacial distribution gap. However, now, the species also occurs there. Most populations are found between 1,300 and 1,800 m a.s.l. The upper elevational range limit is at approximately 2,200 m a.s.l. and the lower one at 1,000 m a.s.l.

Habitat Typical habitats of the Club-legged Grasshopper are mountain pastures, Alpine grasslands and dwarf-shrub heathlands. The microhabitats exhibit a short turf and a high share of bare soil or rocks. Further characteristics of the habitats are permeable soils and sunny conditions.

Population trend and threats Historic records of the Club-legged Grasshopper from Germany are very incomplete. Thus far, no extinctions of populations, as were shown for lower elevations in Austria, have been documented. In Germany, the species seems to benefit from climate change, as the recent colonisation of the Berchtesgaden Alps underlines. Dry and warm conditions in midsummer (July) are assumed to favour population growth. Potential causes of population declines are intensive grazing and abandonment of grazing followed by encroachment of woody plants. However, currently, the species is not considered threatened since one-third of the populations occur in primary grasslands that are naturally treeless.

JOHANNES VOITH

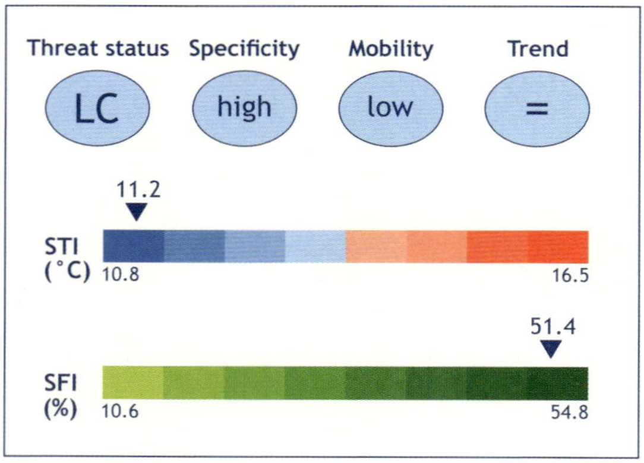

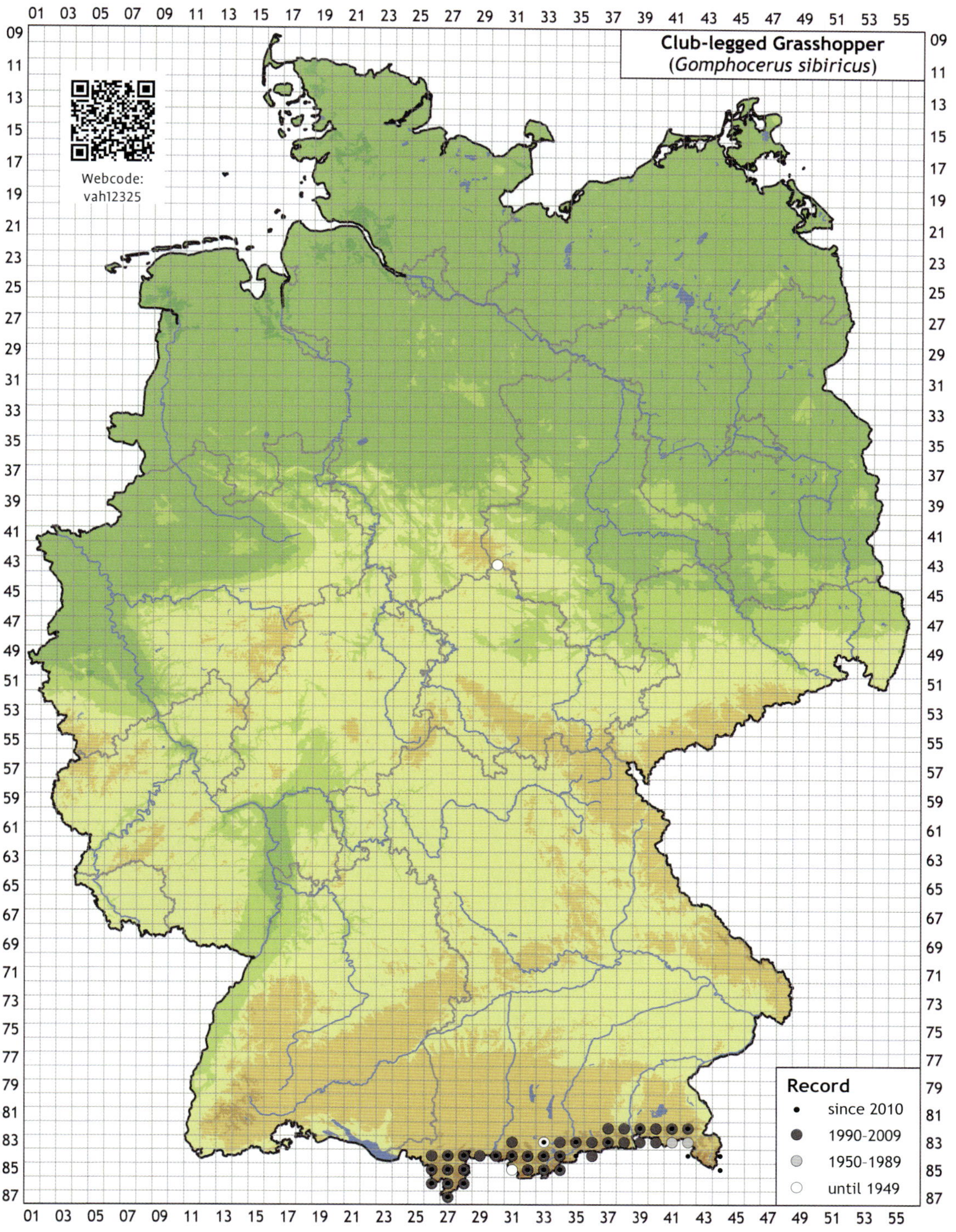
Club-legged Grasshopper
(Gomphocerus sibiricus)
Webcode:
vah12325
Record
since 2010
1990-2009
1950-1989
until 1949

At the Stiegelesfelsen (Baden-Württemberg: Upper Danube Valley), the Rufous Grasshopper inhabits scrubby calcareous grasslands.
(Photo credit: Dominik Poniatowski).

Male Rufous Grasshopper (*Gomphocerippus rufus*).
(Photo credit: Dominik Poniatowski)

Rufous Grasshopper – *Gomphocerippus rufus* (Linnaeus, 1758)

Distribution The Rufous Grasshopper is widespread in southern Germany, especially below 800 m a.s.l. The northwestern edge of its largely closed range is roughly formed by the axis of the low mountain ranges Eifel, Rothaar and Harz. Further southeast, the range extends over the edge of the Thuringian Basin, the middle Saale Valley and the region around Gera to the river Elbe near Dresden. The species is largely absent in the North German Lowland. There are only a few isolated outposts, including Lower Saxony, Saxony-Anhalt and Brandenburg.

Habitat The Rufous Grasshopper is a thermophilic species that prefers sun-exposed fringe biotopes with a heterogeneous herb layer. The habitats are also characterised by a high proportion of woody plants. Typical habitats include abandoned grasslands and semi-dry or dry grasslands with shrubs. These are often located on the edge of forests or near woodlands. Within forests, the Rufous Grasshopper can be observed along pathways and at clearings. Occasionally, secondary biotopes such as mining sites and quarries also serve as a habitat for the species.

Population trend and threats A small-scale expansion has recently been observed, especially on the northwestern and northern edges of the species' range in Germany. The populations are considered stable nationwide. Nevertheless, fringe biotopes and heterogeneous forest edges should be conserved and promoted as many other species also benefit from such habitats.

DOMINIK PONIATOWSKI

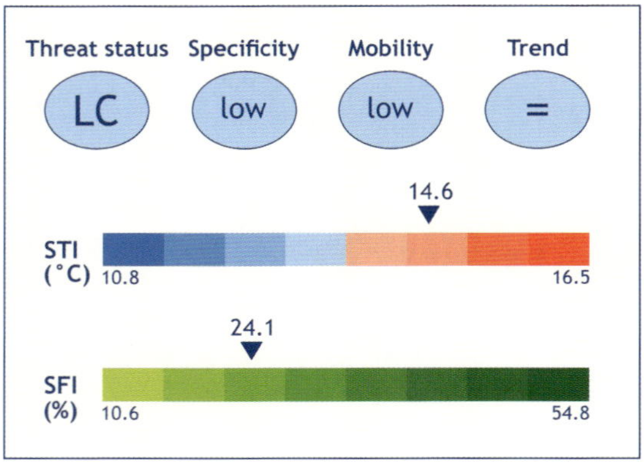

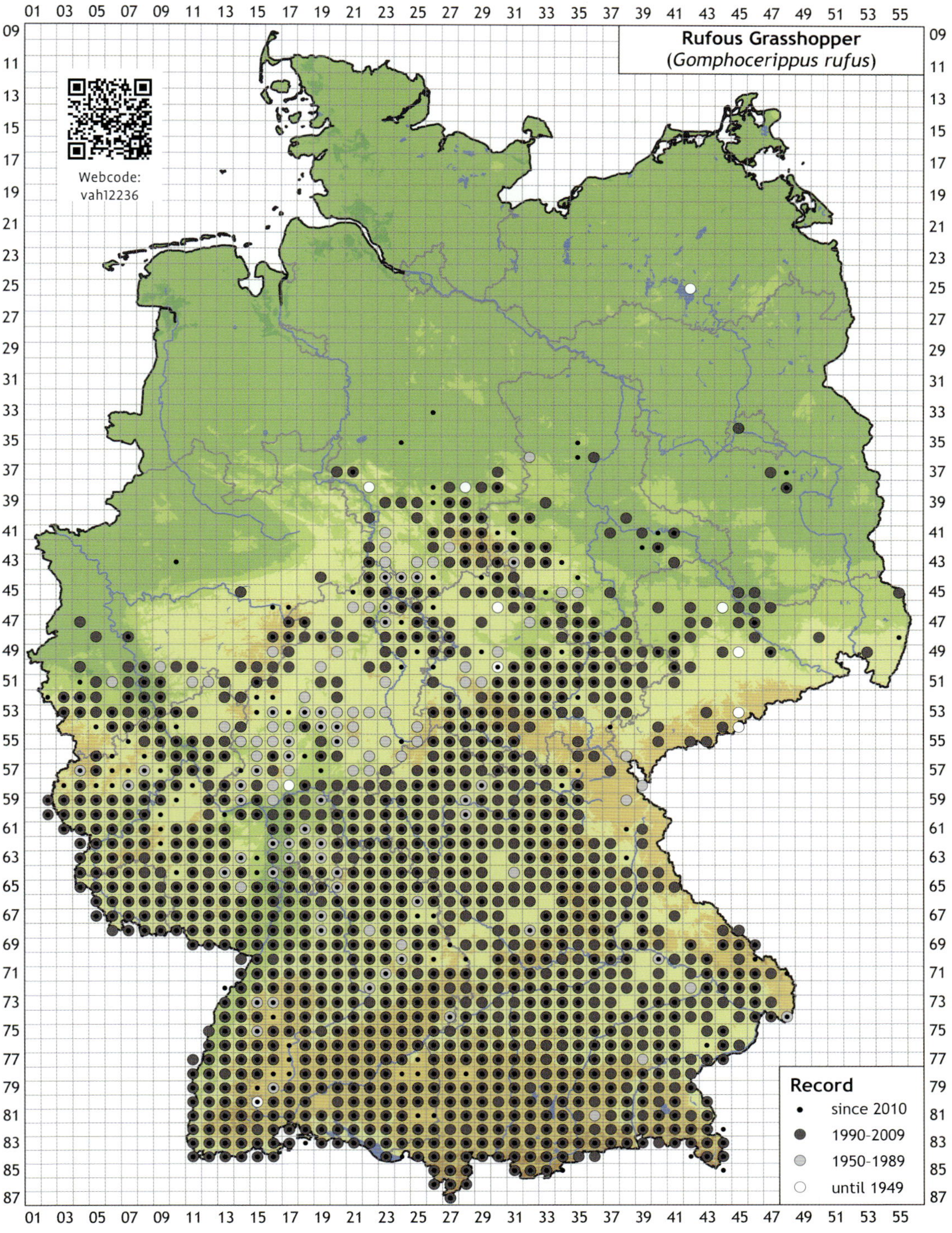
Rufous Grasshopper
(*Gomphocerippus rufus*)
Webcode:
vah12236
Record
since 2010
1990-2009
1950-1989
until 1949
01 03 05 07 09 11 13 15 17 19 21 23 25 27 29 31 33 35 37 39 41 43 45 47 49 51 53 55
09 11 13 15 17 19 21 23 25 27 29 31 33 35 37 39 41 43 45 47 49 51 53 55 57 59 61 63 65 67 69 71 73 75 77 79 81 83 85 87

Female Mottled Grasshopper (*Myrmeleotettix maculatus*).
(Photo credit: Dominik Poniatowski)

The habitats of the Mottled Grasshopper are characterised by sparse vegetation and dry soils; thus, grasslands dominated by the Grey Hair-grass (*Corynephorus canescens*) are very typical (Mecklenburg-Western Pomerania: Ueckermünde Heath).
(Photo credit: Dominik Poniatowski)

Mottled Grasshopper – *Myrmeleotettix maculatus* (Thunberg, 1815)

Distribution The Mottled Grasshopper occurs in all German federal states but has a scattered distribution. The strongholds are the sandy areas of the coastal dunes, the NW and NE German moraine landscape of the Saale glaciation, the Middle Franconian Basin and the Upper Rhine Valley as well as the low mountain ranges consisting of acidic bedrock.

Habitat The Mottled Grasshopper is geo- and xerophilic. Hence, suitable habitats are characterised by sparse vegetation and low water capacity. Among the most important habitats are nutrient-poor grasslands on sandy soils (particularly dominated by tussocks of *Corynephorus canescens*) and heathlands as well as nutrient-poor grasslands on acidic bedrock (partly also on calcareous substrate), woodland clearings and windthrows. During its life cycle, this grasshopper undergoes a microhabitat shift: bare ground is used for oviposition and hosts young nymphs, as do moss-rich microhabitats. By contrast, older nymphs and adults live in taller and denser vegetation.

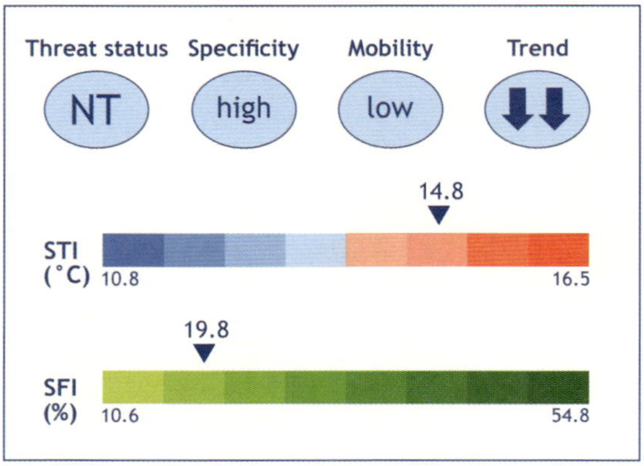

Population trend and threats Land-use change has caused severe population losses of the Mottled Grasshopper in Germany. Particularly the cessation of traditional land use of nutrient-poor grasslands and heathlands (burning, grazing, sod-cutting) and the subsequent succession or afforestation of the habitats were responsible for this. Today, the decline is still ongoing and the Mottled Grasshopper ranks among those native grasshopper species that exhibit the strongest losses. Besides the already mentioned causes, atmospheric nitrogen deposition and abandonment of military training areas are further drivers of recent population decrease.

ANDREAS ZEHM

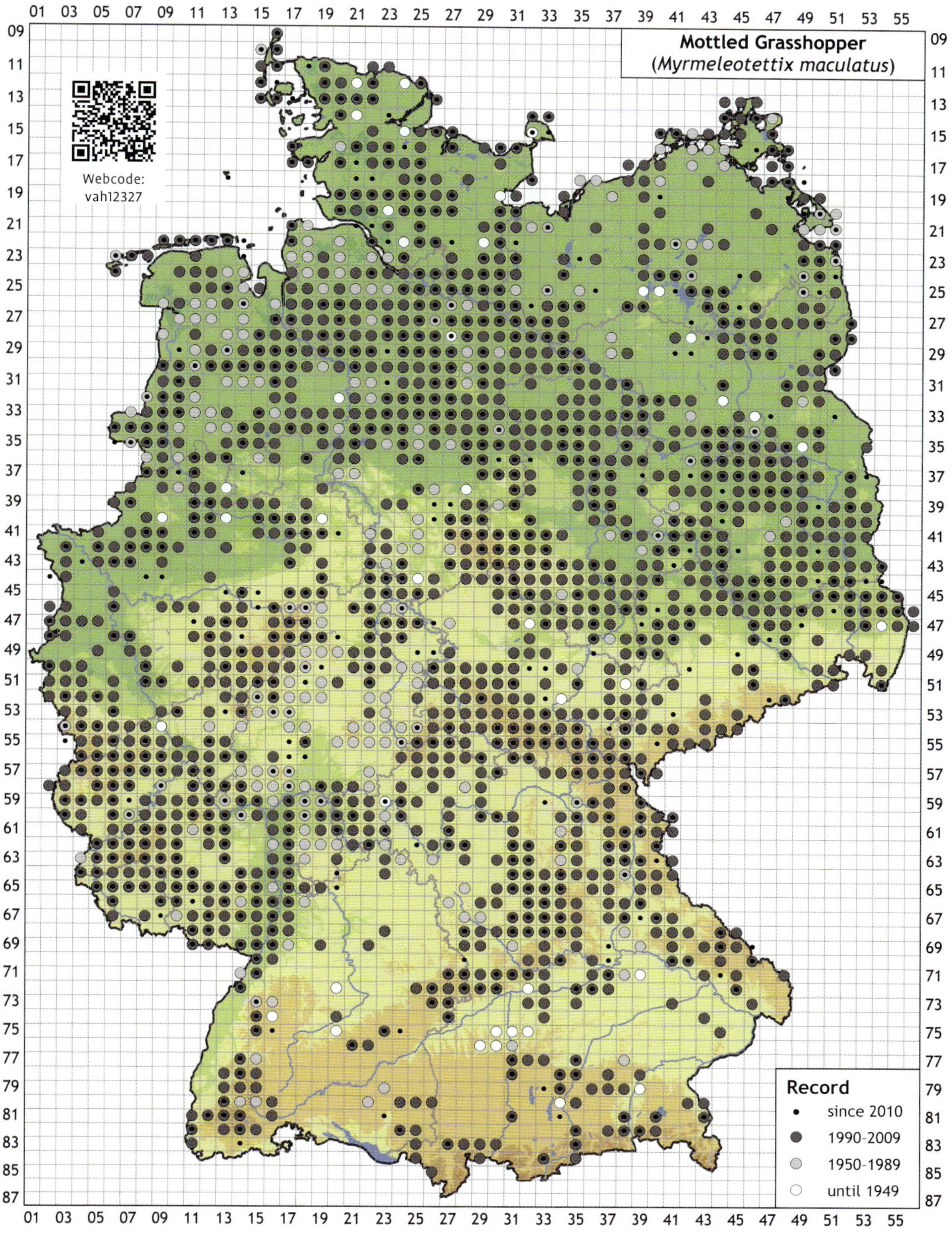
Mottled Grasshopper
(*Myrmeleotettix maculatus*)
Webcode:
vah12327
Record
since 2010
1990-2009
1950-1989
until 1949

The large, heterogeneous cattle pastures in the Black Forest are the most important stronghold of the Large Mountain Grasshopper in Germany (Baden-Württemberg: southern Black Forest).
(Photo credit: Marcel Kettermann)

Male Large Mountain Grasshopper (*Stauroderus scalaris*).
(Photo credit: Thomas Fartmann)

Large Mountain Grasshopper – *Stauroderus scalaris* (Fischer von Waldheim, 1846)

Distribution The occurrence of the Large Mountain Grasshopper in Germany is currently restricted to the Southwestern Uplands. The two distribution centres are the Black Forest and mid-Swabian Jura. Most current records are from higher elevations above 600 m a.s.l. More recently, single dispersing individuals have been found far away from existing populations. Former populations in the North German Lowland had already disappeared by the mid-20th century. By contrast, in the Allgäu Alps, the Large Mountain Grasshopper was recorded up until the late 1990s and it is possibly still present there.

Habitat The Large Mountain Grasshopper only inhabits landscapes that are characterised by a high share of nutrient-poor grasslands that exhibit a low intensity of land use. Correspondingly high is the SFI at 54.5 %. While the species in the Black Forest occurs in large-scale nutrient-poor pastures, it colonises both nutrient-poor pastures and meadows in the Swabian Jura. Overall, heterogeneous grasslands that offer mosaics of bare ground, short turf and taller vegetation are preferred.

Population trend and threats Former records from outside the two current distribution centres provide evidence that the Large Mountain Grasshopper was also more widespread in the North German lowland in historic times. In some parts of the Black Forest, increasing isolation of the populations as a result of abandonment, afforestation and intensified land use has been observed. However, overall, the German populations are currently considered stable.

JENS SACHTELEBEN

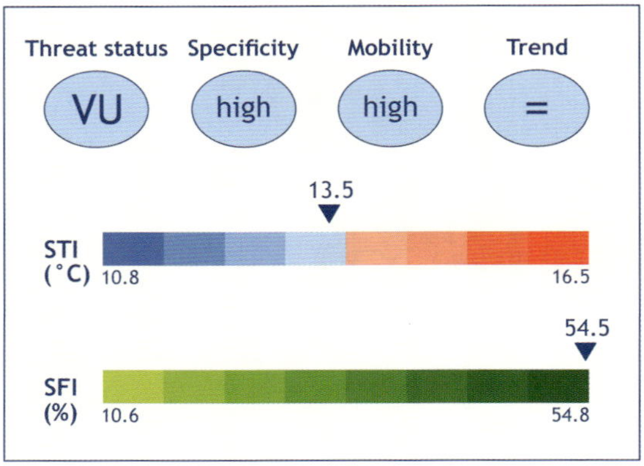

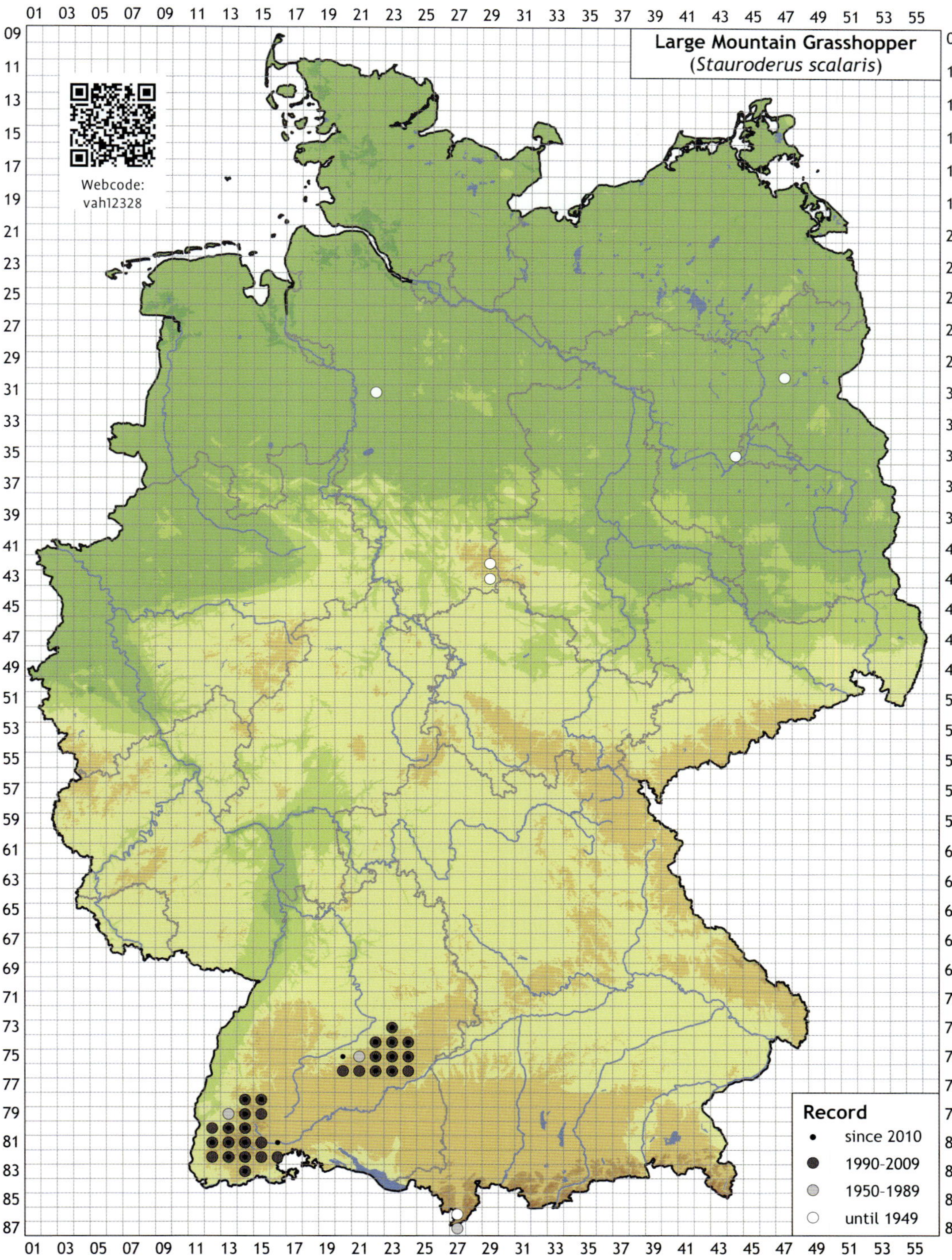
Large Mountain Grasshopper
(*Stauroderus scalaris*)
Webcode:
vah12328
Record
since 2010
1990-2009
1950-1989
until 1949

Nutrient-poor waysides with taller vegetation and patches of bare ground are frequently colonised by the Locomotive Grasshopper (Brandenburg: Ruppiner Land).
(Photo credit: Dominik Poniatowski).

Female Locomotive Grasshopper (*Chorthippus apricarius*).
(Photo credit: Thomas Fartmann)

Locomotive Grasshopper – *Chorthippus apricarius* (Linnaeus, 1758)

Distribution The distribution centre of the Locomotive Grasshopper is in the more continental eastern half of Germany. By contrast, in the western half, the Locomotive Grasshopper is relatively rare. However, local distribution hotspots occur in parts of the Northwest German Lowland – there also in more oceanic regions – and in the more continental northern Upper Rhine Valley.

Habitat The slightly thermophilic Locomotive Grasshopper primarily colonises fresh to dry sites on loose, often sandy soils. Sometimes, it also occurs on periodically moist soils of drained moorland. The sun-exposed habitats have at least partly a high vegetation cover. In the case of a dense vegetation, at least some patches of bare ground are present, for example in the form of animal burrows. Among the main habitats of the Locomotive Grasshopper are ecotones such as field margins, forest edges and road verges. In addition, it regularly inhabits ruderal and semi-ruderal habitats such as abandoned arable fields and grasslands as well as mesic (in particular cattle pastures) and dry grasslands characterised by low intensity of land use.

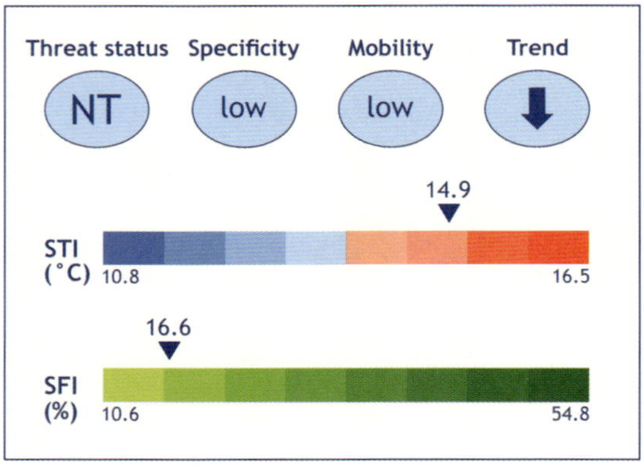

Population trend and threats The Locomotive Grasshopper is moderately common in Germany, but the populations are currently declining. Due to climate change, the species colonises new habitats locally solely at the range margin in the federal states of Lower Saxony and Schleswig-Holstein. The main drivers of decline are the destruction of ecotones, fallow land and ruderal habitats as well as the intensification of management in grasslands and road verges. The abandonment of low-intensity land use and atmospheric nitrogen deposition favour a homogeneous vegetation structure and, hence, also contribute to habitat deterioration.

CHRISTIAN WINKLER

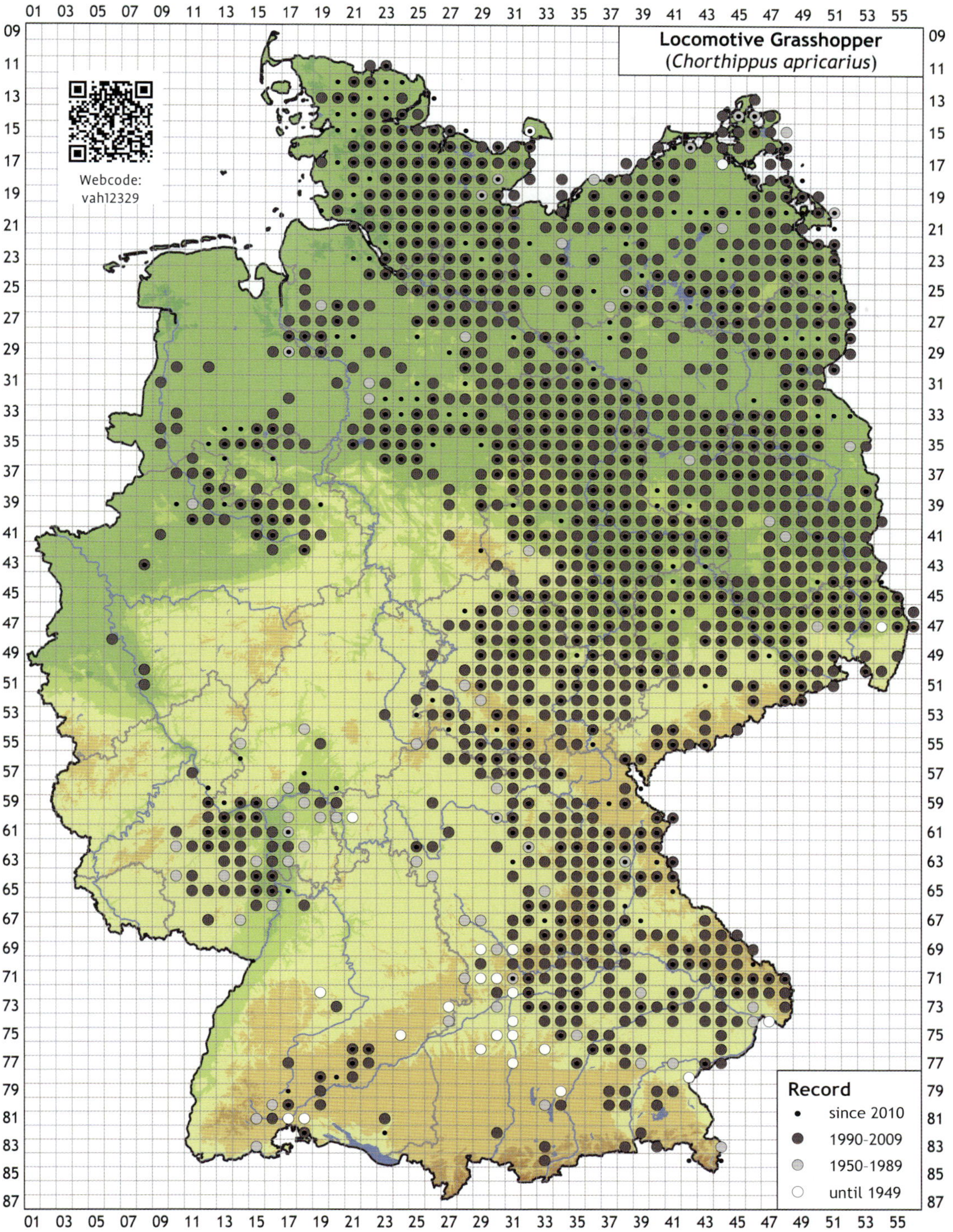
Locomotive Grasshopper
(Chorthippus apricarius)
Webcode:
vah12329
Record
since 2010
1990-2009
1950-1989
until 1949
01 03 05 07 09 11 13 15 17 19 21 23 25 27 29 31 33 35 37 39 41 43 45 47 49 51 53 55
09 11 13 15 17 19 21 23 25 27 29 31 33 35 37 39 41 43 45 47 49 51 53 55 57 59 61 63 65 67 69 71 73 75 77 79 81 83 85 87

Male Gravel Grasshopper (*Chorthippus pullus*). (Photo credit: Thomas Fartmann)

Alluvial gravel banks with sparse stands of the grass *Calamagrostis pseudophragmites* are typical habitats of the Gravel Grasshopper in southern Bavaria (Upper Isar Valley).
(Photo credit: Thomas Fartmann)

Gravel Grasshopper – *Chorthippus pullus* (Philippi, 1830)

Distribution The Gravel Grasshopper is among the rarest grasshopper species in Germany. It only occurs locally in southern Bavaria (Alpine Foreland and Alps), Saxon Switzerland and the Northeast German Lowland (Berlin, Lower and Upper Lusatia).

Habitat Habitats of the Gravel Grasshopper typically exhibit early-successional stages rich in bare soil, which receive a lot of solar irradiation. Accordingly, on sunny days, temperatures are high. The bare ground is used for oviposition. In southern Bavaria, alluvial gravel banks of dynamic rivers that exhibit some refuge habitats above the flooded areas and long-lasting screes are characteristic habitats. By contrast, in eastern Germany, particularly light woodlands as well as dry grasslands and heathlands on sandy soils are colonised.

Population trend and threats Due to the dependence on early-successional stages, natural (or anthropogenic) disturbance is of vital importance for the long-term survival of local populations of the Gravel Grasshopper. During the last century, many German populations of the species became extinct: In Bavaria, especially hydraulic engineering at the Alpine rivers has destroyed the habitats or suppressed their essential natural dynamics. In eastern Germany, particularly succession and afforestation have caused habitat loss. The remaining populations of the Gravel Grasshopper are mostly isolated, exhibit a low abundance and still continue to decline due to succession, particularly in eastern Germany.

ANDREAS ZEHM

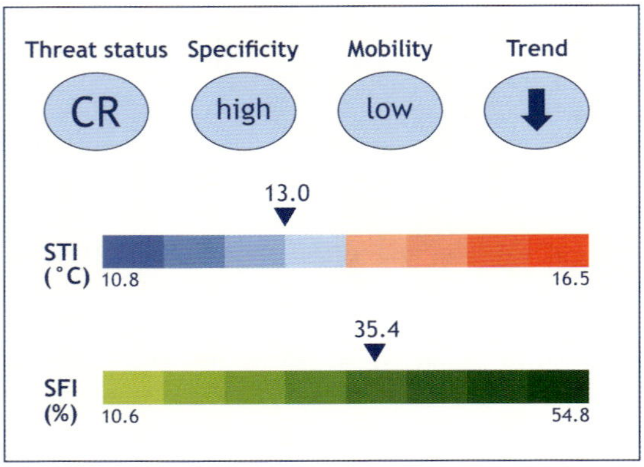

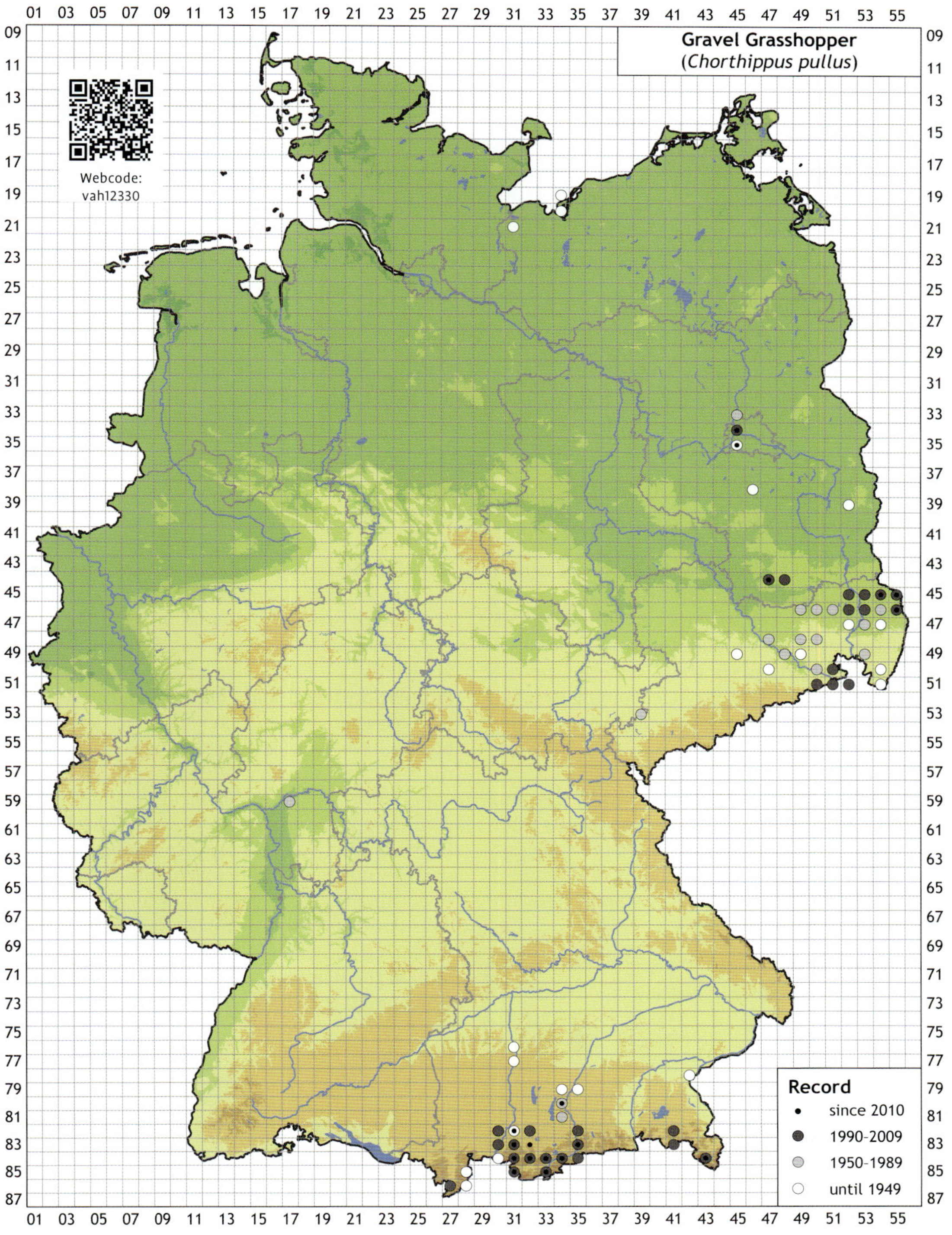
Gravel Grasshopper
(*Chorthippus pullus*)
Webcode:
vah12330
Record
since 2010
1990-2009
1950-1989
until 1949

Thermophilic oak forests, such as this one in the Kellerwald-Edersee National Park (Hesse), are a typical habitat of the Heath Grasshopper.
(Photo credit: Gregor Stuhldreher)

Female Heath Grasshopper (*Chorthippus vagans*).
(Photo credit: Dominik Poniatowski)

Heath Grasshopper – *Chorthippus vagans* (Eversmann, 1848)

Distribution The Heath Grasshopper is rare in Germany. Many of its populations are small and restricted to regional heat islands. At 15.1 °C, the STI of the species is correspondingly high. The distribution strongholds are in southwestern Germany and in Franconia. Clusters of local populations are known to exist in Lower Saxony and some parts of Eastern Germany.

Habitat As a xero-thermophilic species, the Heath Grasshopper colonises mainly dry and warm locations. Light oak and pine forests interspersed with rocks and stones or open sandy soils are preferred. The sunny ecotones of dry grasslands, heathlands, rocky steppes and screes are also considered typical habitats as long as they are at least partially covered by trees. Occasionally, the species can also be found in abandoned quarries, on south-facing slopes along pathways or in vineyards.

Population trend and threats In the course of land-use change, many habitats of the Heath Grasshopper have been lost in Germany. Especially from the mid-20th century onwards, the species has experienced severe population declines. The remaining populations are currently relatively stable. However, their habitats are also disappearing, particularly due to shrub encroachment and afforestation. Climate change might by a further threat. As a result of increasing droughts, many grasses start withering early in the year. This leads to a lack of food in the habitats of the Heath Grasshopper. To what extent the genetic isolation of many populations threatens the survival of the species is still uncertain.

DOMINIK PONIATOWSKI

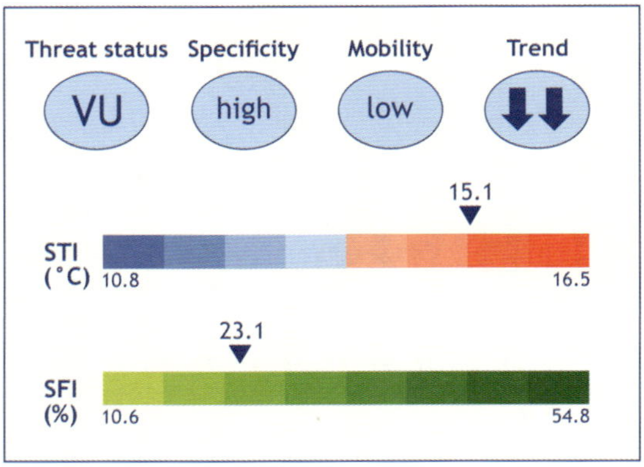

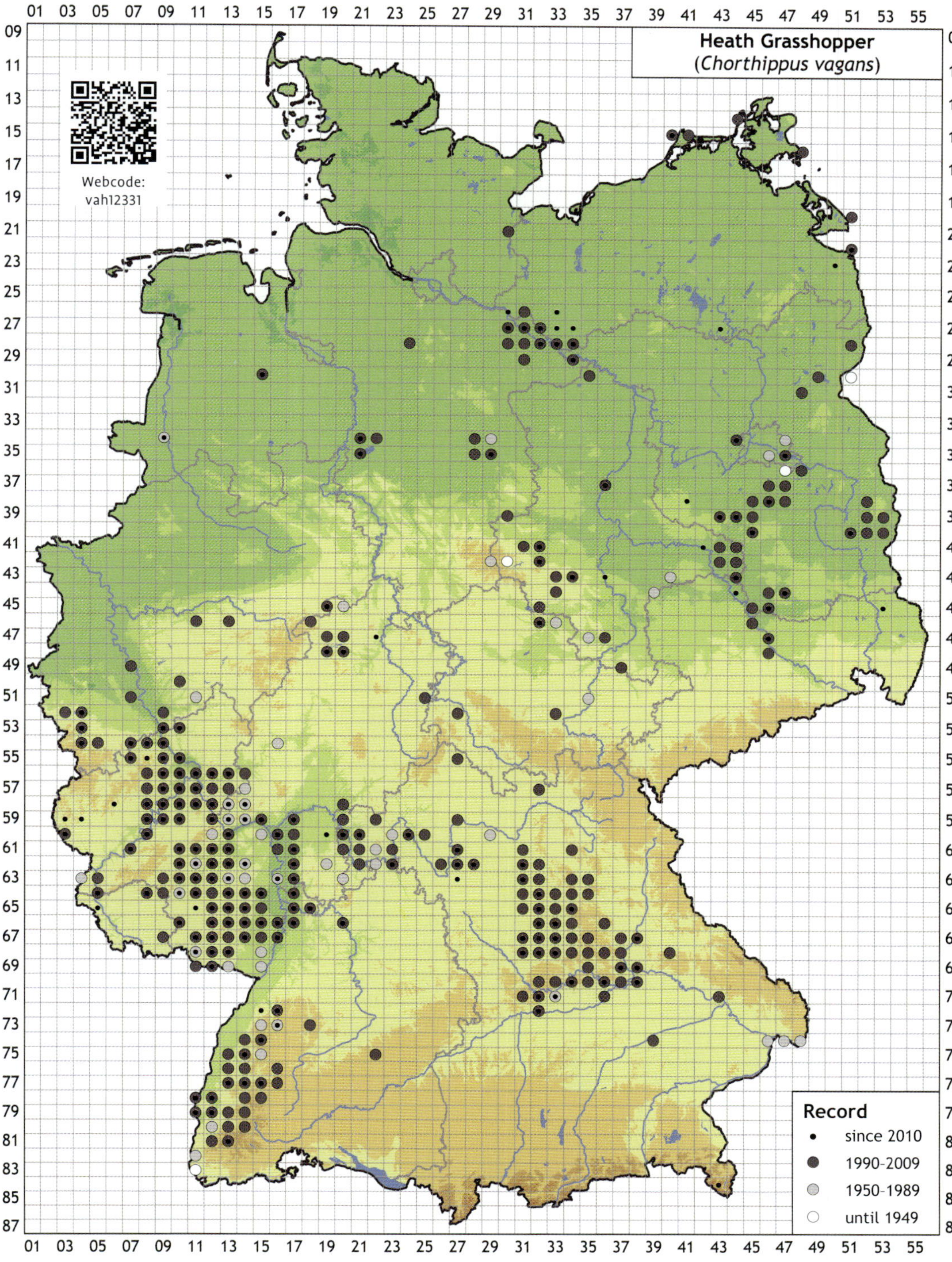
Heath Grasshopper
(*Chorthippus vagans*)
Webcode:
vah12331
Record
since 2010
1990-2009
1950-1989
until 1949

In those landscapes of the Northeast German Lowland that are characterised by very warm summers, the Lesser Grasshopper also occurs in tall-growing sandy grasslands (Brandenburg: Märkische Schweiz).
(Photo credit: Thomas Fartmann)

Female Lesser Grasshopper (*Chorthippus mollis*).
(Photo credit: Dominik Poniatowski)

Lesser Grasshopper – *Chorthippus mollis* (Charpentier, 1825)

Distribution In Germany, the Lesser Grasshopper is typical of lower elevations characterised by warm summers. Records above 600 m a.s.l. are rare. This is reflected by an above average STI of 15.1 °C. Particularly in the Northeast German Lowland, but also in the Northwest German Lowland, the species is widespread and common. Further strongholds are the Thuringian Basin, the northern Upper Rhine Valley, the Franconian Basin and a belt from the eastern Swabian Jura to the southern Franconian Jura.

Habitat As an xero-thermophilic pioneer species, the Lesser Grasshopper prefers sun-exposed habitats with low-growing, sparse vegetation on sandy or stony substrates. In very warm regions, however, the species also occurs in dry habitats with tall-growing vegetation. Dry grassland is a particularly typical habitat. Moreover, dry fallows, heathlands, mining areas, path or roadside verges and forest edges are regularly colonised. Under optimal environmental conditions, population density can be very high and the Lesser Grasshopper is then often the dominant grasshopper species.

Population trend and threats In the long term, a moderate population decline of the Lesser Grasshopper in Germany is to be assumed due to large-scale losses of dry, warm habitats such as sandy and calcareous grasslands or dwarf-shrub heathlands. Currently, the populations are considered to be stable. In some cases, due to climate change, even local expansions of the mobile species have been observed.

LISA HOLTMANN

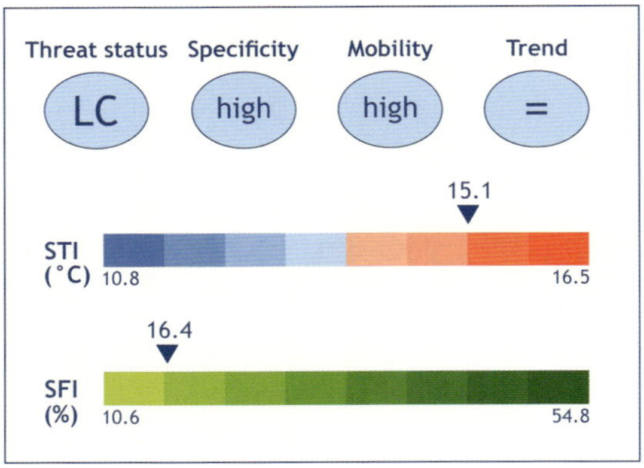

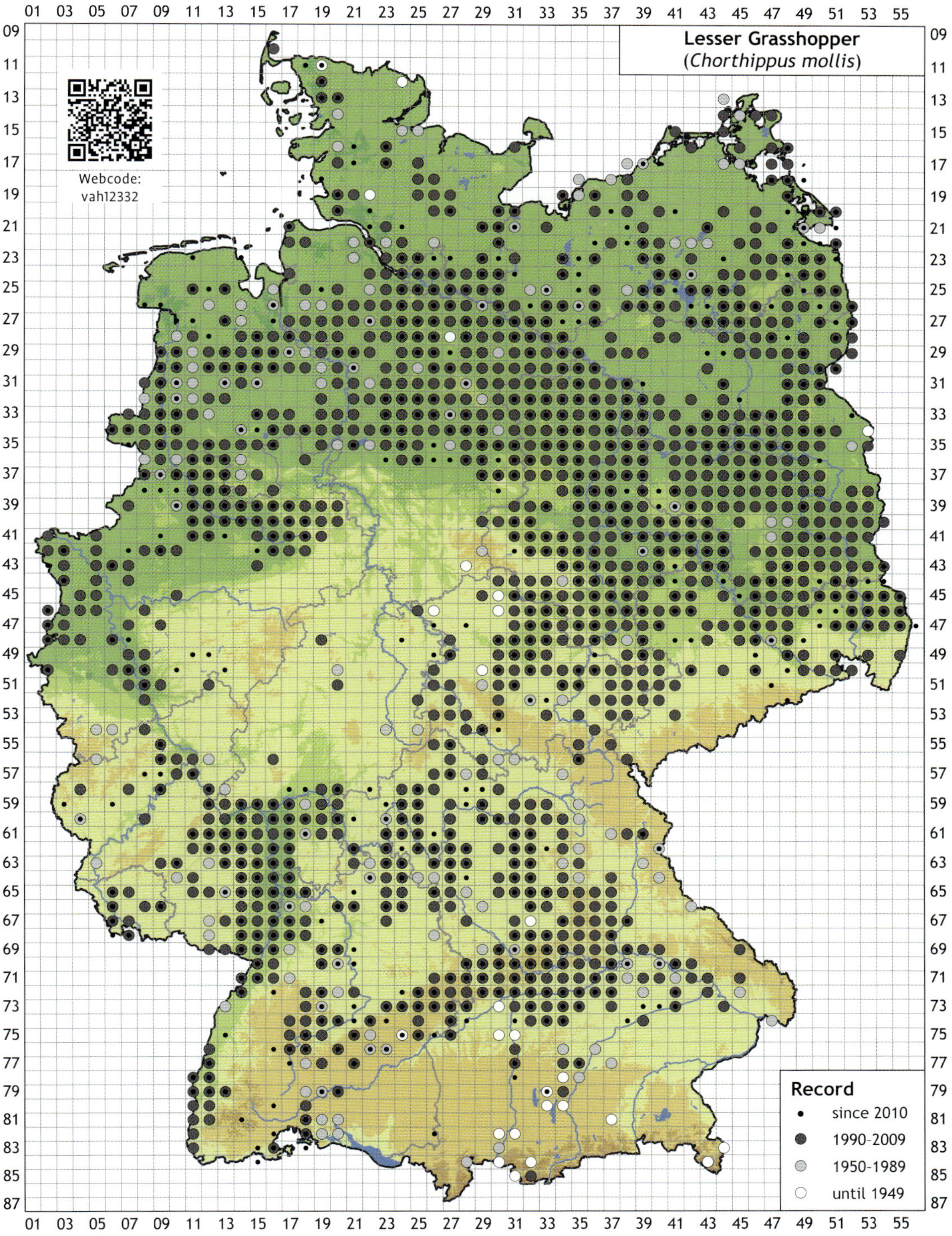
Lesser Grasshopper
(*Chorthippus mollis*)
Webcode:
vah12332
Record
since 2010
1990-2009
1950-1989
until 1949

Sparsely vegetated waysides on sandy soils are regularly colonised by populations of the Common Field Grasshopper (Lower Saxony: Ems-Hunte-Geest). (Photo credit: Dominik Poniatowski)

Female Common Field Grasshopper (*Chorthippus brunneus*). (Photo credit: Thomas Fartmann)

Common Field Grasshopper – *Chorthippus brunneus* (Thunberg, 1815)

Distribution The Common Field Grasshopper is among the most widely distributed and common grasshopper species in Germany. It occurs regularly in all biogeographical regions from the sea level to the Alps at 1,600 m a.s.l.

Habitat The geophilic Common Field Grasshopper inhabits mainly dry and warm habitats – often on sandy soils – that are characterised by some bare soil. The eggs exhibit a high drought resistance and are mainly deposited into open, dry and sunlit soil. Typical habitats include sparsely vegetated arable fallows, urban brownfields, clear-cuts, dry grasslands, heathlands, mining areas, ruderal habitats or waysides. Due to their high capability of flight, the adults are very mobile and can also be observed in habitats with closed swards.

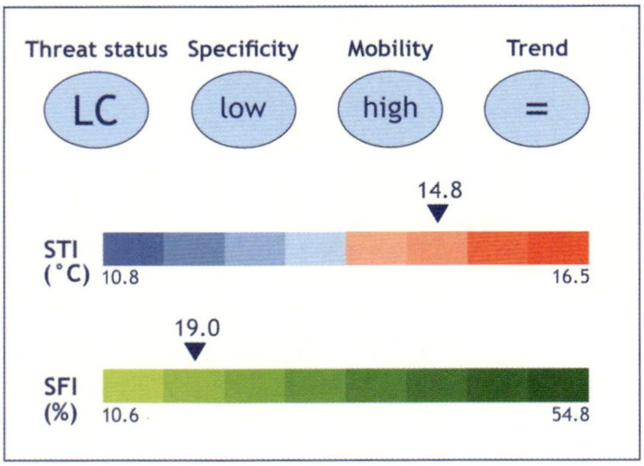

Population trend and threats The long- and short-term trends of the Common Field Grasshopper in Germany are stable. However, regionally, the developments may differ: both local increases and decreases have been detected. In contrast to many other grasshopper species that depend on early-successional stages, the Common Field Grasshopper copes quite well with our highly dynamic landscapes. One reason for this is the relatively low area requirement of the species. Moreover, due to its high mobility, the Common Field Grasshopper is considered a typical pioneer species that is able to colonise even small and isolated habitats rapidly. Recently, in the low mountain ranges, many new habitats emerged due to windthrow events and bark beetle calamities. Furthermore, the Common Field Grasshopper generally benefits from warmer conditions as a result of climate change.

FLORIAN FUMY

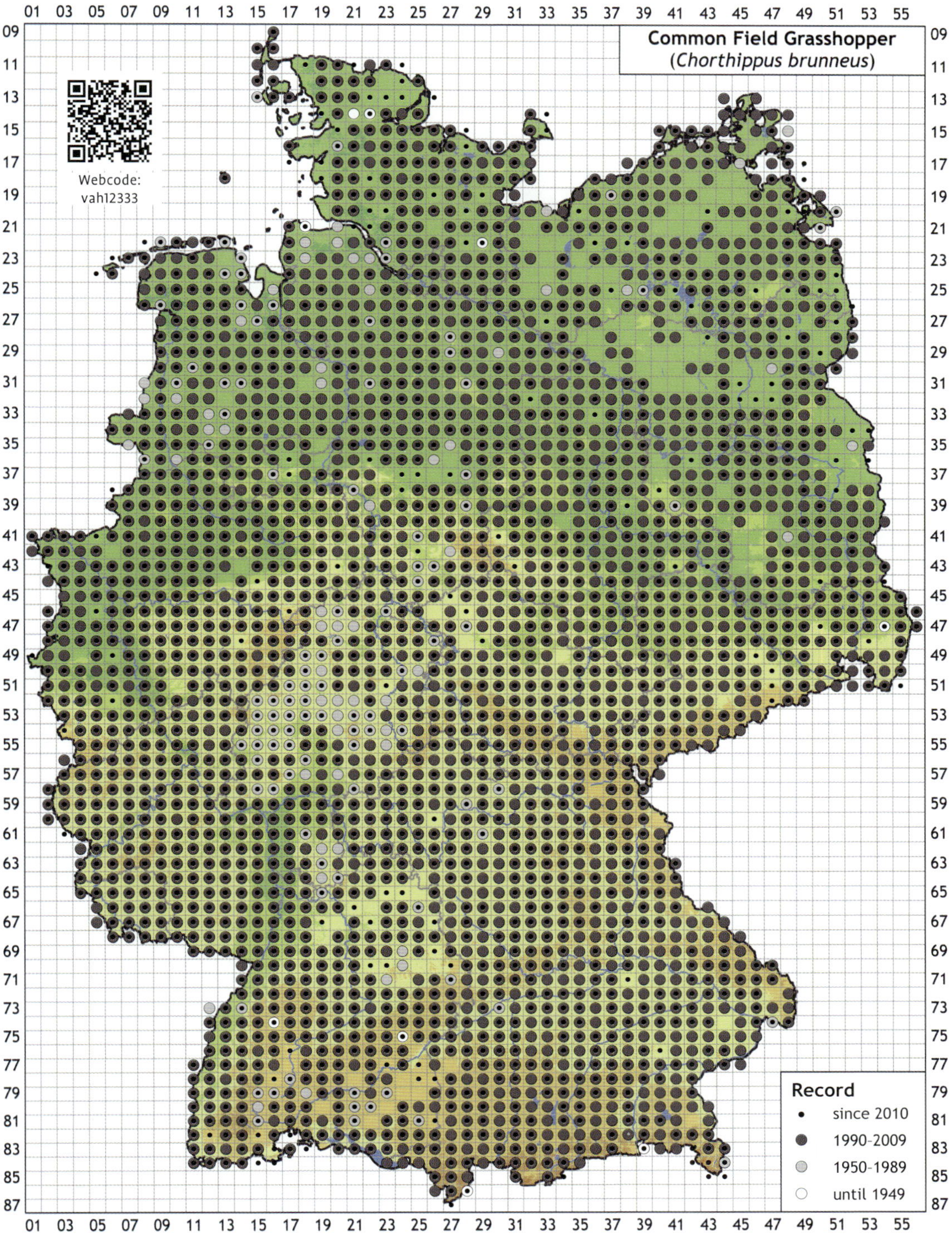
Common Field Grasshopper
(Chorthippus brunneus)
Webcode:
vah12333
Record
since 2010
1990-2009
1950-1989
until 1949

Nutrient-poor pastures with low land-use intensity, here with Przewalski's horses, are among the habitats of the Bow-winged Grasshopper (Bavaria: Franconian Basin). (Photo credit: Dominik Poniatowski)

Female Bow-winged Grasshopper (*Chorthippus biguttulus*). (Photo credit: Dominik Poniatowski)

Bow-winged Grasshopper – *Chorthippus biguttulus* (Linnaeus, 1758)

Distribution The Bow-winged Grasshopper occurs nationwide in all seven major biogeographical regions and is one of the most common grasshopper species in Germany. Only very few areas along the North Sea coast as well as the highest elevations of the Harz Mountains and Alps are unoccupied.

Habitat The basis for the wide distribution and high frequency of the Bow-winged Grasshopper is its high ecological plasticity. The species colonises sunny, moderately dry, open habitats that are characterised by low to intermediate nutrient contents. The mesophilic eggs are deposited in dry to fresh, mostly open soil. Accordingly, the Bow-winged Grasshopper prefers sparser vegetation and avoids wet habitats. Typical habitats include nutrient-poor and improved grasslands or heathlands as well as abandoned grasslands, clear-cuts, forest edges, ruderal habitats and waysides.

Population trend and threats Both the historical and recent population trend of the Bow-winged Grasshopper in Germany is considered stable. As a highly mobile habitat generalist, it has suffered less from the consequences of land-use change than many other grasshopper species. Currently, there is no sign of any significant threat. Due to the high mobility of the Bow-winged Grasshopper, local population losses can may be compensated for by the rapid colonisation of new habitats. On a regional scale, it is even assumed that populations will increase due to climate change.

JONAS BRÜGGESHEMKE

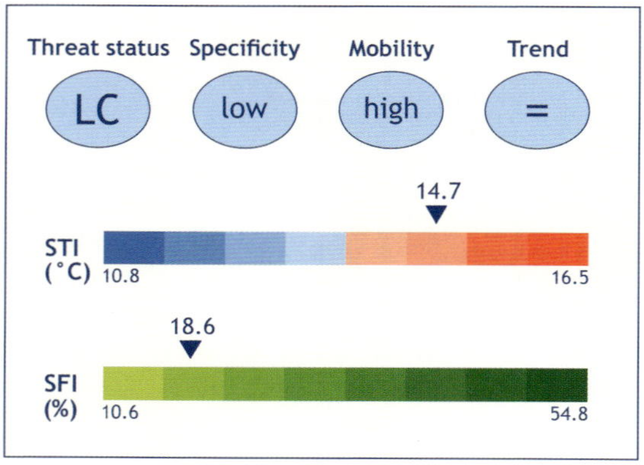

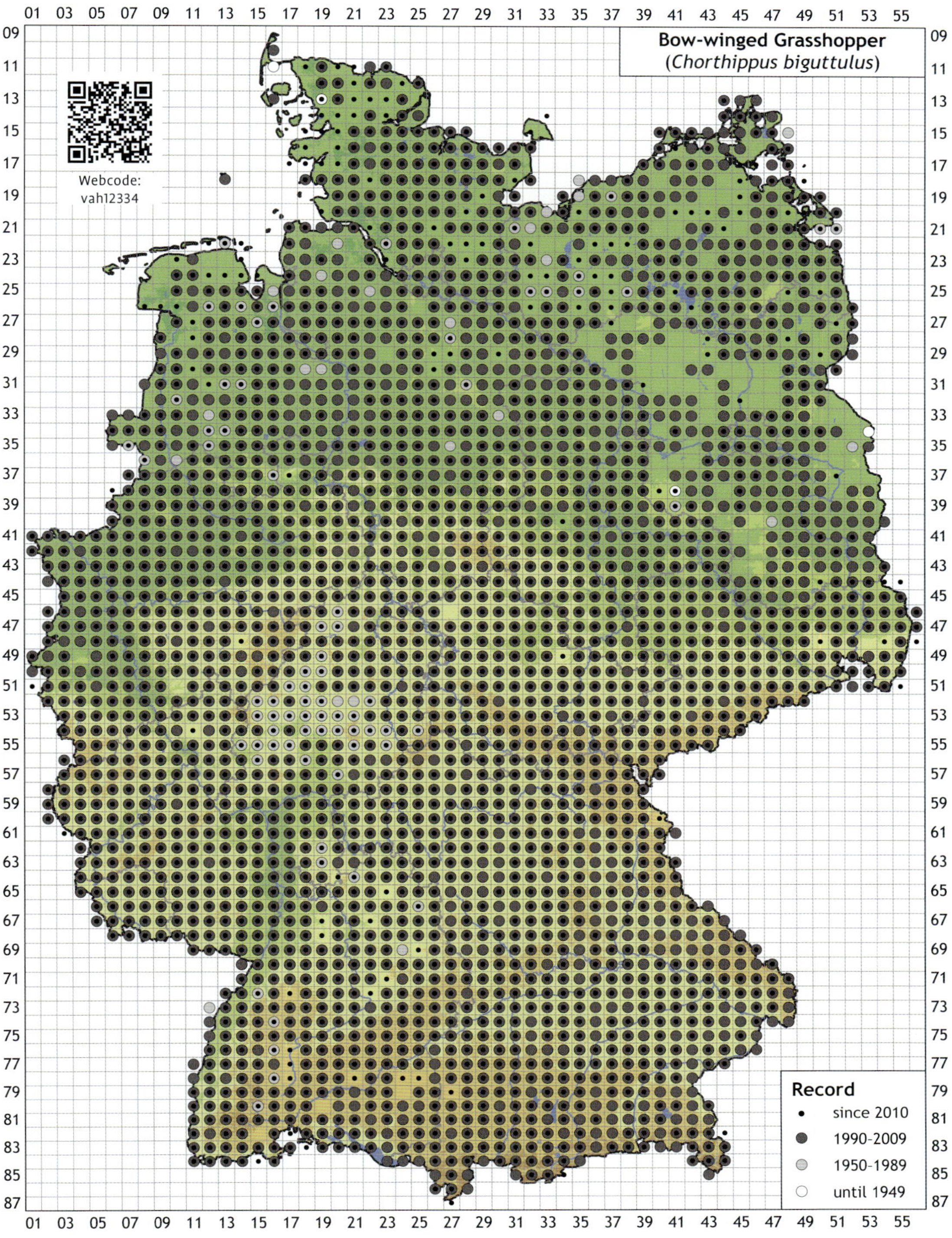
Bow-winged Grasshopper
(*Chorthippus biguttulus*)
Webcode:
vah12334
Record
since 2010
1990-2009
1950-1989
until 1949

Eisentraut's Bow-winged Grasshopper prefers sunny, sparsely vegetated, rocky habitats (Bavaria: Berchtesgaden Alps). (Photo credit: Sebastian König)

Male Eisentraut's Bow-winged Grasshopper (*Chorthippus eisentrauti*). (Photo credit: Sebastian König)

Eisentraut's Bow-winged Grasshopper – *Chorthippus eisentrauti* (Ramme, 1931)

Distribution Eisentraut's Bow-winged Grasshopper is endemic to the Alps and mainly distributed in southern parts of the Alpine arc. In Germany, it occurs exclusively in the southern part of the Bavarian Alps. These populations are part of a disjunct distribution in the Northern Limestone Alps. The occurrence of this species in Germany has only recently been confirmed in the Berchtesgaden Alps. Despite its geographically limited distribution, it can be found at elevations between 600 m a.s.l. at the shore of Lake Königssee and 2,100 m a.s.l. at the Watzmann.

Habitat In Germany, the stenotopic Eisentraut's Bow-winged Grasshopper inhabits naturally open habitats. The habitats of this helio- and thermophilic species are sparsely vegetated and often situated along steep, sunny, south-facing slopes. Characteristic habitats include boulders, outcrops, rocky grasslands, rocky slopes, screes and sparsely vegetated gravel banks of dynamic streams. The diverse vegetation within these partly extreme habitats ranges from swards of Blue Moor Grass to scree communities. Despite the species' macroclimatically cold and rainy distribution area, all these habitats are characterised by a warm and dry local microclimate.

Population trend and threats Historical reports from the Berchtesgaden Alps, although unconfirmed later, suggest a continuous occurrence of Eisentraut's Bow-winged Grasshopper in Germany. Since the species' habitats are located in protected areas and not actively managed, it can be assumed that the populations are at low risk. Habitat loss due to forest succession and newly created habitats due to avalanches or mudslides dynamically maintain a balance. However, due to the specific habitat requirements, many populations are spatially isolated. Warm, dry weather condition during summer have proven to be beneficial for population growth, especially at high elevations. Nevertheless, it is unclear how the climatically driven spread of the closely related Bow-winged Grasshopper will affect the populations since negative interactions could possibly take place despite distinct habitat requirements.

SEBASTIAN KÖNIG

Threat status	Specificity	Mobility	Trend
R	high	high	=

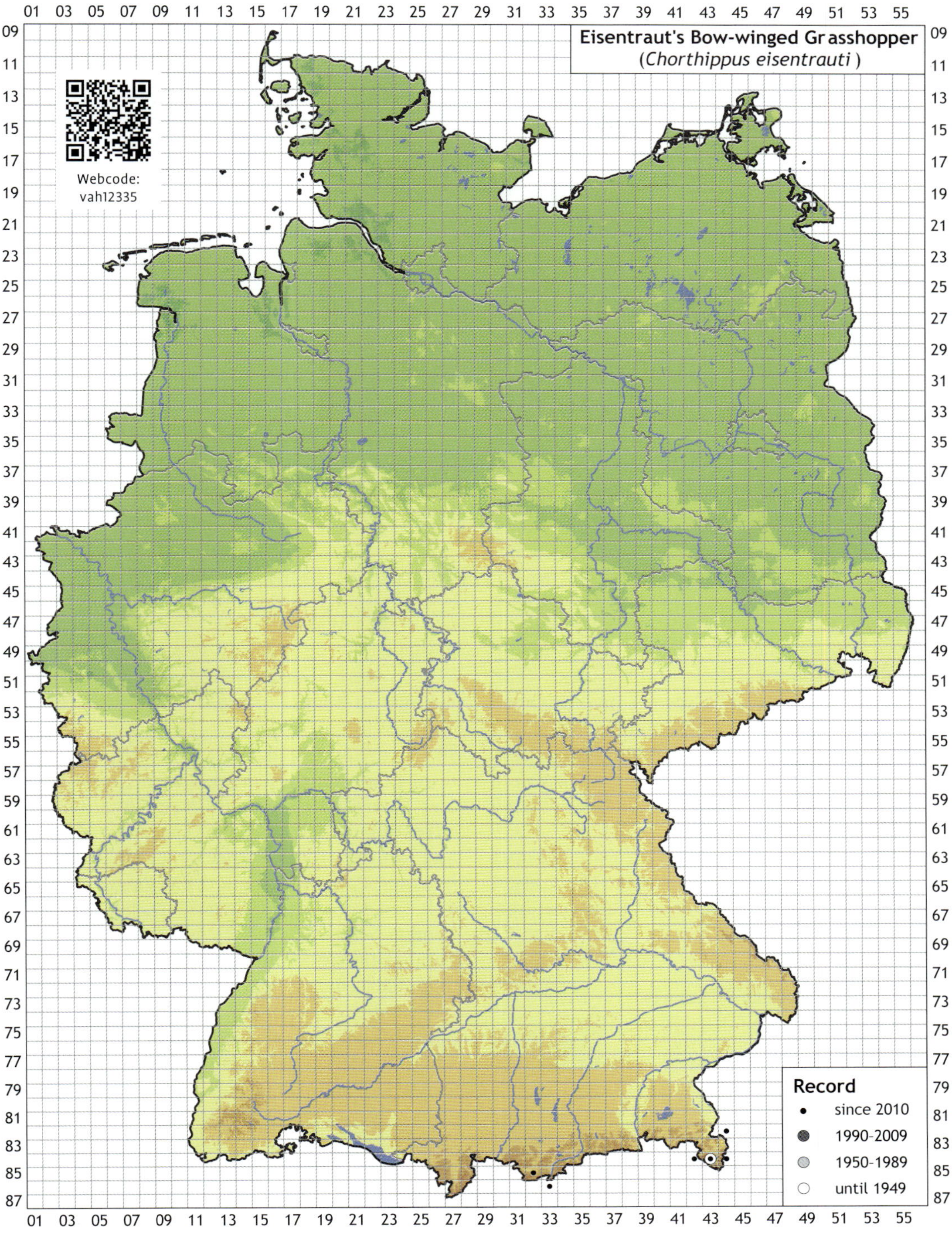
Eisentraut's Bow-winged Grasshopper
(*Chorthippus eisentrauti*)
Webcode:
vah12335
Record
since 2010
1990-2009
1950-1989
until 1949

Fresh to moist pastures with intermediate to high land-use intensity are the preferred habitats of the Lesser Marsh Grasshopper in the North German Lowlands (North Rhine-Westphalia: Westphalian Bay). (Photo credit: Felix Helbing)

Female Lesser Marsh Grasshopper (*Chorthippus albomarginatus*). (Photo credit: Dominik Poniatowski)

Lesser Marsh Grasshopper – *Chorthippus albomarginatus* (De Geer, 1773)

Distribution In Germany, the slightly thermophilic (STI = 14.8 °C) Lesser Marsh Grasshopper is widespread and common in the lowlands. By contrast, it is rare at higher elevations of the low mountain ranges, Alpine Foreland and Alps. However, due to climate change, the species has expanded its elevational range upwards. Such range shifts have been observed in the Eifel Mountains, Rothaar Mountains, Black Forest and Alpine Foreland. Nevertheless, above 700 m a.s.l., the Lesser Marsh Grasshopper is still rare.

Habitat Although the Lesser Marsh Grasshopper generally inhabits a wide range of open habitats, it is typical of fresh to moist meadows and pastures. The eggs are mesophilic and deposited into the soil or slightly above it into plant material. It is one of the few species that tolerates high grassland management intensity. In some cases, the species is even more abundant on grasslands with intermediate to high land-use intensity than in those with less intensive land use. Dense but not too high swards are preferred. Due to its tolerance for short-term flooding, the Lesser Marsh Grasshopper is also particularly common in floodplains and the edges of bogs and lakes. The species is very mobile and can therefore colonise small and fragmented habitats. Path verges and field margins may not only serve as dispersal corridors, but also as permanent habitats.

Population trend and threats In Germany, due to climate change, the Lesser Marsh Grasshopper will further expand its range to higher elevations of the mountain ranges. Since the species also has a high tolerance against intensive grassland management, it is currently considered of least concern.

STEFFEN KÄMPFER

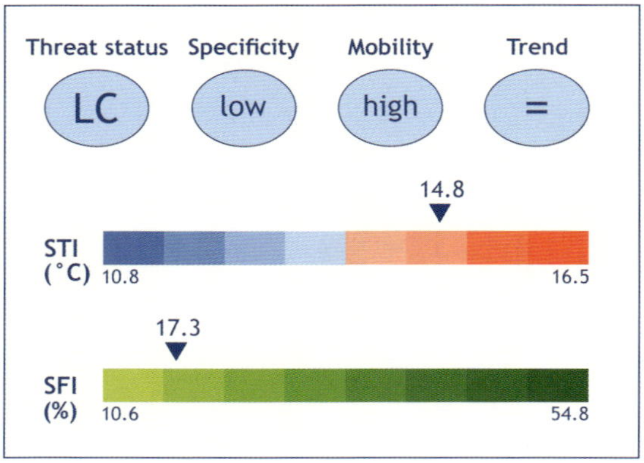

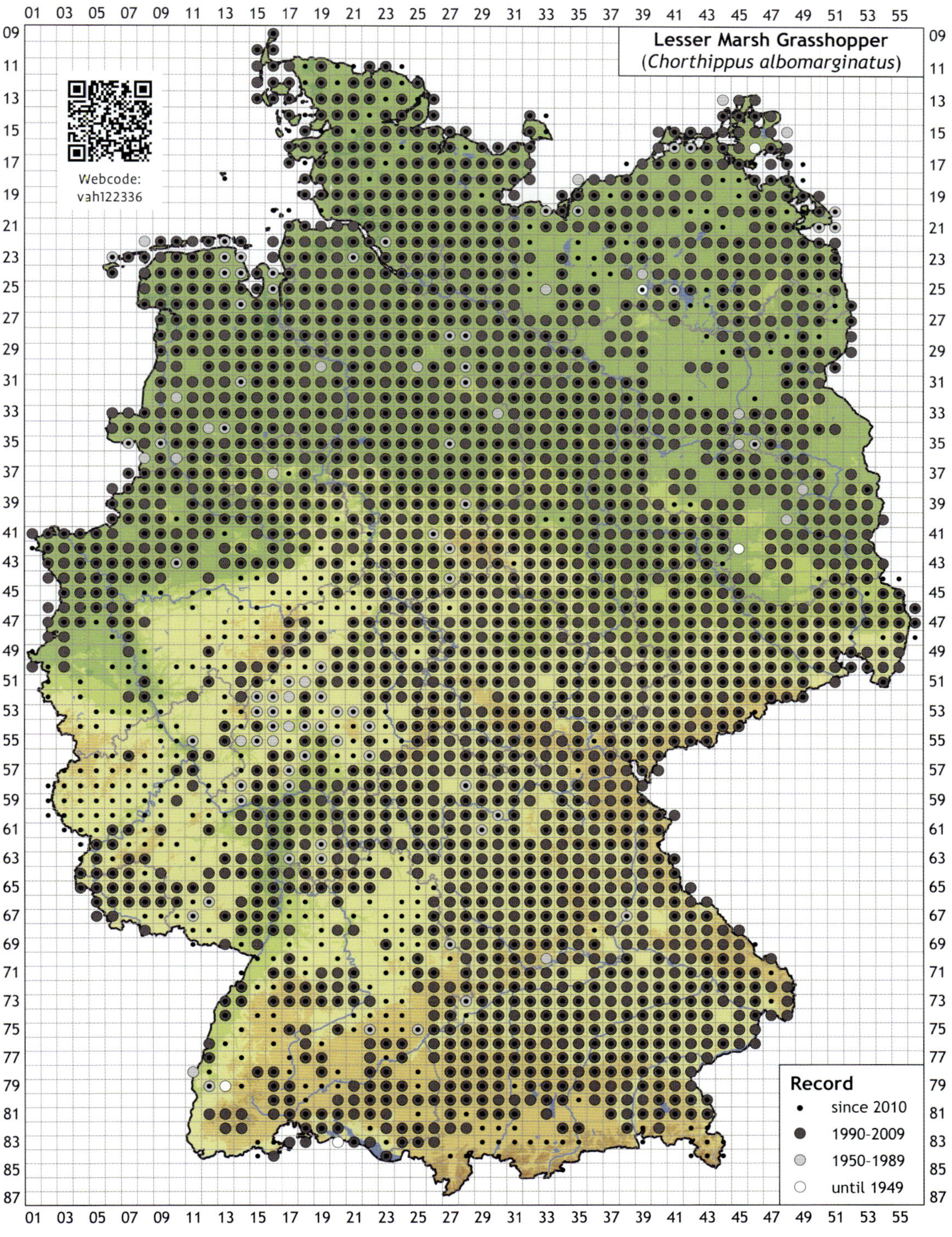
Lesser Marsh Grasshopper
(*Chorthippus albomarginatus*)
Webcode:
vah122336
Record
since 2010
1990-2009
1950-1989
until 1949
01 03 05 07 09 11 13 15 17 19 21 23 25 27 29 31 33 35 37 39 41 43 45 47 49 51 53 55
09 11 13 15 17 19 21 23 25 27 29 31 33 35 37 39 41 43 45 47 49 51 53 55 57 59 61 63 65 67 69 71 73 75 77 79 81 83 85 87

Male Steppe Grasshopper (*Chorthippus dorsatus*).
(Photo credit: Dominik Poniatowski)

Wet grasslands exhibiting low land-use intensity is among the typical habitats of the Steppe Grasshopper (Mecklenburg-Western Pomerania: Ostseeküstenland).
(Photo credit: Wolfgang Wranik).

Steppe Grasshopper – *Chorthippus dorsatus* (Zetterstedt, 1821)

Distribution The Steppe Grasshopper occurs in all German federal states. However, in the northwest of the country, which is characterised by cooler summers, many biogeographical regions have only recently been colonised. This is true for parts of the Western Uplands and Northwest German Lowland, but the species is still missing there at the highest elevations and coastal areas that exhibit the most oceanic climate.

Habitat The Steppe Grasshopper is a typical inhabitant of grasslands. The eggs are hygrophilic, but the nymphs and adults prefer a dry and warm microclimate. Accordingly, wet grasslands are the main habitat of the species and exhibit the highest population densities. However, it also colonises mesic and semi-dry grasslands such as sandy and calcareous grasslands. Where mosaics of wet and dry habitats occur in close proximity, nymphs or adults may migrate from the wet or fresh reproduction sites to the drier and warmer places during the course of the summer. Overall, a low land-use intensity and, hence, taller swards play a crucial role for the occurrence of the Steppe Grasshopper.

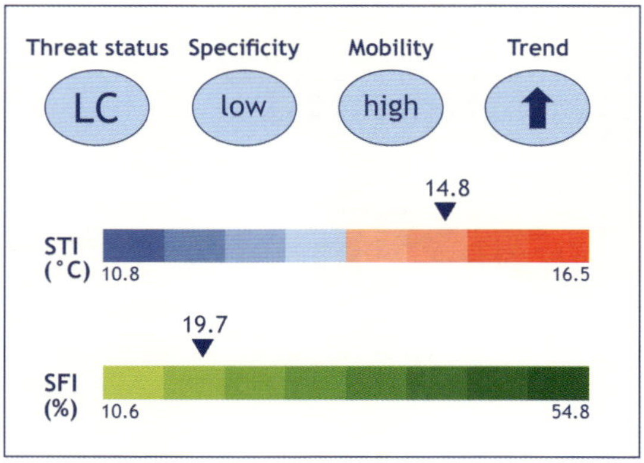

Population trend and threats Due to the conversion of grasslands to other habitats, drainage and intensification of land use, many habitats of the Steppe Grasshopper have regionally disappeared throughout Germany. Today, as a result, the species often occurs exclusively outside managed grasslands, for example in roadside verges or abandoned grasslands. Nonetheless, the Steppe Grasshopper has recently expanded its range in northwestern Germany as a response to climate change. The still available grassy, linear structures in our landscapes may have fostered this process. Overall, the species is currently considered of least concern.

WOLFGANG WRANIK

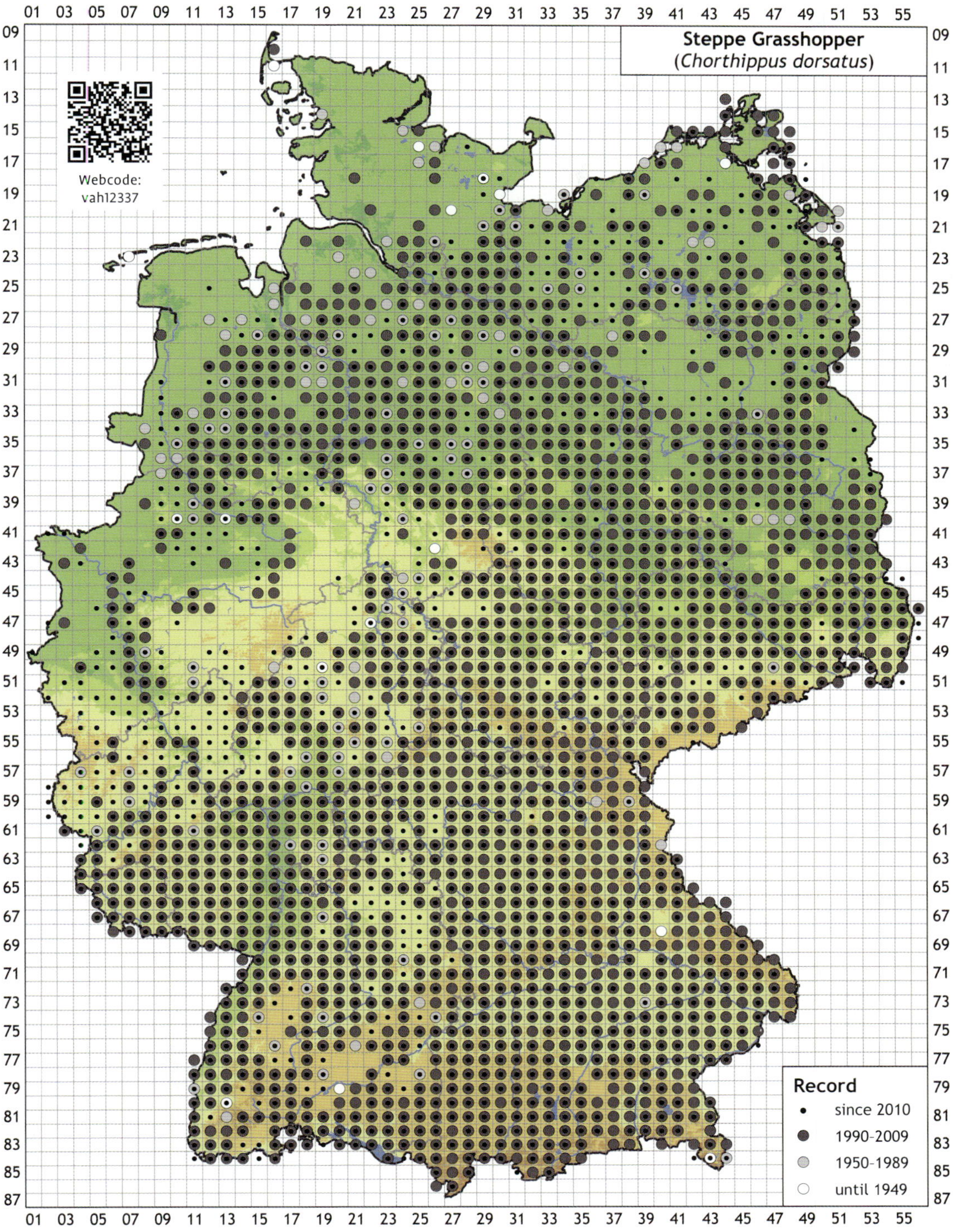
Steppe Grasshopper
(*Chorthippus dorsatus*)
Webcode:
vah12337
Record
since 2010
1990-2009
1950-1989
until 1949

Female Meadow Grasshopper (*Pseudochorthippus parallelus*).
(Photo credit: Dominik Poniatowski)

The Meadow Grasshopper regularly colonises mesic meadows and, in case of too intensive management, often uses adjacent verges as refugia (Saxony: Ore Mountains). (Photo credit: Günter Köhler)

Meadow Grasshopper – *Pseudochorthippus parallelus* (Zetterstedt, 1821)

Distribution Germany is the European range centre of the Eurasian Meadow Grasshopper. Here, it occurs nationwide from the North German Lowland through the uplands to the alpine zone of the Alps (Nebelhorn/Allgäu, 2,100 m a.s.l.). The genetically inherited green-brown polymorphism can, especially across elevation gradients, result in a thermoregulatory caused morph selection. Hence, with increasing elevation of the mountain ranges, brown individuals often become dominant. Normally, the Meadow Grasshopper is short-winged and has a low mobility. However, density stress among nymphs may result in long-winged morphs capable of flight that contribute to local dispersal.

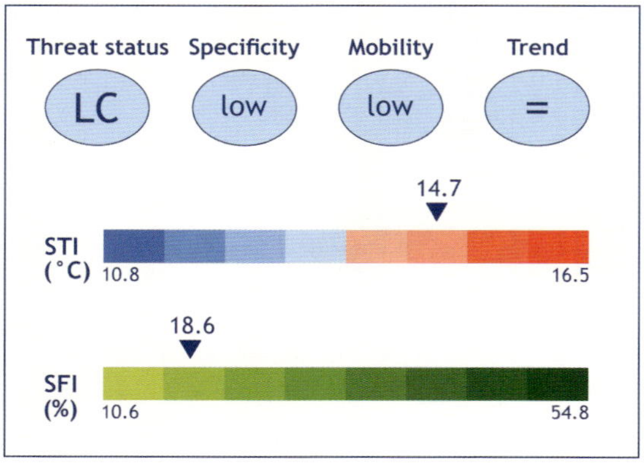

Habitat The Meadow Grasshopper colonises nearly all open habitats that are characterised by a quite closed grass and herb layer of intermediate height. Accordingly, the SFI is relatively low at 18.6%. High population densities, with a tendency for local gradations, may particularly occur in fresh to humid, nutrient-rich meadows and pastures. The Meadow Grasshopper even tolerates highly intensive grassland management. Although, in lower densities, it also inhabits semi-dry and wet habitats. The species is even able to persist in tiny patches that exhibit small populations.

Population trend and threats The Meadow Grasshopper is among the most common species in all German federal states and at all elevational zones. The long-term trend is stable. By contrast, in some regions, populations have recently disappeared or abundance has declined, mainly owing to intensive grassland management but also to drying up of the habitats as a consequence of climate change.

GÜNTER KÖHLER

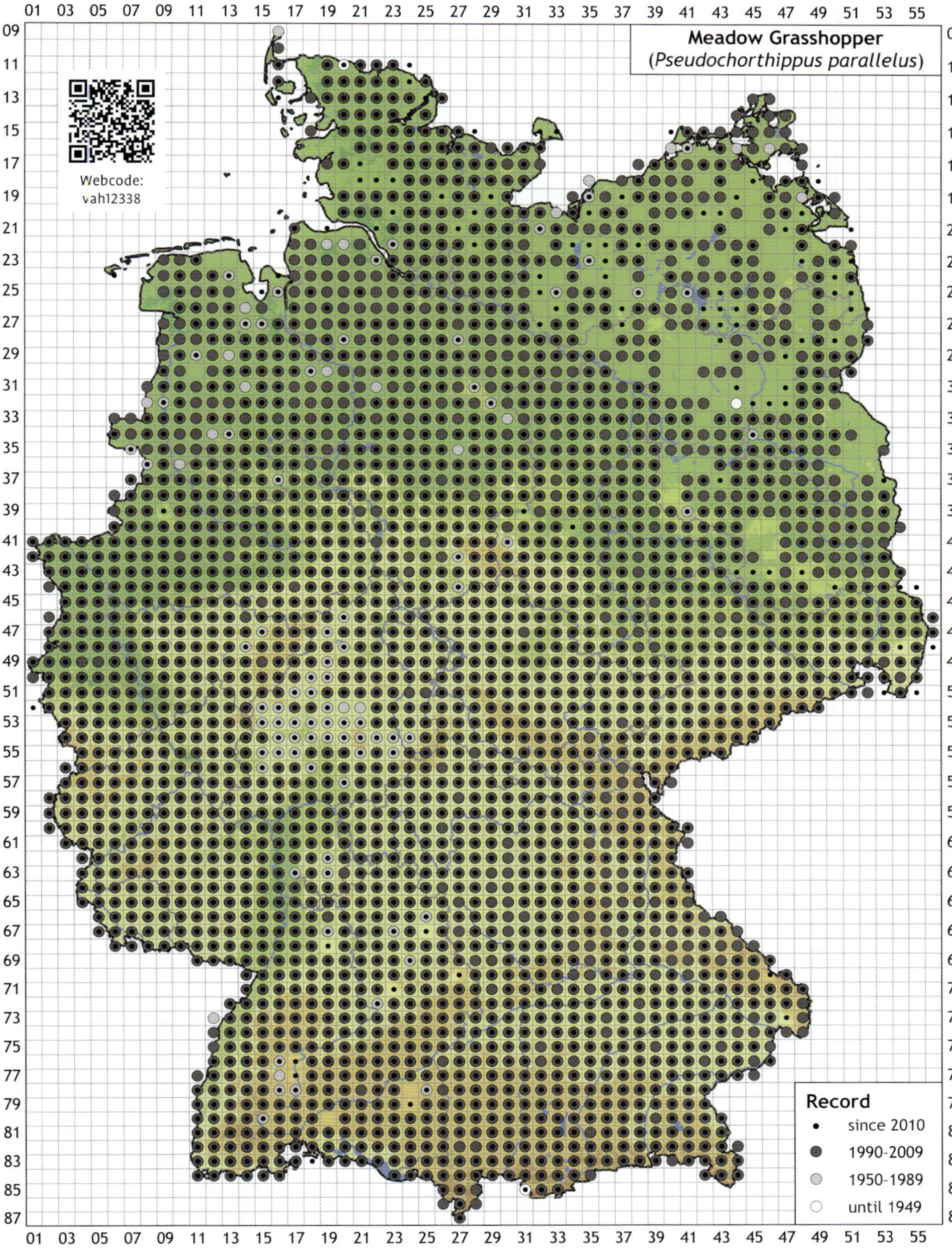
Meadow Grasshopper
(Pseudochorthippus parallelus)
Webcode:
vah12338
Record
since 2010
1990-2009
1950-1989
until 1949

Calcareous fens, which are mown late in the growing season, often exhibit abundant populations of the Marsh Grasshopper (Schleswig-Holstein: Holstein Switzerland). (Photo credit: Arne Drews)

Female Marsh Grasshopper (*Pseudochorthippus montanus*). (Photo credit: Dominik Poniatowski)

Marsh Grasshopper – *Pseudochorthippus montanus* (Charpentier, 1825)

Distribution The Marsh Grasshopper occurs in all seven major biogeographical regions in Germany. It mainly inhabits the planar to montane zone and reaches its upper elevational range limit at about 1,700 m a.s.l. The distribution hotspots are those regions that are characterised by the highest precipitation, particularly the Black Forest, Pre-Alpine Hills and Moorland as well as the Alps.

Habitat The Marsh Grasshopper colonises permanently moist to very wet, open sites. In dry regions, the dependence on high and stable water levels is very pronounced. The low desiccation resistance of the eggs and the high humidity requirements for successful embryonic development explain why the species is highly hygrophilic. Moist and wet grassland or fens, which are, for example, used as litter meadows or low-intensity summer pastures, rank among the characteristic habitats. Tall swards are generally avoided. Hence, an abandonment of land use and subsequent litter accumulation rapidly result in population declines.

Population trend and threats The Marsh Grasshopper is currently still moderately common in Germany. However, both for the long- and short-term, it is assumed that significant population declines have occurred. Due to land-use change since the mid-20th century, many habitats have disappeared due to drainage and abandonment or intensification of land use. Today, climate change has become another threat for the species. The Marsh Grasshopper is usually flightless; long-winged individuals capable of flight occur only very rarely. Since the habitats are mostly also highly isolated, local extinctions due to droughts cannot be compensated in the following years by recolonizations, when hydrological conditions may have improved again.

ARNE DREWS

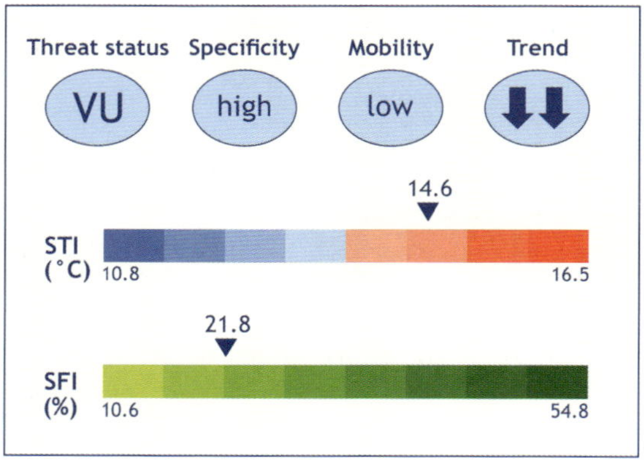

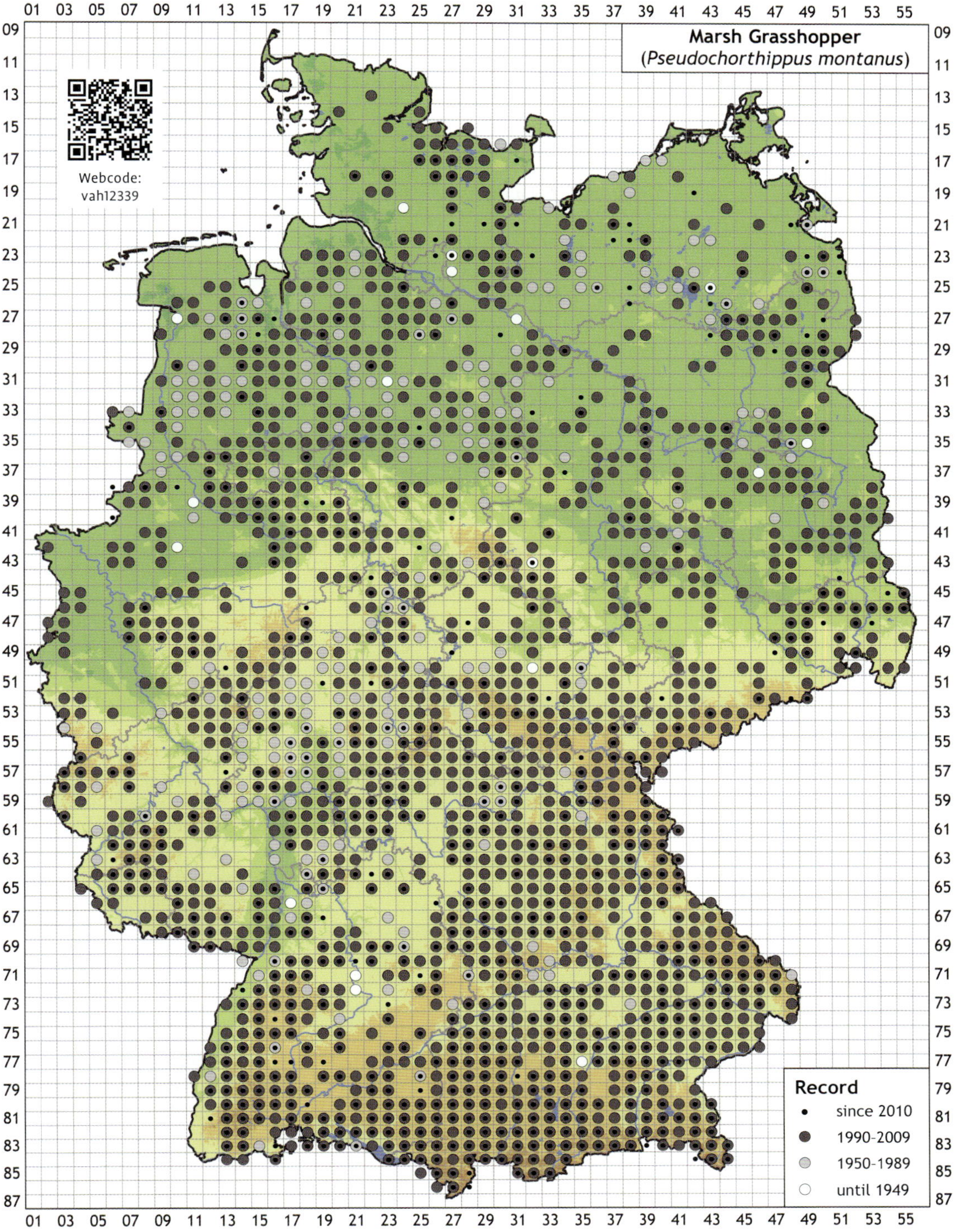
Marsh Grasshopper
(*Pseudochorthippus montanus*)
Webcode:
vah12339
Record
since 2010
1990-2009
1950-1989
until 1949

Service

Key publications

Praying Mantis (*Mantis religiosa*) – p. 56: Beckerbauer et al. (2024), Berg et al. (2011), Detzel (1998), Ehrmann (2011), Ehrmann & Reinhard (2011), Linn (2015), Mutz (2024), Pfeifer (2012), Pfeifer et al. (2011), Poniatowski et al. (2024a), Schädler et al. (2019), Schwarz et al. (2017), Stärz et al. (2010), Stenger (2020), Stübing et al. (2019b).

Common Sickle Bush-cricket (*Phaneroptera falcata*) – p. 58: Detzel et al. (2022), Fartmann et al. (2012), Fischer et al. (2020), Grein (2007, 2010), Hafner & Zimmermann (2019), Handke et al. (2011), Kronshage (1993), Maas et al. (2002), Poniatowski et al. (2018a, 2020).

Southern Sickle Bush-cricket (*Phaneroptera nana*) – p. 60: Aßmann & Ritt (2021), Boczki (2007), Coray (2003), Detzel et al. (2022), Pfeifer (2006, 2012), Pfeifer et al. (2011), Poniatowski et al. (2018a, 2020), Stübing et al. (2019a).

Speckled Bush-cricket (*Leptophyes punctatissima*) – p. 62: Detzel (1998), Detzel et al. (2022), Deura & Hartley (1990), Fartmann (1997a), Fischer et al. (2020), Froehlich (1994), Ingrisch (1978a, b, 1979b, 1987), Maas et al. (2002), Poniatowski et al. (2020, 2024a), Schlumprecht & Waeber (2003), Wallaschek et al. (2004).

Striped Bush-cricket (*Leptophyes albovittata*) – p. 64: Bolz (1998), Detzel (1998), Detzel et al. (2022), Fischer et al. (2020), Grein (2010), Maas et al. (2002), Pfeifer (2012), Pfeifer et al. (2011), Poniatowski et al. (2024a), Samu & Wischhof (2000), Schlumprecht & Waeber (2003), Wallaschek et al. (2004).

Common Saw Bush-cricket (*Barbitistes serricauda*) – p. 66: Gottwald et al. (2002), Grein (1991, 2010), Heller (1988), Ingrisch & Köhler (1998), Klaus (2010), Laußmann (1995), Poniatowski et al. (2024a), Richter (2008), Schlumprecht & Waeber (2003), Unruh & Klaus (2005).

Eastern Saw Bush-cricket (*Barbitistes constrictus*) – p. 68: Fischer et al. (2020), Gottwald et al. (2002), Harz (1987), Ingrisch (1988), Köhler (2001), Laußmann (1995), Maas et al. (2002), Poniatowski et al. (2024a), Schlumprecht & Waeber (2003), Stadler (1997).

Krauss's Bush-cricket (*Isophya kraussii*) – p. 70: Breuer et al. (2017), Detzel (1998), Heller (1988), Heller et al. (2004), Kästner (2016), Köhler (2001, 2013), Köhler et al. (2019), Krech & Köhler (2021), Pfeifer et al. (2011), Schlumprecht & Waeber (2003), Schulze (1998), Strätz & Schlumprecht (1999), Thorn et al. (2022), Wallaschek (2013), Wallaschek et al. (2004).

Large Saw-tailed Bush-cricket (*Polysarcus denticauda*) – p. 72: Beinlich (1993), Birkwald et al. (2022), Detzel et al. (2022), Dürer & Weid (1990), Hawlitschek (2018), Hermann et al. (2022), Kleiner (2013), Köhler (2017, 2021), Krech & Köhler. (2021), Poniatowski et al. (2024a), Rothhaupt (1994), Smettan (1991), Trautner (2024), Voith et al. (2016).

Oak Bush-cricket (*Meconema thalassinum*) – p. 74: Brauner et al. (2021), Detzel et al. (2022), Fartmann (1997a), Fischer et al. (2020), Ingrisch (1978a, b; 1979a, b), Maas et al. (2002), Ogan et al. (2022), Pfeifer et al. (2011, 2019), Poniatowski et al. (2024a), Schlumprecht & Waeber (2003).

Southern Oak Bush-cricket (*Meconema meridionale*) – p. 76: Brauner et al. (2021), Detzel (1998), Fischer et al. (2020), Gottfried & Kästner (2009), Helversen (1969), Köhler et al. (2022c), Maas et al., (2002), Pfeifer (2012), Pfeifer et al. (2011), Poniatowski et al. (2018a, 2024a), Schlumprecht & Waeber (2003), Sczepanski & Jacobi (2005).

Long-winged Conehead (*Conocephalus fuscus*) – p. 78: Brose & Peschel (1998), Fartmann (1997a, 2004), Fartmann et al. (2021b), Fischer et al. (2020), Fuhrmann (2019), Fumy et al. (2020), Grein (2007, 2010), Ingrisch (1988), Ingrisch & Köhler (1998), Maas et al. (2002), Poniatowski et al. (2020, 2024a), Schlumprecht & Waeber (2003).

Short-winged Conehead (*Conocephalus dorsalis*) – p. 80: Detzel et al. (2022), Fartmann et al. (2021b), Grein (2010), Haupt (2005a), Köhler (2021), Lehmann & Haacks (2006), Ogan et al. (2022), Pfeifer et al. (2019), Poniatowski et al. (2018a, 2024a, b), Poschmann et al. (2009), Schlumprecht & Waeber (2003), Voith et al. (2016), Wallaschek (2020), Winkler & Haacks (2019).

Large Conehead (*Ruspolia nitidula*) – p. 82: Detzel et al. (2022), Drukker et al. (2020), Kimmich et al. (2021), Newson et al. (2023), Paulus (2021),

Poniatowski et al. (2018a, 2020), Renker et al. (2020), Sutton et al. (2017), Treiber (2016).

Great Green Bush-cricket (*Tettigonia viridissima*) – p. 84: Fartmann (1997a), Fartmann et al. (2021b, 2022b), Fischer et al. (2020), Grein (2010), Ingrisch & Boekholt (1983), Ingrisch (1986, 1988), Löffler et al. (2019), Poniatowski & Fartmann (2008a), Poniatowski et al. (2018a), Sänger (1977), Schirmel & Fartmann (2007a, b, 2013).

Upland Green Bush-cricket (*Tettigonia cantans*) – p. 86: d'Agostion & Vacher (2021), Fartmann et al. (2021b, 2022a), Fischer et al. (2020), Grein (2010), Hafner & Zimmermann (2019), Ingrisch (1977, 1986, 1988), Ingrisch & Boekholt (1983), Poniatowski et al. (2018a).

Eastern Green Bush-cricket (*Tettigonia caudata*) – p. 88: Bellebaum (2003), Fartmann (1997b), Haupt (1995b, 1997), Ingrisch (1986, 1988), Krištín et al. (2007), Nadig (1991), Schirmel & Fartmann (2007b, 2013), Wallaschek et al. (2004), Zuna-Kratky et al. (2009, 2017).

Common Wart-biter (*Decticus verrucivorus*) – p. 90: Fartmann (1997a), Glück & Ingrisch (1989), Ingrisch (1978a, 1979b, 1986, 1988), Ingrisch & Boekholt (1983), Poniatowski et al. (2020, 2024a), Schuhmacher & Fartmann (2003a, b), Schlumprecht & Waeber (2003), Schwarz & Fartmann (2022), Wallaschek et al. (2004), Wünsch et al. (2012).

Heath Bush-cricket (*Gampsocleis glabra*) – p. 92: Clausnitzer & Clausnitzer (2005), Fischer et al. (2020), Grein (2010), Grzędzicka & Vahed (2020), Hawlitschek et al. (2023), Hochkirch et al. (2023), Knipper (1932), Poniatowski et al. (2020, 2024a), Schäfer (2013), Schäfer & Hennigs (2020), Schlumprecht & Waeber (2003), Wallaschek (2005), Wallaschek et al. (2004), Zuna-Kratky et al. (2017).

Common Grey Bush-cricket (*Platycleis albopunctata*) – p. 94: Detzel (1998), Fartmann et al. (2012), Fischer et al. (2020), Gottschalk (1996, 1997), Grein (2010), Hafner & Zimmermann (2019), Hamann & Weber (2012), Ingrisch (1988), Poniatowski et al. (2018a, 2020), Schlumprecht & Waeber (2003), Thorn et al. (2022), Wünsch et al. (2010, 2012).

Brown-spotted Bush-cricket (*Tessellana tessellata*) – p. 96: Betzin & Neugebauer (2020), Coray (1993), Detzel (1998), Detzel et al. (2022), Fischer et al. (2020), Fusek (2005), Heitz & Hermann (1993), Hochkirch et al. (2023), Krämer et al. (2010), Maas et al. (2002), Poniatowski et al. (2024a).

Steppe Bush-cricket (*Montana montana*) – p. 98: Fartmann et al. (2001), Haupt (1995b, 1997), Holuša & Kočárek (2006), Kämpf & Fartmann (2014), Lehmann et al. (2016), Poniatowski et al. (2024a), Zuna-Kratky et al. (2009, 2017).

Bog Bush-cricket (*Metrioptera brachyptera*) – p. 100: Löffler & Fartmann (2017), Ogan et al. (2022), Poniatowski & Fartmann (2007, 2008a, b, 2009, 2010, 2011b), Poniatowski et al. (2018a, 2020, 2024a), Schwarz & Fartmann (2022), Thorn et al. (2022).

Two-coloured Bush-cricket (*Bicolorana bicolor*) – p. 102: Detzel (1998), Fartmann (1997a), Fartmann et al. (2012), Grein (2010), Ingrisch (1981, 1988), Köhler (2001), Nadig (1986b), Ogan et al. (2022), Poniatowski et al. (2018a, 2024a), Renker (1995), Samietz (1994), Schlumprecht & Waeber (2003), Wranik & Kleeberg (2011).

Roesel's Bush-cricket (*Roeseliana roeselii*) – p. 104: Fartmann et al. (2021b, 2022b), Hochkirch & Damerau (2009), Ingrisch (1988), Löffler et al. (2019), Poniatowski & Fartmann (2005, 2009, 2011b, c), Poniatowski et al. (2012, 2018a, 2024a), Schwarz & Fartmann (2022).

Alpine Dark Bush-cricket (*Pholidoptera aptera*) – p. 106: Fischer et al. (2020), Nadig (1991), Rutschmann & Roesti (2024), Schlumprecht & Waeber (2003), Treiber & Stephan (1995), Zuna-Kratky et al. (2017).

Common Dark Bush-cricket (*Pholidoptera griseoaptera*) – p. 108: Detzel (1998), Fartmann et al. (2021b, 2022b), Fischer et al. (2020), Fumy et al. (2020), Hochkirch (2001), Ingrisch (1988), Kleinert (1992), Löffler et al. (2019), Maas et al. (2002), Ogan et al. (2022), Poniatowski et al. (2018a), Thorn et al. (2022).

Western Saddle Bush-cricket (*Ephippiger diurnus*) – p. 110: Detzel (1998), Pfeifer et al. (2011, 2019), Rühlke (2021).

Neglected Cave-cricket (*Troglophilus neglectus*) – p. 112: Brunk et al. (2003), Christian & Moog (2023), Fischer et al. (2020), Karaman et al. (2011), Kiefer et al. (2000), Moritz et al. (2022), Pfeifer et al. (2011), Poniatowski et al. (2024a), Schlumprecht & Waeber (2003), Zinke (2000), Zuna-Kratky et al. (2017).

European Mole Cricket (*Gryllotalpa gryllotalpa*) – p. 114: Detzel et al. (2022), Fartmann (1997a), Fischer et al. (2020), Hahn (1958), Maas et al.

(2002), NLWKN (2011), Pfeifer (2007, 2012), Pfeifer et al. (2019), Poniatowski et al. (2018a), Schlumprecht & Waeber (2003), Winkler & Haacks (2019), Wöss et al. (2020), Wranik et al. (1996).

Common Ant-cricket (*Myrmecophilus acervorum*) – p. 116: Bönsel & Möller (2008), Detzel et al. (2022), Fischer et al. (2020), Hölldobler (1947), Hörren et al. (2019), Iorgu et al. (2023), Junker (1997), Kettermann et al. (2019), Klaus (2022), Schlick-Steiner et al. (2003), Schlumprecht & Waeber (2003), Schmidt & Schmitt (2011), Tillmanns (2007).

Italian Cricket (*Oecanthus pellucens*) – p. 118: Ingrisch & Köhler (1998), Kästner & Reinhardt (2018), Köhler et al. (2023b), Meßner & Kladny (2017), Pfeifer (2012), Pfeifer et al. (2011), Poniatowski et al. (2018a, 2020), Sander (1992, 1995), Stübing et al. (2019a).

Wood Cricket (*Nemobius sylvestris*) – p. 120: Detzel (1998), Detzel et al. (2022), Fartmann (1997b), Fischer et al. (2020), Fumy et al. (2020), Köhler & Harzdorf (2011), Köhler & Samietz (2006), Maas et al. (2002), Pfeifer et al. (2011), Poniatowski et al. (2020, 2024a), Schlumprecht & Waeber (2003).

Marsh Cricket (*Pteronemobius heydenii*) – p. 122: Detzel (1998), Detzel et al. (2022), Maas et al. (2002), Poniatowski et al. (2020), Schlumprecht & Waeber (2003), Voith et al. (2016).

Field Cricket (*Gryllus campestris*) – p. 124: Detzel (1998), Fartmann (1997a), Fumy et al. (2020), Grein (2010), Maas et al. (2002), Ogan et al. (2022), Pfeifer (2012), Poniatowski et al. (2020, 2024a), Remmert (1979, 1992), Schlumprecht & Waeber (2003), Schuhmacher & Kelm (2021), Wallaschek et al. (2004).

House Cricket (*Acheta domesticus*) – p. 126: Federhen (1955), Kevan (1995), Pfeifer et al. (2011), Poniatowski et al. (2024a), Roe et al. (1980), Schlumprecht & Waeber (2003), Weidner (1972).

Bordeaux Cricket (*Eumodicogryllus bordigalensis*) – p. 128: Boczki (2007), Brauner & Ristow (2022), Pfeifer (2012), Pfeifer et al. (2011), Poniatowski et al. (2018a), Stübing et al. (2019a), van Elst & Schulte (1995).

Eastern Stripe-headed Cricket (*Modicogryllus frontalis*) – p. 130: Buchweitz & Trautner (1997), Detzel et al. (2022), Poniatowski et al. (2024a), Trautner & Buchweitz (2002), Trautner & Rietze (2020), Zechner et al. (2000).

Slender Groundhopper (*Tetrix subulata*) – p. 132: Detzel (1998), Distel et al. (2010), Fischer et al. (2020), Grein (2010), Haupt (1997), Hochkirch et al. (2000b), Krütgen (2012), Maas et al. (2002), Poniatowski et al. (2018a, 2020, 2024a), Schlumprecht & Waeber (2003), Schielzeth (2000), Schwarz & Fartmann (2022).

Cepero's Groundhopper (*Tetrix ceperoi*) – p. 134: Detzel et al. (2022), Garbelmann et al. (2021), Grein (2010), Gröning et al. (2007), Poniatowski et al. (2024a), Venne (2008), Wallaschek et al. (2004), Worschech (2007).

Alpine Groundhopper (*Tetrix tuerki*) – p. 136: Devriese (1996), Fischer et al. (2020), Fuhrmann (2015), Höfler et al. (2020), Janssen et al. (1996), Maas et al. (2002), Nadig (1986a), Pfeuffer (2020), Poniatowski et al. (2024a), Schlumprecht & Waeber (2003), Voith et al. (2016).

Common Groundhopper (*Tetrix undulata*) – p. 138: Detzel (1998), Fartmann (1997a), Fischer et al. (2020), Grein (2010), Köhler (2001), Laußmann (1993), Maas et al. (2002), Pfeifer et al. (2011), Poniatowski et al. (2020, 2024a), Reinhardt et al. (2005), Schielzeth (2000), Schlumprecht & Waeber (2003), Wallaschek et al. (2004), Wranik et al. (2008).

Two-spotted Groundhopper (*Tetrix bipunctata*) – p. 140: Detzel (1998), Detzel et al. (2022), Fartmann et al. (2022a), Fischer et al. (2020), Fumy et al. (2020), Helbing et al. (2014a), Köhler (2001), Korn & Stübing (2020b), Pfeifer et al. (2011), Poniatowski et al. (2020, 2024a), Reinhardt et al. (2005), Schlumprecht & Waeber (2003), Schulte (2003), Wallaschek et al. (2004).

Long-horned Groundhopper (*Tetrix tenuicornis*) – p. 142: Detzel (1998), Detzel et al. (2022), Fartmann (1997a), Fischer et al. (2020), Köhler (2001, 2020), Maas et al. (2002), Pfeifer et al. (2011), Poniatowski & Fartmann (2008a), Poniatowski et al (2020, 2024a), Schlumprecht & Waeber (2003), Schulte (2003), Wallaschek et al. (2004), Wranik et al. (2008).

Italian Locust (*Calliptamus italicus*) – p. 144: Bönsel (2022), Brose (1997), Detzel (1998), Fischer et al. (2020), Franz & Döring (2019), Pfeifer (2012), Pfeifer et al. (2020), Roderus (2020), Rutschmann & Roesti (2024), Schlumprecht & Waeber (2003), Sobczyk & Köhler (2024), Sobczyk & Trampenau (2011), Wranik (2024), Zechner & Koschuh (2000), Zuna-Kratky et al. (2017).

Common Mountain Grasshopper (*Podisma pedestris*) – p. 146: Detzel (1998), Hemp (2000a, b), Hemp & Hemp (1996), Lakeberg (2000), Poniatowski et al. (2024a), Schlumprecht & Waeber (2003), Zinner et al. (2000), Zuna-Kratky et al. (2017).

Green Mountain Grasshopper (*Miramella alpina*) – p. 148: Andreä (2020), d'Agostino & Vacher (2021), Detzel (1998), Fumy et al. (2020), Hafner & Zimmermann (2019), Holuša (1999), Kimmich et al. (2021), Neumayer (2020), Poniatowski et al. (2018a, 2024a), Schlumprecht & Waeber (2003), Zuna-Kratky et al. (2017).

Leek Grasshopper (*Mecostethus parapleurus*) – p. 150: Detzel (1998), Detzel et al. (2022), Helbing et al. (2014b), Kimmich et al. (2021), Korn & Stübing (2020a), Maas et al. (2002), Pfeifer et al. (2011, 2019), Poniatowski et al. (2018a, 2020), Schlumprecht & Waeber (2003), Voith et al. (2016), Zimmermann & Hafner (2011).

Large Marsh Grasshopper (*Stethophyma grossum*) – p. 152: Distel et al. (2010), Fartmann et al. (2021b), Fischer et al. (2020), Fumy et al. (2020), Maas et al. (2002), Poniatowski et al. (2018a, 2020, 2024a, b), Schwarz & Fartmann (2022), Thorn et al. (2022), Trautner & Hermann (2008), Wallaschek (2020).

Rattle Grasshopper (*Psophus stridulus*) – p. 154: Bönsel (2004), Buchweitz (1993), Detzel (1998), Detzel et al. (2022), Köhler (2001, 2021), Poniatowski et al. (2020, 2024a), Rada et al. (2017), Schlumprecht & Waeber (2003), Thorn et al. (2022), Trampenau & Sobczyk (2011), Voith et al. (2016), Vossen & Piper (1996), Weibart & Fischer (2006).

Blue-winged Grasshopper (*Oedipoda caerulescens*) – p. 156: Anger et al. (2020), Detzel (1998), Detzel et al. (2022), Fartmann (1997a), Fartmann et al. (2012), Fischer et al. (2020), Heidinger et al. (2010), Ingrisch (1980), Maas et al. (2002), Maes et al. (2006), Ogan et al. (2022), Pfeifer et al. (2011), Poniatowski et al. (2020), Straube (2013), Thorn et al. (2022).

Red-winged Grasshopper (*Oedipoda germanica*) – p. 158: Detzel (1991, 1998), Detzel et al. (2022), Franz & Döring (2019), Hiller et al. (2020), Köhler & Wagner (2000), Maier (2003), Reinhardt & Köhler (2002), Sändig (2025), Wagner (2000), Zöller (1995).

Speckled Buzzing Grasshopper (*Bryodemella tuberculata*) – p. 160: Dey et al. (2021), Grein (2010), Harz (1957), Höfler et al. (2020), Maas et al. (2002), Pfeuffer (2020), Poniatowski et al. (2024a), Reich (1991a, 1991b, 2006), Schlumprecht & Waeber (2003), Stelter et al. (1997), Voith et al. (2016).

Blue-winged Sand Grasshopper (*Sphingonotus caerulans*) – p. 162: Detzel et al. (2022), Drees (2003), Fischer et al. (2020), Hafner & Zimmermann (2019), Kettermann & Fartmann (2018), Kronshage (2009), Maas et al. (2002, 2011), Poniatowski et al. (2018a, 2020), Straube (2013), Wranik & Lange (2007), Zahn (2020).

Migratory Locust (*Locusta migratoria*) – p. 164: Harz (1957), Kirschbaum (1861), Köhler (2001), Krausch (1967), Poniatowski et al. (2024a), Röber (1951), Rudy (1925), Weidner (1953), Zacher (1917).

Slender Green-winged Grasshopper (*Aiolopus thalassinus*) – p. 166: Buse & Griebeler (2011), Detzel (1998), Detzel et al. (2022), Fischer et al. (2020), Ott (2014), Pfeifer (2012), Pfeifer et al. (2011), Poniatowski et al. (2018a, 2020, 2024a), Seehausen (2016), Stapf et al. (2021).

River Blue-legged Grasshopper (*Epacromius tergestinus*) – p. 168: Baur et al. (2006), Carron et al. (2001), Fischer et al. (2020), Hochkirch et al. (2016), Maas et al. (2002), Nadig (1991), Pfeuffer (2015), Poniatowski et al. (2024a), Sardet et al. (2015), Schlumprecht & Waeber (2003), Zuna-Kratky et al. (2017).

Large Banded Grasshopper (*Arcyptera fusca*) – p. 170: Detzel (1991, 1998), Detzel et al. (2022), Döler & Detzel (2008), Faber (1953), Fischer et al. (2020), Knörzer (1942), Krištín (2000), Maas et al. (2002), Nadig (1986a, b, 1991), Voith et al. (2016), Zuna-Kratky (2017).

Eastern Banded Grasshopper (*Arcyptera microptera*) – p. 172: Bellmann et al. (2019), Fischer et al. (2020), Harz (1957), Kenyeres et al. (2008), Knipper (1932), Maas et al. (2002), Pfeifer et al. (2011), Poniatowski et al. (2024a), Schlumprecht & Waeber (2003), Zuna-Kratky et al. (2017).

Large Gold Grasshopper (*Chrysochraon dispar*) – p. 174: Detzel (1998), Distel et al. (2010), Fumy et al. (2020), Grein (2010), Hochkirch (1997), Löffler et al. (2019), Ogan et al. (2022), Poniatowski et al. (2018a, 2020), Schlumprecht & Waeber (2003).

Small Gold Grasshopper (*Euthystira brachyptera*) – p. 176: Detzel (1998), Detzel et al. (2022), Fischer et al. (2020), Helbing et al. (2014a), Köhler (2001), Landeck & Wiedemann (1998), Löffler

& Fartmann (2017), Maas et al. (2002), Poniatowski et al. (2020, 2024a), Reinhardt (1998), Schlumprecht & Waeber (2003), Schwarz & Fartmann (2022), Thorn et al. (2022).

Common Green Grasshopper (*Omocestus viridulus*) – p. 178: Detzel et al. (2022), Fartmann et al. (2021b), Fumy et al. (2020), Gardiner (2010), Grein (2010), Köhler (2021), Löffler et al. (2019), Organ et al. (2022), Poniatowski et al. (2024a), Pfeifer et al. (2011, 2019), Schlumprecht & Waeber (2003), Voith et al. (2016), Wallaschek (2020), Winkler & Haacks (2019).

Woodland Grasshopper (*Omocestus rufipes*) – p. 180: Brose & Peschel (1998), Detzel (1998), Detzel et al. (2022), Grein (2010), Hochkirch (1997), Köhler (2021), Maas et al. (2002), Ogan et al. (2022), Pfeiffer et al. (2019), Poniatowski et al. (2024a), Schlumprecht & Waeber (2003), Voith et al. (2016), Wallaschek (2020), Winkler & Haacks (2019), Wranik & Kleeberg (2011).

Orange-tipped Grasshopper (*Omocestus haemorrhoidalis*) – p. 182: Detzel (1998), Dolek (1994), Fartmann (1997a), Fartmann et al. (2012), Froehlich (1990), Ingrisch (1987), Köhler et al. (2022a), Poniatowski et al. (2020, 2024a), Sänger (1977), Schlumprecht & Waeber (2003), Wallaschek et al. (2004).

Stripe-winged Grasshopper (*Stenobothrus lineatus*) – p. 184: Behrens & Fartmann (2004a, b), Detzel (1998), Dolek (1994), Fartmann (1997a), Fartmann et al. (2012), Fumy et al. (2020), Ingrisch (1983a), Johannesen et al. (1999), Poniatowski & Fartmann (2008a), Poniatowski et al. (2024a), Samietz (1996, 1998), Schlumprecht & Waeber (2003), Wallaschek (1995).

Black-spotted Toothed Grasshopper (*Stenobothrus nigromaculatus*) – p. 186: Behrens & Fartmann (2004a, b), Detzel (1991, 1998), Detzel et al. (2022), Dolek (1994), Köhler (2021), Ingrisch (1981), Maas et al. (2002), Pfeifer et al. (2011), Poniatowski et al. (2024a), Sändig (2024), Schlumprecht & Waeber (2003, 2022), Weipert & Köhler (2022).

Lesser Mottled Grasshopper (*Stenobothrus stigmaticus*) – p. 188: Behrens & Fartmann (2004a, b), Detzel (1998), Fumy et al. (2020), Herrmann (1995), Poniatowski et al. (2024a), Schlumprecht & Waeber (2003), Wallaschek (1995).

Pygmy Toothed Grasshopper (*Stenobothrus crassipes*) – p. 190: Autorenteam (2022), Köhler (1985, 2001, 2009, 2020), Köhler & Berger (2017), Köhler et al. (2023a), Meineke & Menge (1993), Schädler (2009), Wallaschek (2013, 2014, 2020), Wallaschek et al. (2004), Weipert & Köhler (2022).

Club-legged Grasshopper (*Gomphocerus sibiricus*) – p. 192: Göhler (2005), Illich & Winding (1999), Schlumprecht & Waeber (2003), Voith et al. (2016), Zuna-Kratky et al. (2017).

Rufous Grasshopper (*Gomphocerippus rufus*) – p. 194: Detzel (1998), Fumy et al. (2020), Helbing et al. (2014a), Neumayer (2020), Ogan et al. (2022), Poniatowski et al. (2024a), Schlumprecht & Waeber (2003).

Mottled Grasshopper (*Myrmeleotettix maculatus*) – p. 196: Detzel (1991, 1998), Göhler (2005), Hemp & Zehm (1995), Rutschmann & Roesti (2024), Schirmel (2011), Schlumprecht & Waeber (2003), Wünsch et al. (2010), Zehm (1997, 2003), Zuna-Kratky et al. (2017).

Large Mountain Grasshopper (*Stauroderus scalaris*) – p. 198: Carlsson & Kindvall (2001), Detzel (1989, 1998), Detzel & Sachs (2001), Detzel et al. (2022), Fumy et al. (2020), Grein (2010), Maas et al. (2002), Poniatowski et al. (2024a), Schlumprecht & Waeber (2003), Voith et al. (2016).

Locomotive Grasshopper (*Chorthippus apricarius*) – p. 200: Detzel et al. (2022), Fartmann et al. (2021b), Grein (2010), Köhler (2021), Nissen (2021), Ogan et al. (2022), Pfeifer et al. (2019), Poniatowski et al. (2018a, 2024a), Schulz (2003), Schlumprecht & Waeber (2003), Voith et al. (2016), Wallaschek (2020), Winkler & Haacks (2019), Worschech (2017).

Gravel Grasshopper (*Chorthippus pullus*) – p. 202: Fischer et al. (2020), Höfler et al. (2020), Lemke et al. (2010), Nadig (1991), Rutschmann & Roesti (2024), Schädler & Stadler (2000), Schlumprecht & Waeber (2003), Wiesener et al. (2004), Zechner & Klapf (2005), Zuna-Kratky et al. (2017).

Heath Grasshopper (*Chorthippus vagans*) – p. 204: Brandt & Schäfer (2004), Detzel (1998), Detzel et al. (2022), Fischer et al. (2020), Ingrisch (1983a), Köhler (2001), Maas et al. (2002), Pfeifer et al. (2011), Poniatowski et al. (2020, 2024a), Schlumprecht & Waeber (2003).

Lesser Grasshopper (*Chorthippus mollis*) – p. 206: Detzel (1998), Detzel et al. (2022), Fartmann (1997a), Fartmann et al. (2012), Fischer et al. (2020), Grein (2010), Maas et al. (2002), Poniatowski et al. (2020, 2024a), Schlumprecht &

Waeber (2003), Thorn et al. (2022), Wallaschek (1995), Wallaschek et al. (2004).

Common Field Grasshopper (*Chorthippus brunneus*) – p. 208: Bruckhaus (1992), Detzel (1998), Fartmann (1997a), Fartmann et al. (2012, 2021b), Fumy et al. (2020, 2021), Ingrisch (1983a, 1984), Ingrisch & Boekholt (1983), Laußmann (1993), Ogan et al. (2022), Poniatowski et al. (2020), Schlumprecht & Waeber (2003), Thorn et al. (2022).

Bow-winged Grasshopper (*Chorthippus biguttulus*) – p. 210: Detzel et al. (2022), Fartmann et al. (2021b), Fumy et al. (2020), Ingrisch (1983a), Ingrisch & Boekholt (1983), Köhler (2020), Ogan et al. (2022), Poniatowski & Fartmann (2008a), Poniatowski et al. (2020, 2024a), Schlumprecht & Waeber (2003), Thorens (1989), Thorn et al. (2022), Voith et al. (2016), Winkler & Haacks (2019).

Eisentraut's Bow-winged Grasshopper (*Chorthippus eisentrauti*) – p. 212: Baur et al. (2006), Cremer et al. (1998), Harz (1957), Hawlitschek et al. (2022), Illich et al. (2023), Ingrisch (1995), Ingrisch & Bassangova (1995), König et al. (2024), Koschuh (2012), Malkmus (2003), Schmidt et al. (2024), Zuna-Kratky et al. (2017).

Lesser Marsh Grasshopper (*Chorthippus albomarginatus*) – p. 214: Detzel (1998), Distel et al. (2010), Fischer et al. (2020), Glück & Ingrisch (1989), Hochkirch et al. (2000a), Ingrisch (1981a, 1983), Ogan et al. (2022), Pfeifer (2012), Poniatowski et al. (2018a, 2020, 2024a), Rietze (1994), Schlumprecht & Waeber (2003), Schwarz & Fartmann (2022).

Steppe Grasshopper (*Chorthippus dorsatus*) – p. 216: Detzel (1998), Fartmann (1997a), Fartmann et al. (2012, 2021b, 2022b), Fischer et al. (2020), Ingrisch (1980, 1983a), Löffler & Fartmann (2017), Ogan et al. (2022), Pfeifer (2012), Poniatowski et al. (2018a), Schlumprecht & Waeber (2003), Schwarz & Fartmann (2022), Thorn et al. (2022).

Meadow Grasshopper (*Pseudochorthippus parallelus*) – p. 218: Behrens & Fartmann (2004c), Detzel (1998), Köhler (2001, 2002, 2008, 2013), Köhler & Renker (2004, 2006), Köhler & Wiesner (2010), Köhler et al. (2016), Manzke (1995), Meineke (1994), Pfeifer et al. (2011), Schlumprecht & Waeber (2003), Wallaschek et al. (2004), Wiesner et al. (2011).

Marsh Grasshopper (*Pseudochorthippus montanus*) – p. 220: Distel et al. (2010), Fartmann et al. (2021b), Fischer et al. (2020), Fumy et al. (2020), Ingrisch (1983a, b), Maas et al. (2002), Poniatowski et al. (2018a, 2020, 2024a, b), Schlumprecht & Waeber (2003), Schwarz & Fartmann (2022), Thorn et al. (2022), Wallaschek (2020).

References

Andreä, J. (2020): Nicht nur in den Alpen zu Hause: Nachweis der Alpinen Gebirgsschrecke, *Miramella alpina* (Kollar, 1833) (Acrididae) im Nationalpark Bayerischer Wald. Articulata 35: 129–131.

Anger, F., Buse, J., Förschler, M. I. (2020): Neue Vorkommen der Blauflügeligen Ödlandschrecke *Oedipoda caerulescens* (L., 1758) im Grindenschwarzwald. Mitt. entomol. Ver. Stuttgart 55: 25–30.

Appelt, M., Poethke, H.-J. (1997): Metapopulation dynamics in a regional population of the blue-winged grasshopper (*Oedipoda caerulescens*; Linnaeus 1758). J. Insect Conserv. 1: 205–214.

Aßmann, O., Ritt, R. (2021): Beobachtungen der Vierpunktigen Sichelschrecke (*Phaneroptera nana*, Fieber 1853) im Landkreis Passau. Bayer. Wald 34: 64–66.

Autorenteam (2022): Zwerggrashüpfer (*Stenobothrus crassipes*). In: Thüringer Ministerium für Umwelt, Energie und Naturschutz (Hrsg.): Geschützte und gefährdete Arten in Thüringen. Vorkommen, Erhaltungszustand und Maßnahmen. Erfurt: 186–187.

Badenhausser, I., Amouroux, P., Lerin, J., Bretagnolle, V. (2009): Acridid (Orthoptera: Acrididae) abundance in western European grasslands: sampling methodology and temporal fluctuations. J. Appl. Entomol. 133: 720–732.

Bamann, T. (2018): Die Heuschrecken (Orthoptera) der Niedermoore im württembergischen Allgäu. Articulata 33: 21–46.

Baur, B., Baur, H., Roesti, C., Roesti, D. (2006). Die Heuschrecken der Schweiz. Haupt, Bern.

Bazelet, C., Samways, M. J. (2011): Identifying grasshopper bioindicators for habitat quality assessment of ecological networks. Ecol. Indic. 11: 1259–1269.

Beckerbauer, S., Poorthuis, L., Franzen, A., Kriegs, J. O. (2024): Die Ausbreitung der Europäischen Gottesanbeterin *Mantis religiosa* (Linnaeus, 1758) in NRW, Deutschland und Europa: Citizen Science-Projekte machen die Mechanismen der Ausbreitung einer wärmeliebenden Art sichtbar. Abh. Westf. Mus. Naturkde. 106: 185–196.

Behrens, M., Fartmann, T. (2004a): Die Heuschreckengemeinschaften isolierter Schieferkuppen der Medebacher Bucht (Südwestfalen/Nordhessen). Tuexenia 24: 303–327.

Behrens, M., Fartmann, T. (2004b): Habitatpräferenzen und Phänologie der Heidegrashüpfer *Stenobothrus lineatus, Stenobothrus nigromaculatus* und *Stenobothrus stigmaticus* in der Medebacher Bucht (Südwestfalen/Nordhessen). Articulata 19: 141–165.

Behrens, M., Fartmann, T. (2004c): Sind hohe Populationsdichten die Ursache der Makropterie beim Gemeinen Grashüpfer (*Chorthippus parallelus*, Caelifera: Acrididae)? Articulata 19: 91–102.

Beinlich, B. (1993): Zum Vorkommen der Wanstschrecke *Polysacrus denticauda* in Südwest-Thüringen. Articulata 2: 125–128.

Bellebaum, J. (2003): Kartierung und Schutz des Östlichen Heupferdes *Tettigonia caudata* am Beispiel der Uckermark. Articulata 18: 209–214.

Bellmann, H. (1985): Heuschrecken: beobachten – bestimmen. Neumann-Neudamm. Melsungen.

Bellmann, H. Rutschmann, F., Roesti, C., Hochkirch, A. (2019): Der Kosmos Heuschreckenführer – Die Heuschrecken Mitteleuropas und die wichtigsten Arten Südosteuropas. Kosmos-Verlag, Stuttgart.

Benzler, A., Fuchs, D., Hünig, C. (2015): Methodik und erste Ergebnisse des Monitorings der Landwirtschaftsflächen mit hohem Naturwert in Deutschland. Beleg für aktuelle Biodiversitätsverluste in der Agrarlandschaft. Nat. Landsch. 90: 309–316.

Berg, M. K., Schwarz, C. J., Mehl, J. E. (2011): Die Gottesanbeterin *Mantis religiosa*. Westarp Wissenschaften, Hohenwarsleben.

Betzin, A., Neugebauer, H. (2020): Erste Nachweise der Braunfleckigen Beißschrecke *Tessellana tessellata* (Charpentier, 1825) in Nordbaden. Carolinea 78: 29–35.

BfN (Bundesamt für Naturschutz) (Hrsg.) (2023): Einheitlicher Methodenleitfaden „Insektenmonitoring“ mit weiterentwickelter Methodik für die Erfassung von Insekten und Umweltvariablen (Kapitel 7, Bearbeitungsstand: Januar 2023). https://www.bfn.de. (Zugriff am 30.11.2023).

Birkwald, T., Gombert, J., Hoßfeld, C., Renninger, L., Scheffler, D. (2022): Aktuelle Verbreitung der Wanstschrecke *Polysarcus denticauda* im Bereich des Grünen Bandes zwischen Rhön und Grabfeld. Articulata 37: 59–66.

Boczki, R. (2007): Arealerweiterungen zweier Langfühlerschrecken (Orthoptera: Ensifera): neu in Hessen: Südliche Grille, *Eumodicogryllus bordigalensis* (Latreille, 1804) (Gryllinae), neu in Hessen und Nord-Baden-Württemberg: die Vierpunkt Sichelschrecke, *Phaneroptera nana* (Fieber, 1853) (Phaneropterinae). Articulata 22: 235–248.

Bolz, R. (1998): Anmerkungen zum Fraß- und Eiablageverhalten der Gestreiften Zartschrecke *Leptophyes albovittata* (Kollar, 1833). Articulata 13: 101–103.

Bonari, G., Fajmon, K., Malenovský, I., Zelený, D., Holuša, J., Jongepierová, I., Kočárek, P., Konvička, O., Uřičář, J., Chytrý, M. (2017): Management of semi-natural grasslands benefiting both plant and insect diversity: The importance of heterogeneity and tradition. Agric. Ecosyst. Environ. 246: 243–252.

Bönsel, A. (2004): Ethological and morphological adaptations of *Psophus stridulus* Linnaeus 1758 to habitat islands (Caelifera: Acrididae). Beitr. Entomol. 54: 241–253.

Bönsel, A. (2022): Ungewöhnliche Fundorte von *Calliptamus italicus* in Brandenburg. Articulata 37: 139–144.

Bönsel, A., Möller, S. (2008): Die Ameisengrille *Myrmecophilus acervorum* (Panzer, 1799) in MecklenburgVorpommern. Articulata 23: 81–87.

Brandt, T., Schaefer, F. (2004): Der Steppengrashüpfer (*Chorthippus vagans*, Eversmann 1848) an seiner nördlichen Verbreitungsgrenze am Steinhuder Meer, Niedersachsen. Articulata 19: 61–74.

Brauner, O., Ristow, M. (2022): Erste Nachweise der Südlichen Grille *Eumodicogryllus bordigalensis* (Latreille, 1804) in Brandenburg. Articulata 37: 31–46.

Brauner, O., Nordalm, J., Schulz U. (2021): Vorkommen, Verbreitung und Gehölzhabitatnutzung der Südlichen Eichenschrecke (*Meconema meridionale* Costa, 1860) und der Gemeinen Eichenschrecke (*Meconema thalassinum* (De Geer, 1773)) in Eberswalde. Veröff. Naturkundemus. Potsdam 7: 17–33.

Breuer, M., Simons, I. H., Gerlach, T., Reinhardt, K. (2017): Die Größe und Dichte einer Population der Plumpschrecke (*Isophya kraussii*) in der bayerischen Rhön. Articulata 32: 125–133.

Brocksieper, R. (1978): Der Einfluß des Mikroklimas auf die Verbreitung der Laubheuschrecken, Grillen und Feldheuschrecken im Siebengebirge und auf dem Rodderberg bei Bonn (Orthoptera: Saltatoria). Decheniana, Beih. 21: 1–141.

Brose, U. (1997): Untersuchungen zur Ökologie von *Calliptamus italicus* (Linnaeus, 1758) unter Berücksichtigung von Habitatpräferenzen, Populationsaufbau und Ausbreitungsverhalten. Articulata 12: 19–33.

Brose, U., Peschel, R. (1998): Neue Nachweise von *Conocephalus discolor* Thunberg, 1815, *Chrysochraon dispar* (Germar, 1831), *Oedipoda caerulescens* (Linnaeus, 1758) und *Platycleis albopunctata* (Goeze, 1778) an der nördlichen Verbreitungsgrenze. Articulata 13: 191–195.

Bruckhaus, A., (1992): Ergebnisse zur Embryonalentwicklung bei Feldheuschrecken und ihre Bedeutung für den Biotop- und Artenschutz. Articulata, Beih. 2: 1–112.

Brunk, I., Dieke, M., Krawzcynski, R., Landeck, I. (2003): Ein weiterer Nachweis der Bedornten Höhlenschrecke (*Troglophilus neglectus* Krauss 1879) aus dem Elbsandsteingebirge (Orthoptera, Rhaphidophoridae, Troglophilinae). Cottbus.

Buchweitz, M. (1993): Zur Ökologie der Rotflügeligen Schnarrschrecke (*Psophus stridulus* L. 1758) unter besonderer Berücksichtigung der Mobilität, Populationsstruktur und Habitatwahl. Articulata 8: 39–62.

Buchweitz, M., Trautner, J. (1997): In vino veritas? Zum Vorkommen der Östlichen Grille *Modicogryllus frontalis* (Fieber, 1844) im Jagsttal (Baden-Württemberg). Articulata 12: 201–209.

Bugla, B., Poschlod, P. (2006): Biotopverbund für die Migration von Pflanzen – Förderung von Ausbreitungsprozessen statt „statischen“ Korridoren und Trittsteinen. Das Fallbeispiel „Pflanzenarten der Sandmagerrasen“ in Bamberg, Bayern. In: Reck, H., Hänel, K., Böttcher, M., Winter, A. (Hrsg.): Lebensraumkorridore für Mensch und Natur. Naturschutz Biol. Vielfalt 17: 101–117.

Buse J., Griebeler, E. M. (2011): Incorporating classified dispersal assumptions in predictive distribution models – A case study with grasshoppers and bush-crickets. Ecol. Model. 222: 2130–2141.

Butchart, S. H., Walpole, M., Collen, B., Van Strien, A., Scharlemann, J. P., Almond, R. E., Baillie, J. E., Bomhard, B., Brown, C., Bruno, J. (2010): Global biodiversity: indicators of recent declines. Science 328 (5982): 1164–1168.

Cardoso, P., Barton, P. S., Birkhofer, K., Chichorro, F., Deacon, C., Fartmann, T., Fukushima, C. S., Gaigher, R., Habel, J., Hallmann, C. A., Hill, M., Hochkirch, A., Kwak, M. L., Mammola, S., Noriega, J. A., Orfinger, A. B., Pedraza, F., Pryke,

J. S., Roque, F. O., Settele, J., Simaika, J. P., Stork, N. E., Suhling, F., Vorster, C., Samways, M. J. (2020): Scientists' warning to humanity on insect extinctions. Biol. Conserv. 242: 108426.

Carlsson, A., Kindvall, O. (2001): Spatial dynamics in a metapopulation network: recovery of a rare grasshopper *Stauroderus scalaris* from population refuges. Ecography 24: 452–460.

Carron, G., Sardet, E., Praz, C., Wermeille, E. (2001): *Epacromius tergestinus* (Charpentier, 1825) and other interesting Orthoptera in the floodplains of braided rivers of the Alps. Articulata 16: 27–40.

Cherrill, A. J., Brown, V. K. (1992): Ontogenetic changes in the microhabitat preferences of *Decticus verrucivorus* (Orthoptera: Tettigoniidae) at the edge of its range. Ecography 15: 37–44.

Christian, E., Moog, O. (2023): Die Höhlenheuschrecke *Troglophilus neglectus* Krauss, 1879 (Orthoptera: Rhaphidophoridae) ist in Österreich weiter im Vormarsch. Beitr. Entomofaunist. 24: 184–187.

Cigliano, M. M., Braun, H., Eades, D. C., Otte, D. (2023): Orthoptera Species File. URL: http://orthoptera.speciesfile.org (Zugriff am 18.12.2023).

Clausnitzer, C., Clausnitzer, H.-J. (2005): Die Auswirkung der Heidepflege auf das Vorkommen der vom Aussterben bedrohten Heideschrecke (*Gampsocleis glabra*, Herbst 1786) in Norddeutschland. Articulata 20: 23–25.

Cooke, R., Mancini, F., Boyd, R. J., Evans, K. L., Shaw, A., Webb, T. J., Isaac, N. J. B. (2023): Protected areas support more species than unprotected areas in Great Britain, but lose them equally rapidly. Biol. Conserv. 278: 109884.

Coray, A. (1993): Neuere Funde von *Platycleis tessellata* (Charp.) aus dem Elsass mit Hinweisen zu Lebensraum und Biologie (Ensifera, Tettigoniidae). Articulata 8: 69–81.

Coray, A. (2003): *Phaneroptera nana* (Fieber, 1853) (Ensifera: Phaneropteridae) überwindet den Rhein bei Basel. Articulata 18: 247–250.

Cremer, S., Mayer, F., Helversen, O. von (1998): Two sibling grasshopper biotypes: Differences in song structure but a single genepool. 91. Annu. Meet. Dtsch. Zool. Ges., Zoology 101, Suppl. l: 25.

d'Agostino, R., Vacher, J.-P. (2021): Évolution temporelle de la répartition des Orthoptères dans les massif vosgien: possibles effets des changements climatiques sur la biodiversité. Bull. Soc. d'Hist. Nat. d'Ethnogr. Colmar 77: 10–17.

DWD (Deutscher Wetterdienst) (2024): Climate Data Center. https://www.dwd.de. (Zugriff am 10.01.2024).

Detzel, P. (1989): Verbreitung und Biologie des Gebirgsgrashüpfers (*Stauroderus scalaris* Fischer-Waldheim 1846) in Baden-Württemberg. Mitt. Badisch. Landesver. Naturkde. Naturschutz, N. F. 14: 927–934.

Detzel, P. (1991): Ökofaunistische Analyse der Heuschreckenfauna Baden-Württembergs (Orthoptera). Diss., Univ. Tübingen.

Detzel, P. (1995): Herkunft und Verbreitung der Heuschrecken in Baden-Württemberg. Articulata 10: 107–118.

Detzel, P. (1998): Die Heuschrecken Baden-Württembergs. Ulmer, Stuttgart.

Detzel, P. (2001): Verzeichnis der Langfühlerschrecken (Ensifera) und Kurzfühlerschrecken (Caelifera) Deutschlands. In: Klausnitzer, B. (Hrsg.): Entomofauna Germanica 5. Entomol. Nachr. Ber., Beih. 6: 63–90.

Detzel, P., Sachs, N. (2001): GIS-gestützte Populationsgefährdungsanalyse: Artenschutzprogramm Gebirgsgrashüpfer (*Stauroderus scalaris* Fischer von Waldheim, 1846) auf der Schwäbischen Alb. Nat. Landsch. 76: 491–497.

Detzel, P., Neugebauer, H., Niehues M., Zimmermann, P. (2022): Rote Liste und kommentiertes Verzeichnis der Heuschrecken und Fangschrecken Baden-Württembergs. Stand 31.12.2019. Naturschutz-Prax. Artenschutz 15: 1–179.

Deura, K., Hartley, J. C. (1990): Egg laying behaviour of the bush cricket, *Leptophyes punctatissima*. Entomol. 109: 100–105.

Devriese, H. (1996): Bijdrage tot de systematiek, morfologie en biologie van de West-Palearktische Tetrigidae. Nieuwsbr. Saltabel 15: 2–38.

Dey, L., Simoes, M., Hawlitschek, O., Sergeev, M., Xu, S., Lkhagvasuren, D., Husemann, M. (2021): Analysis of geographic centrality and genetic diversity in the declining grasshopper species *Bryodemella tuberculata* (Orthoptera: Oedipodinae). Biodiv. Conserv. 30: 2773–2796.

Díaz, S., Settele, J., Brondízio, E., Ngo, H. T., Guèze, M., Agard, J., Arneth, A., Balvanera, P., Brauman, K. A., Butchart, S. H. M., Chan, K. M. A., Garibaldi, L. A., Ichii, K., Lui, J., Subramanian, S. M., Midgley, G. F., Miloslavich, P., Molnár, Z., Obura, D., Pfaff, A., Polasky, S., Purvis, A., Razzaque, J., Reyers, B., Roy Showdhury, R.,

Shin, Y. J., Visseren-Hamakers, I. J., Willis, K. J., Zayas, C. N. (2019): Summary for Policymakers of the Global Assessment Report on Biodiversity and Ecosystem Services of the Intergovernmental Science-Policy Platform on Biodiversity and Ecosystem Services. IPBES Secretariat, Bonn.

Dierschke, H., Briemle, G. (2002): Kulturgrasland – Wiesen, Weiden und verwandte Staudenfluren. Ulmer, Stuttgart.

Distel, J., Fartmann, T., Schulte, A., Poniatowski, D. (2010): Die Heuschreckenfauna der Medebacher Bucht (Südwestfalen/Nordhessen). Articulata 25: 199–220.

Döler, H. P., Detzel, P. (2008): Zur Heuschreckenfauna des Naturschutzgebiets Irndorfer Hardt. Analyse und Pflegeempfehlungen. Articulata 23: 1–14.

Dolek, M. (1994): Der Einfluss der Schafbeweidung von Kalkmagerrasen in der Südlichen Frankenalb auf die Insektenfauna (Tagfalter, Heuschrecken). Agrarökol. 10: 1–126.

Dolek, M., Geyer, A. (2002): Conserving biodiversity on calcareous grasslands in the Franconian Jura by grazing: a comprehensive approach. Biol. Conserv. 104: 351–360.

Dolek, M., Radlmair, S., Geyer, A. (1999): Der Einfluss der Nutzung (Weide, Mahd, Brache) voralpiner Moorgebiete auf die Insektenfauna (Tagfalter, Heuschrecken). Schriftenr. Bayer. Landesamt Umwelt 150: 129–136.

Drees, M. (2003): Die Sandschrecke *Sphingonotus caerulans* (L.) nun auch bei Hagen nachgewiesen. Decheniana 156: 309–310.

Drukker, D., van der Arend, I., van Deijk, J., Houkes, J., Maas, D., Schut, D. (2020): De grote spitskop *Ruspolia nitidula* in Nederland (Orthoptera). Ned. Faunist. Meded. 54: 7–16.

Dürer, S., Weid, S. (1990): Zum Vorkommen der Wanstschrecke (*Polysarcus denticauda*) im ehemaligen innerdeutschen Grenzbereich zwischen Bayern und Thüringen. Articulata 5: 39–42.

Dvořák, T., Hadrava J., Knapp, M. (2022): The ecological niche and conservation value of Central European grassland orthopterans: A quantitative approach. Biol. Conserv. 265: 109406.

Eckert, S., Möller, M., Buchholz, S. (2017): Grasshopper diversity of urban wastelands is primarily boosted by habitat factors. Insect Conserv. Div. 10: 248–257.

EEA (European Environment Agency) (2024): https://www.eea.europa.eu. (Zugriff am 10.01.2024).

Ehrmann, R. (2011): *Mantis religiosa religiosa* Linné, 1758 in Deutschland und angrenzenden Ländern (Insecta: Mantodea). Articulata 26: 135–146.

Ehrmann, R., Reinhardt, R. (2011): Fauna der Fangschrecken (Mantodea) Sachsens. Beitr. Insektenfauna Sachs. 9: 82–96.

Ellenberg, H., Leuschner, C. (2010): Vegetation Mitteleuropas mit den Alpen. 6. Aufl. Ulmer, Stuttgart.

Ellenberg, H., Weber, H. E., Düll, R., Wirth, V., Werner, W. (2001): Zeigerwerte von Pflanzen in Mitteleuropa. Scripta Geobot. 18: 1–262.

Engelhardt, E. K., Biber, M. F., Dolek, M., Fartmann, T., Hochkirch, A., Leidinger, J., Löffler, F., Pinkert, S., Poniatowski, D., Voith, J., Winterholler, M., Zeuss, D., Bowler, D. E., Hof, C. (2022): Consistent signals of a warming climate in occupancy changes of three insect taxa over 40 years in central Europe. Glob. Change Biol. 28: 3998–4012.

Faber, A. (1953): Laut- und Gebärdensprache bei Insekten. Orthoptera (Geradflügler) I. Ges. Freunde Mitarb. Staatl. Mus. Naturkde., Stuttgart.

Fartmann, T. (1997a): Biozönologische Untersuchungen zur Heuschreckenfauna auf Magerrasen im Naturpark Märkische Schweiz (Ostbrandenburg). In: Mattes, H. (Hrsg.): Ökologische Untersuchungen zur Heuschreckenfauna in Brandenburg und Westfalen. Arb. Inst. Landschaftsökol. 3: 1–62.

Fartmann, T. (1997b): Zur Verbreitung von *Tettigonia caudata* (Charpentier, 1825) und *Nemobius sylvestris* (Bosc, 1792) in Berlin und Brandenburg. Articulata 12: 59–74.

Fartmann, T. (2004): Hydrochorie und warme Jahre – sind das die Gründe für die Ausbreitung der Langflügeligen Schwertschrecke (*Conocephalus fuscus*) in Ostbrandenburg? Articulata 19: 75–90.

Fartmann, T. (2017): Überleben in fragmentierten Landschaften – Grundlagen für den Schutz der Biodiversität Mitteleuropas in Zeiten des globalen Wandels. Naturschutz Landschaftspl. 49 (9): 277–282.

Fartmann, T. (2020): Heuschrecken als Indikatoren für den ökologischen Zustand des Offenlandes in Deutschland. In: Züghart, W., Stenzel, S., Fritsche, B. (Hrsg.): Umfassendes bundesweites Biodiversitätsmonitoring. BfN-Skr. 585: 33–42.

Fartmann, T. (2024): Insect Conservation in Grasslands. In: Pryke, J., Samways, M. J., New, T., Cardoso, P., Gaigher, R. (Hrsg.) Routledge Handbook of Insect Conservation. London: Routledge: 263–274.

Fartmann, T., Mattes, H. (1997): Heuschreckenfauna und Grünland – Bewirtschaftungsmaßnahmen und Biotopmanagement. In: Mattes, H. (Hrsg.): Ökologische Untersuchungen zur Heuschreckenfauna in Brandenburg und Westfalen. Arb. Inst. Landschaftsökol. 3: 179–188.

Fartmann, T., Gunnemann, H., Salm, P., Schröder, E. (2001): Berichtspflichten in Natura 2000-Gebieten. Empfehlungen zur Erfassung der Arten des Anhangs II und Charakterisierung der Lebensraumtypen des Anhangs I der FFH-Richtlinie. Angew. Landschaftsökol. 42: 1–725.

Fartmann, T., Krämer, B., Stelzner, F., Poniatowski, D. (2012): Orthoptera as ecological indicators for succession in steppe grassland. Ecol. Indic. 20: 337–344.

Fartmann, T., Jedicke, E., Stuhldreher, G., Streitberger, M. (2021a): Insektensterben in Mitteleuropa. Ursachen und Gegenmaßnahmen. Ulmer, Stuttgart.

Fartmann, T., Poniatowski, D., Holtmann, L. (2021b): Habitat availability and climate warming drive changes in the distribution of grassland grasshoppers. Agric. Ecosyst. Environ. 320: 107565.

Fartmann, T., Stuhldreher, G., Streitberger, M., Helbing, F. (2021c): Die Bedeutung der Habitatqualität für den Schutz der Insektendiversität – Mikroklima, Phytodiversität, Habitatheterogenität und Totholz sind Schlüsselfaktoren für artenreiche Insektengemeinschaften. Naturschutz Landschaftspl. 53 (7): 2–7.

Fartmann, T., Brüggeshemke, J., Poniatowski, D., Löffler, F. (2022a): Summer drought affects abundance of grassland grasshoppers differentially along an elevation gradient. Ecol. Entomol. 47: 778–790.

Fartmann, T., Poniatowski, D., Holtmann, L. (2022b): Effects of land-use and climate change on grasshopper assemblages differ between protected and unprotected grasslands. Basic Appl Ecol. 63: 83–92.

Fartmann, T., Freienstein, F. M., Helbing, F., Scherer, G., Poniatowski, D. (2024): A box quadrat for standardised sampling of Orthoptera in open habitats: design, handling, applications and baseline data. Glob. Ecol. Conserv. 55: e03217.

Federhen, M. T. (1955): Zur Biologie und Physiologie der Hausgrille (*Gryllus domesticus* L.). Z. Angew. Entomol. 38: 224–244.

Feurdean, A., Ruprecht, E., Molnár, Z., Hutchinson, S. M., Hickler, T. (2018): Biodiversity-rich European grasslands: Ancient, forgotten ecosystems. Biol. Conserv. 228: 224–232.

Finck, P., Heinze, S., Raths, U., Riecken, U., Ssymank, A. (2017): Rote Liste der gefährdeten Biotoptypen Deutschlands. Dritte fortgeschriebene Fassung 2017. Naturschutz Biol. Vielfalt 156: 1–637.

Fischer, J., Steinlechner, D., Zehm, A., Poniatowski, D., Fartmann, T., Beckmann, A., Stettmer, C. (2020): Die Heuschrecken Deutschlands und Nordtirols. Bestimmen – Beobachten – Schützen. 2. Aufl. Quelle & Meyer, Wiebelsheim.

Fischer, S. F., Poschlod, P., Beinlich, B. (1995): Die Bedeutung der Wanderschäferei für den Artenaustausch zwischen isolierten Schaftriften. Beih. Veröff. Naturschutz Landschaftspfl. Baden-Württ. 83: 229–256.

Fischer, S. F., Poschlod, P., Beinlich, B. (1996): Experimental studies on the dispersal of plants and animals on sheep in calcareous grasslands. J. Appl. Ecol. 63: 1206–1221.

Franz, M., Döring, M. (2019): Über das Vorkommen von *Oedipoda germanica* (Latreille, 1804) und *Calliptamus italicus* (Linnaeus, 1758) im Landkreis Bad Kissingen. Articulata 34: 109–126.

French, H. M. (2018): The Periglacial Environment. 4. Aufl., Wiley Blackwell, Chichester (UK).

Froehlich, C. (1990): Verbreitung und Gefährdungssituation der Heuschrecken (Insecta: Saltatoria) im Regierungsbezirk Koblenz. Fauna Flora Rheinland-Pfalz 6 (1): 5–200.

Froehlich, C. (1994): Analyse der Habitatpräferenzen von Heuschreckenarten (Orthoptera: Saltatoria) in einem Mittelgebirgsraum unter Berücksichtigung regionaler Differenzierungen. Articulata, Beih. 4: 1–176.

Fuhrmann, K. (2015): Türks Dornschrecke *Tetrix tuerki* (Krauss, 1876) an der Ammer (Orthoptera: Tetrigidae). Nachrichtenbl. Bay. Ent. 64 (1/2): 50–52.

Fuhrmann, K. (2019): Die Heuschrecke *Conospehalus fuscus* (Insecta: Orthoptera) erreicht Westniedersachsen. Drosera 2014: 37–39.

Fumy, F., Fartmann, T. (2023): Low-intensity land use fosters species richness of threatened butterflies and grasshoppers in mires and grasslands. Glob. Ecol. Conserv. 41: e02357.

Fumy, F., Löffler, F., Samways, M. J., Fartmann, T. (2020): Response of Orthoptera assemblages to environmental change in a low-mountain range differs among grassland types. J. Environ. Manag. 256: 109919.

Fumy, F., Kämpfer, S., Fartmann, T. (2021): Land-use intensity determines grassland Orthoptera assemblage composition across a moisture gradient. Agric. Ecosyst. Environ. 315: 107424.

Fumy, F., Schwarz, C., Fartmann, T. (2023): Intensity of grassland management and landscape heterogeneity determine species richness of insects in fragmented hay meadows. Glob. Ecol. Conserv. 47: e02672

Fusek, A. (2005): Der Einfluss unterschiedlicher Strukturen auf die Heuschreckenfauna entlang eines Höhengradienten in den spanischen Pyrenäen. Dipl.-Arb., Univ. Münster.

Garbelmann, K., Hawlitschek, O., Dey, L.-S., Fuhrmann, K., Schuhmacher, O., Husemann, M. (2021): Die Heuschreckenfauna von Pevestorf (Lüchow-Dannenberg, Niedersachsen) und Umgebung basierend auf Sammlungsdaten von mehr als 10 Jahren. Articulata 36: 77–90.

Gardiner, T. (2010): Precipitation and habitat degradation influence the occurrence of the common green grasshopper *Omocestus viridulus* in south-eastern England. J. Orthoptera Res. 19: 315–326.

Gardiner, T., Hill, J. (2006): A comparison of three sampling techniques used to estimate the population density and assemblage diversity of Orthoptera. J. Orthoptera Res. 15: 45–51.

Gardiner, T., Dover, J. (2008): Is microclimate important for Orthoptera in open landscapes? J. Insect Conserv. 12: 705–709.

Gardiner, T., Hill, J., Chesmore, D. (2005). Review of the methods frequently used to estimate the abundance of Orthoptera in grassland ecosystems. J. Insect Conserv. 9: 151–173.

Gibson, D. J. (Hrsg.) (2009): Grasslands and Grassland Ecology. Oxford University Press, Oxford.

Glück, E., Ingrisch, S. (1989): Heuschrecken und andere Geradflügler des Federseebeckens. Veröffentl. Naturschutz Landschaftspfl. Bad.-Württ. 64/65: 289–321.

Göhler, U. (2005): Habitatpräferenzen der Keulenschrecken *Gomphocerus sibiricus* und *Myrmeleotettix maculatus* auf Allmendweiden der spanischen Pyrenäen (Provinz Huesca). Dipl.-Arb. Univ. Münster.

Gottfried, T., Kästner, A. (2009): Erstnachweise der Südlichen Eichenschrecke (*Meconema meridionale* [Costa, 1860]) in Sachsen und Sachsen-Anhalt (Saltatoria). Sächs. Entomol. Z. 4: 3–9.

Gottschalk, E. (1996): Population Vulnerability of the Grey Bush Cricket *Platycleis albopunctata* (Goeze, 1778) (Ensifera: Tettigoniidae). In: Settele, J., Margueles, C., Poschlod, P., K. Henle (Hrsg.): Species Survival in Fragmented Landscapes. Kluwer Academic Publishers, Dordrecht (NL): 324–328.

Gottschalk, E. (1997): Habitatbindung und Populationsökologie der Westlichen Beißschrecke (*Platycleis albopunctata*, Goeze 1778) (Orthoptera: Tettigoniidae): Eine Grundlage für den Schutz der Art. Diss., Univ. Göttingen.

Gottwald, J., Richter, C., Wörner, M. (2002): Habitatwahl, Nahrungswahl und Entwicklung von *B. serricauda* und *B. constrictus* Brunner von Wattenwyl, 1878 (Phaneropterinae). Articulata 17: 51–78.

Grein, G. (1991): Zum Vorkommen der Laubholz-Säbelschrecke *Barbitistes serricauda* (Fabricius, 1794) in Niedersachsen. Articulata 6: 35–46.

Grein, G. (2007): Zur Ausbreitung von *Phaneroptera falcata* (Poda, 1761) und *Conocephalus fuscus* (Fabricius, 1793) in Niedersachsen. Articulata 22: 91–98.

Grein, G. (2010): Fauna der Heuschrecken (Ensifera & Caelifera) in Niedersachsen. Naturschutz Landschaftspf. Niedersachs. 46: 1–183.

Grime, J. P., Hodgson, J. G., Hunt, R. (2007): Comparative Plant Ecology. 2. Aufl. Castlepoint Press, Dalbeattie (UK).

Gröning, J., Krause, S., Hochkirch, A. (2007): Habitat preferences of an endangered insect species, Cepero's ground-hopper (*Tetrix ceperoi*). Ecol. Res. 22: 767–773.

Grzędzicka, E., Vahed, K. (2020): Habitat requirements of the endangered heath bush-cricket *Gampsocleis glabra* (Orthoptera, Tettigoniidae) in an isolated population. J. Insect Conserv. 24: 935–945.

Guggemoos, T. (2023): Erstfund der Atlantischen Bergschrecke *Antaxius pedestris* (Fabricius, 1787) in den Bayerischen Kalkalpen. Articulata 38: 123–126.

Hafner, A., Zimmermann, P. (2019): Fang- und Heuschrecken der Naturschutzgebiete im Landkreis Rastatt und im Stadtkreis Baden-Baden. Naturschutz Landschaftspfl. Baden-Württ. 79: 205–264.

Hahn, E. (1958): Untersuchungen über die Lebensweise und Entwicklung der Maulwurfsgrille (*Gryllotalpa vulgaris* Latr.) im Lande Brandenburg. Beitr. Entomol. 8 (3/4): 334–365.

Hamann, M., Weber, G. (2012): Bat-Detektor ermöglicht Neufunde bei Heuschrecken – Neufunde der Westlichen Beißschrecke (*Platycleis albopunctata albopunctata*) im Ballungsraum Rhein-Ruhr. Nat. NRW 3: 25–29.

Handke, K., Horstkotte, J., Kleinekuhle, J. (2011): Neue Funde der Sichelschrecke *Phaneroptera falcata* (Poda, 1761) in Nordwestdeutschland. Articulata 26: 163–167.

Hansen, L. O., Olsen, K. M. (2010): Orthoptera, Dermaptera, Blattodea. In: Kålås, J. A., Viken, Å., Henriksen, S., Skjelseth, S. (Hrsg.): The 2010 Norwegian Red List for Species. Norwegian Biodiversity Information Centre: 235–240.

Harásek, M., Klinkovská, K., Chytrý, M. (2023): Vegetation change in acidic dry grasslands in Moravia (Czech Republic) over three decades: Slow decrease in habitat quality after grazing cessation. Appl. Veg. Sci. 26: e12726.

Harthun, M. (1999): Zur Bedeutung der Biberwiesen in der mitteleuropäischen Urlandschaft. In: Gerken, B., Görner, M. (Hrsg.): Europäische Landschaftsentwicklung mit großen Weidetieren. Geschichte, Modelle und Perspektiven. Nat. Kulturlandsch. 3: 146–155.

Harvey, J. A., Tougeron, K., Gols, R., Heinen, R., Abarca, M., Abram, P. K., Basset, Y., Berg, M., Boggs, C., Brodeur, J., Cardoso, P., de Boer, J. G., De Snoo, G. R., Deacon, C., Dell, J. E., Desneux, N., Dillon, M. E., Duffy, G. A., Dyer, L. A., Ellers, J., Espíndola, A., Fordyce, J., Forister, M. L., Fukushima, C., Gage, M. J. G., García-Robledo, C., Gely, C., Gobbi, M., Hallmann, C., Hance, T., Harte, J., Hochkirch, A., Hof, C., Hoffmann, A. A., Kingsolver, J. G., Lamarre, G. P. A., Laurance, W. F., Lavandero, B., Leather, S. R., Lehmann, P., Le Lann, C., López-Uribe, M. M., Ma, C.-S., Ma, G., Moiroux, J., Monticelli, L., Nice, C., Ode, P. J., Pincebourde, S., Ripple, W. J., Rowe, M., Samways, M. J., Sentis, A., Shah, A. A., Stork, N., Terblanche, J. S., Thakur, M. P., Thomas, M. B., Tylianakis, J. M., Van Baaren, J., Van de Pol, M., Van der Putten, W. H., Van Dyck, H., Verberk, W. C. E. P., Wagner, D. L., Weisser, W. W., Wetzel, W. C., Woods, H. A., Wyckhuys, K. A. G., Chown, S. L. (2023): Scientists' warning on climate change and insects. Ecol. Monogr. 93 (1): e1553

Harz, K. (1957): Die Geradflügler Mitteleuropas. Fischer, Jena.

Harz, K. (1960): Geradflügler oder Orthopteren (Blattodea, Mantodea, Saltatoria, Dermaptera). In: Dahl, F. (Hrsg.): Die Tierwelt Deutschlands und der angrenzenden Meeresteile nach ihren Merkmalen und nach ihrer Lebensweise. Fischer, Jena.

Harz, K. (1987): Vordringen von *Barbitistes constrictus* Br. nach Westen. Articulata 3 (1): 6.

Haupt, H. (1995a): Zum Eiablageverhalten der Kurzflügligen Schwertschrecke (*Conocephalus dorsalis* Latreille, 1804). Articulata 10: 97–100.

Haupt, H. (1995b): Faunistische Beobachtungen an Heuschrecken (Orthoptera: Saltatoria) im Unteren Odertal bei Schwedt (Brandenburg) mit einem Wiederfund von *Platycleis montana* (Kollar, 1833). Articulata 10: 161–175.

Haupt, H. (1997): Analyse von Habitatfaktoren der Heuschreckenfauna des Offenlandes im deutsch-polnischen Überflutungsraum der Unteren Oder und seiner Randgebiete (Orthoptera: Saltatoria). Dipl.-Arb., Univ. Bonn.

Hawlitschek, O. (2018): Ein neuer Nachweis der Wanstschrecke (*Polysarcus denticauda*) in Bayern. Articulata 33: 65–68.

Hawlitschek, O., Edgardo, M., Ortiz, E. M., Noori, S., Webster, K. C., Husemann, M., Pereira, R. J. (2022): Transcriptomic data reveals nuclear-mitochondrial discordance in Gomphocerinae grasshoppers (Insecta: Orthoptera: Acrididae). Mol. Phylogenet. Evol. 170: 107439.

Hawlitschek, O., Bruns, C., Dey, L.-S., Nuhlícková, S., Felix, R., Van Kleef, H., Nakel, J., Husemann, M. (2023): The genomics of isolated populations of *Gampsocleis glabra* (Orthoptera: Tettigoniidae) in Central and Western Europe. Insects 14 (12): 946.

Heidinger, I. M., Hein, S., Bonte, D. (2010): Patch connectivity and sand dynamics affect dispersal-related morphology of the blue-winged grasshopper *Oedipoda caerulescens* in coastal grey dunes. Insect Conserv. Div. 3: 205–212.

Heitz, S., Hermann, G. (1993): Wiederfund der Braunfleckigen Beißschrecke (*Platycleis tessellata*) (Charpentier 1829) in der Bundesrepublik Deutschland. Articulata 8: 83–87.

Helbing, F., Blaeser, T. P., Löffler, F., Fartmann, T. (2014a): Response of Orthoptera communities to succession in alluvial pine woodlands. J. Insect Conserv. 18: 215–224.

Helbing, F., Löffler, F., Thoms, A., Fartmann, T. (2014b): Erstfund der Lauchschrecke (*Mecostethus parapleurus*, Hagenbach, 1822) in den mittleren Bayerischen Kalkalpen. Articulata 29: 75–78.

Helbing, F., Fartmann, T., Morkel, C., Poniatowski, D. (2023): Rapid response of vascular plants and insects to restoration of montane grasslands. Front. Ecol. Evol. 11: 1148266.

Heller, K.-G. (1988): Bioakustik der europäischen Laubheuschrecken. Verlag Josef Margraf, Weikersheim.

Heller, K.-G., Korsunovskaya, O., Ragge, D. R., Vedenina, V., Willemse, F., Zhantiev, R. D., Frantsevich, L. (1998): Check-list of European Orthoptera. Articulata, Beih. 7: 1–61.

Heller, K.-G., Orci, K. M., Grein, G., Ingrisch, S. (2004): The *Isophya* species of Central and Western Europe (Orthoptera: Tettigonioidea: Phaneropteridae). Tijdschr. Entomol. 147: 237–258.

Helversen, O. von (1969): *Meconema meridionale* (Costa 1860) in der südlichen Oberrhein-Ebene (Orth. Ensifera). Mitt. Dtsch. Entomol. Ges. 28: 19–22.

Hemp, A., Hemp, C. (2000): Die Heuschrecken-Zönosen auf Kalkschutthalden der Nördlichen Frankenalb und ihre Beziehung zur Vegetation. Tuexenia 20: 259–281.

Hemp, C. (2000a): Wiederansiedlung von Rote Liste-Arten: *Podisma pedestris* (L., 1758) (Caelifera: Acrididae, Melanoplinae). Articulata 15: 233–241.

Hemp, C. (2000b): Ein neues Vorkommen von *Podisma pedestris* (L., 1758) in der Nördlichen Frankenalb (Caelifera: Acrididae, Melanoplinae). Articulata 15: 243–244.

Hemp, C., Hemp, A. (1996): *Podisma pedestris* Linne, 1761 (Saltatoria: Acrididae, Catantopinae) in der Hersbrucker Alb. Ber. ANL 20: 277–286.

Hemp, C., Zehm, A. (1995): Eine zweite Larvenschlupfwelle bei Heuschrecken – ein Beitrag zur Populationsbiologie von *Psophus stridulus* und *Myrmeleotettix maculatus* (Orthoptera: Acrididae). Articulata 12: 123–129.

Hermann, S., Detzel, P., Bamann, T., Anthes, N. (2022): Association between mowing regimes and abundance of the Bull Bush Cricket (*Polysarcus denticauda*). Articulata 37: 83–102.

Herrmann, M. (1995): Die Heuschrecken-Gemeinschaften verinselter Trockenstandorte in Nordwestniedersachsen. Articulata 10: 119–139.

Hewitt, G. M. (1990): Divergence and speciation as viewed from an insect hybrid zone. Can. J. Zool. 68: 1701–1715.

Hewitt, G. M. (1999): Post-glacial re-colonization of European biota. Biol. J. Linn. Soc. 68: 87–112.

Hill, J. K., Thomas, C. D., Huntley, B. (1999): Climate and habitat availability determine 20th century changes in a butterfly's range margin. Proc. Royal Soc. B 266: 197–206.

Hiller, G., Elias, D., Köhler, M., Tischew, S. (2020): Ziegenrotationsbeweidung auf verbuschten Kalktrockenrasen. Förderung der Rotflügeligen Ödlandschrecke (*Oedipoda germanica*). Naturschutz Landschaftspl. 52 (11): 518–523.

Hochkirch, A. (1997): Neue Nachweise von *Chrysochraon dispar* (Germar, 1831) in Nordwestdeutschland – Ausbreitung oder Erfassungslücken? Articulata 12: 221–230.

Hochkirch, A. (2001): Rezente Areal- und Bestandsveränderungen bei Heuschrecken Nordwestdeutschlands (Orthoptera, Saltatoria). Verh. Westdt. Entomologentag. 2000: 167–178.

Hochkirch, A., Damerau, M. (2009): Rapid range expansion of a wing-dimorphic bush-cricket after the 2003 climatic anomaly. Biol. J. Linn. Soc. 97: 118–127.

Hochkirch, A., Blank, C., Dieling, H., Dormann, W., Hämker, S., Hoffmann, J., Rahmel, U. (2000a): Wiesen, Weiden und Witterung: Einflussgrößen für *Chorthippus albomarginatus* (Degeer, 1773) in einem nordwestdeutschen Binnendelta (Lkr. Osterholz, Niedersachsen). Articulata 15: 35–48.

Hochkirch, A., Gröning, J., Loos, T., Metzing, C., Reichelt, M. (2000b): Specialized diet and feeding habits as key factors for the habitat requirements of the grasshopper species *Tetrix subulata* (Orthoptera: Tetrigidae). Entomol. Gen. 25: 39–52.

Hochkirch, A., Nieto, A., García Criado, M., Cálix, M., Braud, Y., Buzzetti, F. M., Chobanov, D., Odé, B., Presa Asensio, J. J., Willemse, L., Zuna-Kratky, T., Barranco Vega, P., Bushell, M., Clemente, M. E., Correas, J. R., Dusoulier, F., Ferreira, S., Fontana, P., García, M. D., Heller, K-G., Iorgu, I. S., Ivkovic;, S., Kati, V., Kleukers, R., Krištín, A., Lemonnier-Darcemont, M., Lemos, P., Massa,

B., Monnerat, C., Papapavlou, K. P., Prunier, F., Pushkar, T., Roesti, C., Rutschmann, F., Sirin, D., Skejo, J., Szövényi, G., Tzirkalli, E., Vedenina, V., Barat Domenech, J., Barros, F., Cordero Tapia, P. J., Defaut, B., Fartmann, T., Gomboc, S., Gutiérrez-Rodríguez, J., Holuša, J., Illich, I., Karjalainen, S., Kočárek, P., Korsunovskaya, O., Liana, A., López, H., Morin, D., Olmo-Vidal, J. M., Puskás, G., Savitsky, V., Stalling, T., Tumbrinck, J. (2016): European Red List of Grasshoppers, Crickets and Bush-crickets. Publications Office of the European Union, Luxembourg.

Höfler, J., Leitinger, G., Reich, M., Sonntag, H., Zerbe, S. (2020): Habitatansprüche von gefährdeten Heuschreckenarten alpiner Flussökosysteme. Naturschutz Landschaftspl. 52: 228–235.

Hölldobler, K. (1947); Studien über die Ameisengrille (*Myrmecophila acervorum* Panzer) im mittleren Maingebiet. Mitt. Schweiz. Entomol. Ges. 20 (7): 607–648.

Holuša, J. (1999): The results of faunistic research of grasshoppers (Orthoptera: Caelifera), crickets (Orthoptera: Ensifera) and cockroaches (Dictyoptera: Blattodea) in the Sumava Mts. and in several localities of the Sumavske podhuri foothills. Silva Gabreta 3: 123–140.

Holuša, J., Kočárek, P. (2006): Occurrence of bush-cricket *Platycleis montana* in the Czech Kepuhlic (Orthoptera: Tettigoniidae). Entomol. Probl. 32 (2): 99.

Hörren, T., Bodingbauer, S., Enß, J., Rautenberg, T. (2019): Die Ameisengrille *Myrmecophilus acervorum* (Panzer, 1799) im Ballungsraum Ruhrgebiet und ihre aktuelle Verbreitung in Nordrhein-Westfalen (Orthoptera: Gryllotalpoidea: Myrmecophilidae). Ser. Nat. 1: 1–8.

Hülber, K., Kuttner, M., Moser, D., Rabitsch, W., Schindler, S., Wessely, J., Gattringer, A., Essl, F., Dullinger, S. (2020): Habitat availability disproportionally amplifies climate change risks for lowland compared to alpine species. Glob. Ecol. Conserv. 23: e01113.

Illich, I., Winding, N. (1999): Dynamik von Heuschrecken-Populationen (Orthoptera: Saltatoria) in subalpinen und alpinen Rasen des Nationalparks Hohe Tauern (Österreichische Zentralalpen) von 1990 bis 1997. Wiss. Mitt. Nationalpark Hohe Tauern 5: 63–85.

Illich, I., Gottsberger, B., Zuna-Kratky, T. (2023): Weiter verbreitet als bisher bekannt – der Südalpen-Grashüpfer *Chorthippus eisentrauti* (Ramme, 1931) nördlich des Alpenhauptkammes. Articulata 38: 9–36.

Ingrisch, S. (1977). Beitrag zur Kenntnis der Larvenstadien mitteleuropäischer Laubheuschrecken. Z. angew. Zool. 64: 459–501.

Ingrisch, S. (1978a): Labor- und Freilanduntersuchungen zur Dauer der postembryonalen Entwicklung einiger mitteleuropäischer Laubheuschrecken und ihre Beeinflussung durch Temperatur und Feuchte. Zool. Anz. 200: 309–320.

Ingrisch, S. (1978b): Zum Verhalten mitteleuropäischer Laubheuschrecken (Orthoptera: Tettigoniidae) in Temperatur- und Feuchtegradienten sowie gegenüber visuellen Reizen. Dtsch. Ent. Z. 25: 349–360.

Ingrisch, S. (1979a): Untersuchungen zum Einfluß von Temperatur und Feuchtigkeit auf die Embryogenese einiger mitteleuropäischer Laubheuschecken (Orthoptera: Tettigoniidae). Zool. Beitr., N. F. 25: 343–364.

Ingrisch, S. (1979b): Experimentell-ökologische Freilanduntersuchungen zur Monotopbindung der Laubheuschrecken (Orthoptera, Tettigoniidae) im Vogelsberg. Beitr. Naturkde. Osthessen 15: 33–95.

Ingrisch, S. (1980): Zur Feuchte-Präferenz von Feldheuschrecken und ihren Larven. Verh. Ges. Ökol. 8: 403–410.

Ingrisch, S. (1981): Zur Verbreitung der Orthopteren in Hessen. Mitt. Internat. Entomol. Ver. Frankfurt 6 (2–3): 29–58.

Ingrisch, S. (1983a): Zum Einfluß der Feuchte auf die Schlupfrate und Entwicklungsdauer der Eier mitteleuropäischer Feldheuschrecken. Dtsch. Entomol. Zeitschr., N. F. 30: 1–15.

Ingrisch, S. (1983b): Zum Einfluß der Feuchte auf den Wasserhaushalt der Eier und die Größe des 1. Larvenstadiums bei mitteleuropäischen Feldheuschrecken (Orthoptera: Acrididae). Zool. Anz. 210: 357–368.

Ingrisch, S. (1984): Zur Verbreitung und Vergesellschaftung der Orthopteren in der Nordeifel. Decheniana 137: 79–104.

Ingrisch, S. (1986): The plurennial life cycles of the European Tettigoniidae (Insecta: Orthoptera). 1. The effect of temperature on embryonic development and hatching. Oecologia 70: 606–616.

Ingrisch, S. (1987): Die Geradflügler (Orthopteroidea, Dermaptera und Blattaria) des Mainzer Sandes. Mainzer Naturwiss. Arch. 25: 233–252.

Ingrisch, S. (1988): Wasseraufnahme und Trockenresistenz der Eier europäischer Laubheuschrecken (Orthoptera: Tettigoniidae). Zool. Jb. Physiol. 92: 117–170.

Ingrisch, S. (1995): Evolution of the *Chorthippus biguttulus* group (Orthoptera, Acrididae) in the Alps based on morphology and stridulation. Rev. Suisse Zool. 100: 475–535.

Ingrisch, S., Bassangova, N. (1995): Paarungswahlverhalten von *Chorthippus biguttulus* und *C. eisentrauti* (Orthoptera: Acrididae). Mitt. Schweiz. Entomol. Ges. 68: 1–6.

Ingrisch, S., Boekholt, I. (1983). Zur Wahl des Eiablageplatzes durch mitteleuropäische Saltatoria. Zool. Beitr. 28: 33–46.

Ingrisch, S., Köhler, G. (1998): Die Heuschrecken Mitteleuropas. Westarp Wissenschaften, Magdeburg.

Iorgu, I. Ş., Iorgu, E. I., Stalling, T., Puskás, G., Chobanov, D., Szövényi, G., Moscaliuc, L. A., Motoc, R., Tăuşan, I., Fusu, L. (2023): Ant crickets and their secrets: *Myrmecophilus acervorum* is not always parthenogenetic (Insecta: Orthoptera: Myrmecophilidae). Zool. J. Linn. Soc. 197: 211–228.

Iorio, C., Scherini, R., Fontana, P., Buzzetti, F. M., Kleukers, R. M. J. C., Odé, B., Massa, B. (2019): Grasshoppers & Crickets of Italy. A Photographic Field Guide to all the Species. WBA Handbooks 10. Verona.

IPCC/Intergovernmental Panel on Climate Change (2023): Synthesebericht zum Sechsten IPCC-Sachstandsbericht (AR6). https://www.de-ipcc.de. (Zugriff am 21.11.2023).

Janssen, B., Manderbach, R., Reich, M. (1996): Zur Verbreitung und Gefährdung von *Tetrix tuerki* (Krauss, 1876) in Deutschland. Articulata 11: 81–86.

Jerrentrup, J. S., Wrage-Mönnig, N., Röver, K.-U., Isselstein, J. (2014). Grazing intensity affects insect diversity via sward structure and heterogeneity in a long-term experiment. J. Appl. Ecol. 51: 968–977.

Joern, A. (2005): Disturbance by fire frequency and bison grazing modulate grasshopper assemblages in tallgrass prairie. Ecology 86: 861–873.

Joern, A., Laws, A. N. (2013): Ecological mechanisms underlying arthropod species diversity in grasslands. Annu. Rev. Entomol. 58: 19–36.

Johannesen, J., Samietz, J., Wallaschek, M., Seitz, A., Veith, M. (1999): Patch connectivity and genetic variation in two congeneric grasshopper species with different habitat preferences. J. Insect Conserv. 3: 201–209.

Junker, E. A. (1997): Untersuchungen zur Lebensweise und Entwicklung von *Myrmecophilus acervorum* (Panzer, 1799) (Saltatoria, Myrmecophilidae). Articulata 12: 93–106.

Kämpf, I., Fartmann, T. (2014): Die Heuschreckenfauna des Nationalparks Unteres Odertal und ihre Veränderungen in den letzten 20 Jahren. Nationalpark Jahrb. Unteres Odertal 11: 101–109.

Kästner, T. (2016): Ein neues Vorkommen der Gemeinen Plumpschrecke *Isophya kraussii* Brunner von Wattenwyl, 1878 im Landkreis Meißen (Insecta: Orthoptera: Ensifera). Sächs. Entomol. Z. 8: 218–222.

Kästner, T., Reinhardt, K. (2018): Neue Funde des Weinhähnchens *Oecanthus pellucens* (Scopoli, 1763) im Elbtal um Dresden (Insecta: Ensifera). Sächs. Entomol. Z. 10: 11–14.

Karaman, I., Hammouti, N., Pavićević, D., Kiefer, A., Horvatović, M., Seitz, A. (2011): The genus *Troglophilus* Krauss, 1879 (Orthoptera: Rhaphidophoridae) in the west Balkans. Zool. J. Linn. Soc. 163: 1035–1063.

Karjalainen, S. (2009): The Grasshoppers and Crickets of Finland, Orthoptera. Tammi Kustannusosakeyhtio.

Kenyeres, Z., Nagy, B., Bauer, N. (2008): Distribution and habitat requirements of *Arcyptera microptera* (Fischer von Waldheim, 1833) in Hungary. Articulata 23: 25–36.

Kenyeres, Z., Takács, G., Király, G. (2024): Challenges of Orthoptera conservation in grasslands with land use-determined sizes and structural heterogeneity. Landsc. Ecol. Eng. 20: 441–453.

Kettermann, M., Fartmann, T. (2018): Auswirkungen des globalen Wandels auf Heuschrecken. Besiedelung von Steinbrüchen der Westfälischen Bucht (NW-Deutschland) durch die Blauflügelige Sandschrecke. Naturschutz Landschaftspl. 50 (1): 23–29.

Kettermann, M., Scherer, G., Drung, M., Münsch, T., Poniatowski, D., Fartmann, T. (2019): Verbreitung der Ameisengrille *Myrmecophilus acervorum* (Panzer, 1799) (Saltatoria, Gryllidae) im Diemeltal (Ostwestfalen/Nordhessen). Nachr. Entomol. Ver. Apollo, N. F. 39 (3/4): 156–159.

Kevan, D. K. M. (1955): The home of the House Cricket *Acheta domesticus* (L.) (Orth., Gryllidae). Entomol. Mon. Mag. 91: 263.

Kiefer, A., Buzila, R., Kosuch, J., Seitz, A. (2000): A record of the cave cricket *Troglophilus neglectus* (Orthoptera: Rhaphidophoridae) in South-Western Germany, identified by mitochondrial sequences analysis. In: Onac, B. P., Tamas, T. (Hrsg.): Proc. Joint Meet. Friends Karst, Theor. Appl. Karstology. Presa Universitara, Cluj-Napoca: 168–169.

Kimmich, T., Anger, F., Brandt, D., Zimmermann, P., Buse, J. (2021): Kommentierte Artenliste der Fang- und Heuschrecken im Gebiet des Nationalpark Schwarzwald. Articulata 36: 43–60.

Kindvall, O. (1996): Habitat heterogeneity and survival in a bush cricket metapopulation. Ecology 77: 207–214.

Kindvall, O. (2018): Skärrande gräshoppa – Analys avartens utbredning 2016 med förslag på åtgärder. Meddelande 2018:02. Calluna AB, Länsstyrelsen Kalmar län.

Kirschbaum, C. L. (1861): Bei Wiesbaden gefangene Exemplare von *Pachytylus migratorius* L. Jahrb. Nassau. Ver. Naturkde. 16: 1–305.

Klaiber, J., Altermatt, F., Birrer, S., Chittaro, Y., Dziock, F., Gonseth, Y., Hoess, R., Keller, D., Küchler, H., Luka, H., Manzke, U., Müller, A., Pfeifer, M.A., Roesti, C., Schneider, K., Schlegel, J., Sonderegger, P., Walter, T., Holderegger, R., Bergamini, A. (2017): Fauna Indicativa. WSL Ber. 54: 1–192.

Klaus, D. (2010). Heuschrecken, Schaben und Ohrwürmer. In: Unruh, M. (Hrsg.): Der Zeitzer Forst – Natur und Nutzungsgeschichte einer Landschaft. Druck-zuck Verlag, Halle (Saale): 129–146.

Klaus, D. (2022): Die Ameisengrille, *Myrmecophilus acervorum* (Panzer, 1799) in Sachsen – neuere Nachweise und ergänzende Bemerkungen zu einigen Altfunden (Insecta, Orthoptera, Myrmecophilidae). Mauritiana 41: 1–25.

Kleiner, A. (2013): Mähwiesenmanagement im Zollernalbkreis im Hinblick auf den Artenschutz. Naturschutz Landschaftspfl. Bad.-Württ. 76: 153–178.

Kleinert, H. (1992): Entwicklung eines Biotopbewertungskonzeptes am Beispiel der Saltatoria (Orthoptera). Articulata, Beih. 1: 1–117.

Klinkovská, K., Kučerová, A., Pustková, Š., Rohel, J., Slachová, K., Sobotka, V., Szokala, D., Danihelka, J., Kočí, M., Šmerdová, E., Chytrý, M. (2023): Subalpine vegetation changes in the Eastern Sudetes (1973–2021): Effects of abandonment, conservation management and avalanches. Appl. Veg. Sci. 26: e12711.

Knipper, H. (1932): Beiträge zur deutschen Orthopterenfauna. I. *Arcyptera microptera* F. W. und *Gampsocleis glabra* Herbst bei Darmstadt aufgefunden. Entomol. Rundsch. 49: 233–235.

Knörzer, A. (1942): Grundlagen zur Erforschung der Orthopterenfauna und Dermapterenfauna Südostbayerns. Mitt. Münch. Ent. Ges. 32: 626–648.

Köhler, G. (1985): *Stenobothrus crassipes* (Charp., 1825) (Orthoptera, Acrididae) – Erstnachweis für das Gebiet der DDR. Ent. Nachr. Ber. 29: 217–219.

Köhler, G. (2001): Fauna der Heuschrecken (Ensifera et Caelifera) des Freistaates Thüringen. Naturschutzrep. 17: 1–378.

Köhler, G. (2002): Experimente und Erhebungen zur Flügeligkeit beim Gemeinen Grashüpfer *Chorthippus parallelus* (Zetterstedt) (Caelifera: Acrididae) – ein Beitrag zur Interpretation von Makropterie. Articulata 17: 1–19.

Köhler, G. (2008): Populationsstudien am Gemeinen Grashüpfer, *Chorthippus parallelus* (Zetterstedt), entlang eines Höhengradienten im Mittleren Erzgebirge. Mauritiana 20: 349–370.

Köhler, G. (2009): Zur Lebensgeschichte des Zwerggrashüpfers, *Stenobothrus crassipes* (Charpentier, 1825), im Vergleich zu *Stenobothrus lineatus* (Panzer, [1796]) und *Stenobothrus nigromaculatus* (Herrich-Schäffer, 1840). Articulata 24: 31–47.

Köhler, G. (2013): Der erzgebirgische Zechengrund und seine Heuschrecken (Insecta: Saltatoria). Veröff. Mus. Naturkde. Chemnitz 36: 23–54.

Köhler, G. (2017): Beobachtungen zur Embryonal- und Juvenilentwicklung der Wanstschrecke, *Polysarcus denticauda* (Charpentier, 1825). Articulata 32: 23–38.

Köhler, G. (2020): Checkliste der wild lebenden Heuschrecken (Insecta: Orthoptera) Thüringens. 5., aktualisierte und erweiterte Fassung: Stand: Mai 2020. Checklisten Thüringer Insekten und Spinnentiere 28: 13–28.

Köhler, G. (2021): Rote Liste der Heuschrecken (Insecta: Orthoptera) Thüringens. 5. Fassung, Stand: 05/2020. Naturschutzrep. 30: 118–124.

Köhler, G., Berger, D. (2017): Älteste Belege des Zwerggrashüpfers, *Stenobothrus crassipes* (Charpentier, 1825), vom Kyffhäuser/Thüringen im Museum für Tierkunde (Senckenberg Naturhistorische Sammlungen Dresden). Articulata 32: 87–96.

Köhler, G., Harzdorf, M. (2011): Zur Winteraktivität der Waldgrille, *Nemobius sylvestris* (Bosc, 1792). Articulata 26: 79–86.

Köhler, G., Renker, C. (2004): Zur Heuschreckenfauna (Ensifera, Caelifera) extensiv genutzter Gebirgsmähwiesen im Thüringer Schiefergebirge und im Frankenwald (BIOLOG-Flächen). Veröff. Naturkundemus. Erfurt 23: 115–121.

Köhler, G., Renker, C. (2006): Verteilung, Morphometrie und Fitness der Farbmorphen in Wildpopulationen von *Chorthippus parallelus* (Zetterstedt) (Caelifera: Acrididae). Articulata 21: 59–75.

Köhler, G., Samietz, J. (2006): Semivoltine Entwicklung der Waldgrille, *Nemobius sylvestris* (Bosc), an ihrem mitteldeutschen Arealrand. Articulata 21: 183–194.

Köhler, G., Wagner, G. (2000): Lebensräume der Rotflügligen Ödlandschrecke, *Oedipoda germanica* (Latr.), und ihre Vergesellschaftung mit anderen Heuschreckenarten in Thüringen. Mauritiana 17: 389–416.

Köhler, G., Wiesner, K. R. (2010): Kryptische Reliktpopulationen des Gemeinen Grashüpfers, *Chorthippus parallelus* (Zetterstedt), im Erzgebirge? – Genetische und waldgeschichtliche Argumente. Articulata 25: 151–166.

Köhler, G., Reinhardt, K., Opitz, S. (2001): Unterschiede in Morphometrie und Färbung zwischen zwei benachbarten Populationen der Blauflügligen Ödlandschrecke, *Oedipoda caerulescens* (Linnaeus, 1758) (Caelifera: Acrididae), im Gebiet der Drei Gleichen (Thüringen). Thüring. Faunist. Abh. 8: 83–90.

Köhler, G., Gossner, M. M., Weisser, W. W. (2016): Das Biodiversitätsexploratorium Hainich-Dün und seine Geradflügler (Orthoptera, Dermaptera, Blattoptera) – eine ökofaunistische Übersicht. Vernate 35: 227–250.

Köhler, G., Ebeling, A., Schumacher, J. (2019): Untersuchungen an einer Wildpopulation adulter *Isophya kraussii* Brunner von Wattenwyl, 1878 bei Jena/Thüringen (Orthoptera: Phaneropteridae). Articulata 34: 57–70.

Köhler, G., Krech, M., Korsch, H., Worschech, K. (2022a): Verbreitung und Biotopbindung des Rotleibigen Grashüpfers, *Omocestus haemorrhoidalis* (Charpentier, 1825), in Thüringen (Insecta: Orthoptera: Acrididae). Thüring. Faunist. Abh. 27: 99–127.

Köhler, G., Schweiger, M., Neubert, D. (2022b): Oberflächenstruktur der Tarsen von *Meconema meridionale* (A. Costa, 1860) und *M. thalassinum* (De Geer, 1773) (Ensifera, Meconematidae). Articulata 37: 123–134.

Köhler, G., Worschech, K., Krech, M., Korsch, H. (2022c): Die Südliche Eichenschrecke (*Meconema meridionale* A. Costa, 1860) in Thüringen (Orthoptera: Meconematidae). Mitt. Thüring. Entomologenverb. 29: 9–30.

Köhler, G., Korsch, H., Weipert, J. (2023a): Beiträge zur Fauna des Kyffhäusergebirges, Teil XVI. Der Kyffhäuser und seine Geradflügler (Insecta: Mantodea, Orthoptera, Dermaptera, Blattoptera). Vernate 42.

Köhler, G., Krech, M., Korsch, H., Worschech, K., Lehmeier, S., Fechtler, T. (2023b): Das Weinhähnchen, *Oecanthus pellucens* (Scopoli, 1793), in Thüringen (Orthoptera: Ensifera, Oecanthidae). Mitt. Thüring. Entomologenverb. 30: 2–35.

König, S., Krauss, J., Classen, A., Hof, C., Prietzel, M., Wagner, C., Steffan-Dewenter, C. (2024): Micro- and macroclimate interactively shape diversity, niches and traits of Orthoptera communities along elevational gradients. Div. Distrib. 30 (5): e13810.

Korn, M., Stübing, S. (2020a): Artensteckbrief Lauchschrecke *Mecostethus parapleurus* (Hagenbach, 1822). Hessisches Landesamt für Naturschutz, Umwelt und Geologie.

Korn, M., Stübing, S. (2020b): Artensteckbrief Zweipunkt-Dornschrecke, *Tetrix bipunctata* (Linnaeus, 1758). Hessisches Landesamt für Naturschutz, Umwelt und Geologie.

Koschuh, A. (2012): Der Südalpen-Grashüpfer *Chorthippus eisentrauti* (Ramme, 1931) (Orthoptera: Caelifera) in den östlichen Nordalpen als Erstnachweis für die Steiermark. Joannea Zool. 12: 117–126.

Krämer, B., Poniatowski, D., Villar, L., Fartmann, T. (2010): The Orthoptera communities of sub-mediterranean dry grasslands (Aphyllanthion alliance) in the western Spanish Pyrenees. Articulata 25: 59–72.

Krausch, H.-D. (1967): Heuschreckenplagen im Gubener Land. Gubener Heimatkalender 12: 84–92.

Krech, M., Köhler, G. (2021): Die Wanstschrecke, *Polysarcus denticauda* (Charpentier, 1825) (Insecta: Orthoptera: Phaneropteridae), und ihre Lebensräume an der Raststätte Eichelborn-Nord, 2016–2021. Vernate 40: 215–244.

Krištín, A., Kaňuch, P., Balla, M., Gavlas, V. (2007): On the distribution and ecology of *Gampsocleis glabra* and *Tettigonia caudata* (Orthoptera) in Slovakia. Articulata 22: 53–61.

Kronshage, A. (1993): Neufunde von *Phaneroptera falcata* (Poda) (Saltatoria: Tettigoniidae) am nördlichen Arealrand unter besonderer Berücksichtigung der westfälischen Vorkommen. Nat. Heim. 53 (3): 77–81.

Kronshage, A. (2009): Zum Vorkommen der Blauflügeligen Sandschrecke (*Sphingonotus caerulans*) in Paderborn. Mitt. Naturwiss. Ver. Paderborn, Dezember 2009: 21–27.

Kruess, A., Tscharntke, T. (2002): Grazing intensity and the diversity of grasshoppers, butterflies, and trap-nesting bees and wasps. Conserv. Biol. 16: 1570–1580.

Krütgen, J. (2012): Die Bedeutung wildlebender Huftiere für das Vorkommen von Kurzfühlerschrecken (Caelifera) am Beispiel der Gefleckten Keulenschrecke (*Myrmeleotettix maculatus*) und der Säbeldornschrecke (*Tetrix subulata*). Articulata 27: 67–77.

Lakeberg, H. (2000): Auswirkungen einer Artenschutzmanahme für die Gewöhnliche Gebirgsschrecke *Podisma pedestris* Linné 1758. Articulata 15: 109–112.

Landeck, I., Wiedemann, D. (1998): Die Geradflüglerfauna (Dermaptera, Orthoptera) der Niederlausitzer Bergbaufolgelandschaft – Ein Beitrag zur Ökologie der Arten. Articulata 13: 81–100.

Laußmann, H. (1993): Die Besiedlung neu entstandener Windwurfflächen durch Heuschrecken. Articulata 8: 53–59.

Laußmann, H. (1995): Zum Vorkommen von *Barbitistes constrictus* Br. und *Barbitistes serricauda* (Farb.) in den Fichtenwäldern des nördlichen Tertiär-Hügellandes (Südbayern). Articulata 10: 11–19.

Lehmann, A., Haacks, M. (2006): Vorkommen von *Conocephalus dorsalis* (Latreille, [1804]) in vom Salzwasser beeinflussten Habitaten der schleswig-holsteinischen Nord- und Ostseeküste. Articulata 21: 161–167.

Lehmann, A. W., Klatt, R., Landeck, I., Machatzi, B., Hennigs, S., Brauner, O., Oldorff, S., Lehmann, G. U. C. (2016): Fokusarten für die Erfassung zur Gefährädungsanalyse der Heuschrecken (Orthoptera) in Brandenburg und Berlin. Articulata 31: 23–44.

Lemke, H., Löffler, F., Fartmann, T. (2010): Habitat- und Nahrungspräferenzen des Kiesbank-Grashüpfers (*Chorthippus pullus*) in Südbayern. Articulata 25: 133–149.

Lensink, B. M. (1963): Distributional ecology of some Acrididae (Orthoptera) in the dunes of Voorne, Netherlands. Tijdschr. Entomol. 106: 357–443.

Liebelt, R. (2021): Untersuchungen von Brutvögeln, Tagfaltern, Heuschrecken und Wildbienen im Bereich der Neuen Hute in den Jahren 2017 bis 2020. Dokumentation „20 Jahre Beweidungsprojekte Naturpark Solling-Vogler“: 73–77.

Linn, C. A. (2015): Ausbreitung und Etablierung der Europäischen Gottesanbeterin *Mantis religiosa* in Deutschland unter dem Einfluss des Klima- und Landschaftswandels. Diss., Univ. Mainz.

Löffler, F., Fartmann, T. (2017): Effects of landscape and habitat quality on Orthoptera assemblages of pre-alpine calcareous grasslands. Agric. Ecosyst. Environ. 248: 71 – 81.

Löffler, F., Poniatowski, D., Fartmann, T. (2019): Orthoptera community shifts in response to land-use and climate change – Lessons from a long-term study across different grassland habitats. Biol. Conserv. 236: 315–323.

Löffler, F., Grüneberg, C., Drung, M., Freienstein, F. M., Helbing, F., Holtmann, L., Kämpfer, S., Kettermann, M., Münsch, T., Poniatowski, D., Streitberger, M., Stuhldreher, G., Fartmann, T. (2023): Different environmental conditions in lowlands and uplands highlight challenges for butterfly conservation in Central Europe. Biol. Conserv. 281: 110034.

Ludwig, G., Haupt, H., Gruttke, H., Binot-Hafke, M. (2009): Methodik der Gefährdungsanalyse für Rote Listen. In: Haupt, H., Ludwig, G., Gruttke, H., Binot-Hafke, M., Otto, C., Pauly, A. (Bearb.): Rote Liste gefährdeter Tiere, Pflanzen und Pilze Deutschlands, Band 1: Wirbeltiere. Naturschutz Biol. Vielf. 70: 19–71.

Maas, S., Detzel, P., Staudt, A. (2002): Gefährdungsanalyse der Heuschrecken Deutschlands. Verbreitungsatlas, Gefährdungseinstufung und Schutzkonzepte. Bundesamt für Naturschutz, Bonn-Bad Godesberg.

Maas, S., Detzel, P., Staudt, A. (2011): Rote Liste und Gesamtartenliste der Heuschrecken (Saltatoria) Deutschlands. Naturschutz Biol. Vielf. 70: 577–606.

MacDonald, D., Crabtree, J. R., Wiesinger, G., Dax, T., Stamou, N., Fleury, P., Gutierrez Lazpita, J.,

Gibon, A. (2000): Agricultural abandonment in mountain areas of Europe: Environmental consequences and policy response. J. Environ. Manag. 59: 47–69.

Maes, D., Bonte, D. (2006): Using distribution patterns of five threatened invertebrates in a highly fragmented dune landscape to develop a multispecies conservation approach. Biol. Conserv. 133: 490–499.

Maier, C. (2003): Untersuchungen zur Populationsentwicklung von *Oedipoda germanica* (Latr.) im Naturdenkmal Galgenberg, Main-Tauber-Kreis (Caelifera: Acrididae). Articulata 18: 193–208.

Malkmus, T. (2003): Vorkommen und Habitatvergleich zweier nahe verwandter Feldheuschrecken der *Chorthippus biguttulus*-Gruppe in einer sekundären Kontaktzone. Dipl.-Arb., Univ. Erlangen-Nürnberg.

Manzke, U. (1995): Freilandbeobachtungen zum Abflugverhalten makropterer *Chorthippus parallelus* (Zetterstedt) (Acrididae: Gomphocerinae). Articulata 10: 61–72.

Marini, L., Fontana, P., Battisti, A., Gaston, K. J. (2009): Response of orthopteran diversity to abandonment of semi-natural meadows. Agric. Ecosyst. Environ. 132: 232–236.

Meineke, T. (1994): Ausbreitungsversuche und initiale Populationsstadien von *Chorthippus parallelus* (Zetterstedt, 1821) im Hochharz. Articulata 9: 33–42.

Meineke, T., Menge, K. (1993): *Tetrix ceperoi* (Bolivar, 1887) und andere bemerkenswerte Heuschrecken in Sachsen-Anhalt gefunden (Orthoptera: Tettigoniidae, Tetrigidae, Acrididae). Entomol. Z. 103: 357–380.

Meßner, J., Kladny, M. (2017): Verhalten, Habitatwahl und Ausbreitung des Weinhähnchens *Oecanthus pellucens* (Scopoli, 1763) im Ruhrgebiet. Decheniana 170: 127–143.

Mitschke, A., Sudfeldt, C., Heidrich-Riske, H., Dröschmeister, R. (2005): The new monitoring of common breeding birds in the wider countryside of Germany – monitoring sites, field method and preliminary results. Vogelwelt 126: 127–140.

Moritz, R., Moritz, N., Meyer, S., Nippgen, K. (2022): Neuer Nachweis der Bedornten Höhlenschrecke (*Troglophilus neglectus* Krauss, 1879) außerhalb des bisherigen Verbreitungsmusters in Sachsen – (Orthoptera, Rhaphidophoridae, Troglophilinae). Entomol. Nachr. Ber. 66: 145–147.

Mutz, T. (2024): Zum Nachweis einer über mehrere Jahre reproduzierenden Population der Gottesanbeterin (*Mantis religiosa*) im Kreis Steinfurt, nördliches Nordrhein-Westfalen. Nat. Heim. 84 (3): 101–134.

Nadig, A. (1986a): Oekologische Untersuchungen im Unterengadin: Heuschrecken (Orthoptera). Ergeb. wiss. Unters. schweiz. Nationalpark 12: 103–167.

Nadig, A. (1986b): Massenvermehrung und Makropterie bei *Bicolorana bicolor, Chorthippus parallelus* und *Arcyptera fusca* [Insecta: Orthoptera]. Atti Accad. Roveretana Agiati VI, v. 26 (B): 135–139.

Nadig, A. (1991): Die Verbreitung der Heuschrecken (Orthoptera: Saltatoria) auf einem Diagonalprofil durch die Alpen (Inntal-Maloja-Bregaglia-Lago di Vomo-Furche). Jahresber. Naturforsch. Ges. Graubünden 106 (2): 5–380.

Naeem, S., Duffy, J. E., Zavaleta, E. (2012): The functions of biological diversity in an age of extinction. Science 336 (6087): 1401–1406.

Neumayer, J. (2020): Erfassung von Insekten auf Almen Hummeln, Heuschrecken, Tagfalter und Widderchen im Nationalpark Berchtesgaden. Unveröffentl. Gutachten, Bayer. Landesamt Umwelt.

Newbold, T., Hudson, L. N., Hill, S.L., Contu, S., Lysenko, I., Senior, R. A., Börger, L., Bennett, D. J., Choimes, A., Collen, B. (2015): Global effects of land use on local terrestrial biodiversity. Nature 520: 45–50.

Newson, S. E., Allez, S. L., Coule, E. K., Guille, A. W., Henney, J. M., Higgins, L., McLellan, G. D., Simmons, M. C., Sweet, E., Whitelegg, D., Atkinson, P. W. (2023): Bailiwick Bat Survey: 2022 season report. BTO Research Report 750. BTO, Thetford.

Nissen, H. (2021): Effekte einer abrupten Landnutzungsintensivierung auf den Feldgrashüpfer (*Chorthippus apricarius*) zum Laufzeitende einer extensiven Großweidenutzung. Faunist.-Ökol. Mitt. 10/2021: 19–26.

NLWKN (Niedersächsischer Landesbetrieb für Wasserwirtschaft, Küsten- und Naturschutz) (Hrsg.) (2011): Vollzugshinweise zum Schutz von Wirbellosenarten in Niedersachsen – Wirbellosenarten mit Priorität für Erhaltungs- und Entwicklungsmaßnahmen – Maulwurfsgrille (*Gryllotalpa gryllotalpa*). Niedersächsische Strategie zum Arten- und Biotopschutz, Hannover.

Öckinger, E., Schweiger, O., Crist, T. O., Debinski, D. M., Krauss, J., Kuussaari, M., Petersen, J. D., Pöyry, J., Settele, J., Summerville, K. S., Bommarco, R. (2010): Life-history traits predict species responses to habitat area and isolation: a cross-continental synthesis. Ecol. Lett. 13: 969–979.

Öckinger, E., Lindborg, R., Sjödin, N. E., Bommarco, R. (2012): Landscape matrix modifies richness of plants and insects in grassland fragments. Ecography 35: 259–267.

Ogan, S., Paulus, C., Froehlich, C., Renker, C., Kolwelter, C., Schendzielorz, M., Danielczak, A., Müller, K., Eulering, H., Hochkirch, A. (2022): Re-surveys reveal biotic homogenization of Orthoptera assemblages as a consequence of environmental change. Div. Distrib. 28: 1795–1809.

Oschmann, M. (1973): Untersuchungen zur Biotopbindung der Orthopteren. Faunist. Abh. Staatl. Mus. Tierkde. Dresden 4 (21): 177–206.

Oschmann, M. (1991): Zur Klassifizierung der ökologischen Ansprüche von Schaben (Blattodea) und Heuschrecken (Saltatoria). Faunist. Abh. Staatl. Mus. Tierkde. Dresden 18: 51–57.

Ott, J. (2014): Die Grüne Strandschrecke (*Aiolopus thalassinus*) (Fabricius, 1781) erobert die Westpfalz (Orthoptera: Acrididae). Fauna Flora Rheinland-Pfalz 12: 1523–1526.

Paracchini, M. L., Petersen, J. E., Hoogeveen, Y., Bamps, C., Burfield, I., van Swaay, C. A.M. (2008): High Nature Value Farmland in Europe, an Estimate of the Distribution Patterns on the Basis of Land Cover and Biodiversity Data. Joint Research Centre Scientific and Technical Reports, Office for Official Publications of the European Communities, Luxembourg.

Pauler, C. M., Isselstein, J., Braunbeck, T., Schneider, M. K. (2019): Influence of Highland and production-oriented cattle breeds on pasture vegetation: A pairwise assessment across broad environmental gradients. Agric. Ecosyst. Environ. 284: 106585.

Paulus, C. (2021): Ausbreitung der Großen Schiefkopfschrecke *Ruspolia nitidula* (Scopoli, 1786) im südwestlichen Rheinland-Pfalz. Articulata 36: 31–42.

Pfeifer M. A. (2006): Arealexpansion der Vierpunktigen Sichelschrecke (*Phaneroptera nana*): ein weiterer Nachweis in der nördlichen Oberrheinebene. Pollichia-Kur. 22 (3): 21–22.

Pfeifer, M. A. (2007): Schätzung des Kartierungsstandes und Umfangs der Verbreitung der Maulwurfsgrille (*Gryllotalpa gryllotalpa*) in der Pfalz. Fauna Flora Rheinland-Pfalz 11 (1): 55–70.

Pfeifer, M. A. (2012): Heuschrecken und Klimawandel. Ausbreitung vor allem südlicher Fang- und Heuschreckenarten in Rheinland-Pfalz. Nat. Landsch. 44: 205–212.

Pfeifer, M. A. (2020): Nachweis einer Einschleppung der Griechischen Marmorierten Strauchschrecke – *Eupholidoptera megastyla* (Ramme, 1939) (Saltatoria: Tettigoniidae) – nach Deutschland. Mitt. Pollichia 100: 153–156.

Pfeifer, M. A. (2024): Die Geringelte Beißschrecke *Rhacocleis annulata* Fieber, 1853, ein neues Neozoon in Rheinland-Pfalz sowie aktueller Stand der Verbreitung in Deutschland (Insecta: Saltatoria). Entomol. Z. 134: 71–74.

Pfeifer, M. A., Niehuis, M., Renker, C. (2011): Die Fang- und Heuschrecken in Rheinland-Pfalz. Fauna Flora Rheinland-Pfalz, Beih. 41: 1–678.

Pfeifer, M. A., Renker, C., Hochkirch, A., Braun, M., Braun, U., Schlotmann, F., Weitzel, M., Simon, L. (2019): Rote Liste und Gesamtartenliste der Geradflügler (Heuschrecken, Fangschrecken, Ohrwürmer und Schaben) in Rheinland-Pfalz. Ministerium für Umwelt, Energie, Ernährung und Forsten Rheinland-Pfalz, Mainz.

Pfeifer, M. A., Neugebauer, H., Eller, O. (2020): Zur Bestandsentwicklung der Italienischen Schönschrecke – *Calliptamus italicus* (Linnaeus, 1758) (Insecta: Saltatoria) – in Rheinland-Pfalz und Nordbaden. Mitt. Pollichia 100: 147–151.

Pfeuffer, E. (2015): Zum Habitat der Fluss-Strandschrecke *Epacromius tergestinus ponticus* (Karny, 1907). Ber. Naturwiss. Ver. Schwab. 119: 93–101.

Pfeuffer, E. (2018): Die Blauflügelige Sandschrecke (*Sphingonotus caerulans*) im Unteren Lechtal – einst in der Wildflussaue des Lechs, heute in der Großstadt Augsburg. Articulata 33: 47–55.

Pfeuffer, E. (2020): Biotopvielfalt in Wildflussauen am Beispiel von Heuschreckenhabitaten der einstigen Aue des außeralpinen Lechs. Articulata 35: 133–148.

Poniatowski, D., Fartmann, T. (2005): Die Ökologie von Roesels Beißschrecke (*Metrioptera roeselii*) im Feuchtgrünland der Medebacher Bucht (Südwestfalen). Articulata 20: 85–111.

Poniatowski, D., Fartmann, T. (2007): Kleinräumig heterogen strukturierte Hochheiden in mikroklimatisch günstiger Lage – Lebensräume der Kurzflügeligen Beißschrecke (*Metrioptera brachypte-*

ra) im Quellgebiet der Diemel (Südwestfalen/Nordhessen). Articulata 22: 153–171.

Poniatowski, D., Fartmann, T. (2008a): The classification of insect communities: Lessons from Orthoptera assemblages of semi-dry calcareous grasslands in central Germany. Eur. J. Entomol. 105: 659–671.

Poniatowski, D., Fartmann, T. (2008b): Phänologie und Populationsdynamik der Kurzflügeligen Beißschrecke (*Metrioptera brachyptera*) entlang eines Höhen- und Expositionsgradienten. Articulata 23: 31–41.

Poniatowski, D., Fartmann, T. (2009): Experimental evidence for density-determined wing dimorphism in two bush-crickets. Eur. J. Entomol. 106: 599–605.

Poniatowski, D., Fartmann, T. (2010): What determines the distribution of a flightless bush-cricket (*Metrioptera brachyptera*) in a fragmented landscape? J. Insect Conserv. 14: 637–645.

Poniatowski, D., Fartmann, T. (2011a): Weather-driven changes in population density determine wing dimorphism in a bush-cricket species. Agric., Ecosys. Environ. 145, 5–9.

Poniatowski, D., Fartmann, T. (2011b): Dispersal capability in a habitat specialist bush-cricket: the role of population density and habitat moisture. Ecol. Entomol. 36: 717–723.

Poniatowski, D., Fartmann, T. (2011c): Does wing dimorphism affect mobility in *Metrioptera roeselii*? Eur. J. Entomol. 108: 409–415.

Poniatowski, D., Fartmann, T. (2023): Das deutsche Heuschreckenportal OrthopteraWeb – ein Werkzeug zur Erfassung und Pflege von Beobachtungsdaten im digitalen Zeitalter. Nat. Landsch. 98 (6/7): 322–323.

Poniatowski, D., Heinze, S., Fartmann, T. (2012): The role of macropters during range expansion of a wing-dimorphic insect species. Evol. Ecol. 26: 759–770.

Poniatowski, D., Münsch, T., Helbing, F., Fartmann, T. (2018a): Arealveränderungen mitteleuropäischer Heuschrecken als Folge des Klimawandels. Nat. Landsch. 93 (12): 553–561.

Poniatowski, D., Stuhldreher, G., Löffler, F., Fartmann, T. (2018b): Patch occupancy of grassland specialists: Habitat quality matters more than habitat connectivity. Biol. Conserv. 225: 237–244.

Poniatowski, D., Beckmann, C., Löffler, F., Münsch, T., Helbing, F., Samways, M. J., Fartmann, T. (2020): Relative impacts of land-use and climate change on grasshopper range shifts have changed over time. Glob. Ecol. Biogeogr. 29: 2190–2202.

Poniatowski, D., Detzel, P., Drews, A., Hochkirch, A., Hundertmark, I., Husemann, M., Klatt, R., Klugkist, H., Köhler, G., Kronshage, A., Maas, S., Moritz, R., Pfeifer, M. A., Stübing, S., Voith, J., Winkler, C., Wranik, W., Helbing, F., Fartmann, T. (2024a): Rote Liste und Gesamtartenliste der Heuschrecken und Fangschrecken (Orthoptera et Mantodea) Deutschlands. Naturschutz Biol. Vielf. 170 (7): 1–88.

Poniatowski, D., Weißgräber, V., Drung, M., Freienstein, F. M., Kettermann, M., Scherer, G., Fartmann, T. (2024b): Grassland nature reserves safeguard a high species richness and biomass of grasshoppers. J. Appl. Ecol. 61: 2739–2750.

Poschlod, P. (2017): Geschichte der Kulturlandschaft: Entstehungsursachen und Steuerungsfaktoren der Entwicklung der Kulturlandschaft, Lebensraum- und Artenvielfalt in Mitteleuropa. 2. Aufl., Ulmer, Stuttgart.

Poschlod, P., Bakker, J. P., Bonn, S., Fischer, S. (1996): Dispersal of Plants in Fragmented Land-scapes. In: Settele, J., Margules, C., Poschlod, P., Henle, K. (Hrsg.): Species Survival in Fragmented Landscapes. Kluwer Academic Publishers, Dordrecht: 123–127.

Poschmann, C., Unterberg, U., Poniatowski, D., Fartmann, T. (2009): Ökologie der Kurzflügeligen Schwertschrecke *Conocephalus dorsalis* (Latreille, 1804) im Feuchtgrünland des Münsterlandes (Nordwestdeutschland). Articulata 24: 49–67.

Prangel, E., Kasari-Toussaint, L., Neuenkamp, L., Noreika, N., Karise, R., Marja, R., Ingerpuu, N., Kupper, T., Keerberg, L., Oja, E., Meriste, M., Tiitsaar, A., Ivask, M., Helm, A. (2023): Afforestation and abandonment of semi-natural grasslands lead to biodiversity loss and a decline in ecosystem services and functions. J. Appl. Ecol. 60: 825–836.

Rabitsch, W., Winter, M., Kühn, E., Kühn, I., Götul, M., Essl, F., Gruttke, H. (2011): Auswirkungen des rezenten Klimawandels auf die Fauna in Deutschland. Naturschutz Biol. Vielfalt 98: 1–265.

Rada, S., Spitzer, L., Sipos, J., Kuras, T. (2017): Habitat preferences of the grasshopper *Psophus stridulus*, a charismatic species of submontane pastures. Insect Conserv. Div. 10: 310–320.

Rada, S., Schweiger, O., Harpke, A., Kühn, E., Kuras, T., Settele, J., Musche, M. (2019): Protected areas do not mitigate biodiversity declines: a case study on butterflies. Div. Distrib. 25 (2): 217–224.

Reich, M. (1991a): Struktur und Dynamik einer Population von *Bryodema tuberculata* (Fabricius, 1775) (Saltatoria, Acrididae). Diss., Univ. Ulm.

Reich, M. (1991b): Grasshoppers (Orthoptera, Saltatoria) on alpine and dealpine riverbanks and their use as indicators for natural floodplain dynamics. Regul. Rivers 6: 333–339.

Reich, M. (2006): Linking metapopulation structures and landscape dynamics: Grasshoppers (Saltatoria) in alluvial floodplains. Articulata, Beih. 11: 1–154.

Reinhardt, K. (1998): Zur Eiablagepräferenz der Kleinen Goldschrecke, *Euthystira brachyptera* (Ocskay) – Freiland versus Gewächshaus. Articulata 13: 29–37.

Reinhardt, K., Köhler, G. (2002): Conservation of the red-winged grasshopper, *Oedipoda germanica* (Latr.): the influence of reproductive behaviour. Biol. Conserv. 107: 221–228.

Reinhardt, K., Köhler, G., Maas, S., Detzel, P. (2005): Low dispersal ability and habitat specificity promote extinctions in rare but not in widespread species: the Orthoptera of Germany. Ecography 28: 593–602.

Remmert, H. (1979): Grillen – oder wie groß müssen Naturschutzgebiete sein? Nationalpark 22: 7–9.

Remmert, H. (1992): Ökologie – Ein Lehrbuch. Springer, Berlin.

Renker, C. (1995): Verbreitung der Heuschrecken (Insecta: Saltatoria) im Landkreis Bernkastel-Wittlich. Fauna Flora Rheinland-Pfalz 7: 935–1025.

Renker, C., Groh, K., Pfeifer, M. A., Wiese, A. (2020): Neue Funde der Großen Schiefkopfschrecke – *Ruspolia nitidula* (Scopoli, 1786) (Orthoptera: Tettigoniidae) in Rheinland-Pfalz. Mainzer naturwiss. Arch. 56: 243–248.

Richter, C. (2008). Evolution isolierter Teilpopulationen der Laubholz-Säbelschrecke *Barbitistes serricauda* (Fabricius 1798). Biodiv. Ecol. Ser. A, 3: 1–248.

Rietze, J. (1994): Zum Ausbreitungsverhalten von Feldheuschrecken; Erfahrungen, Methoden und Ergebnisse. Articulata 9: 43–58.

Riggi, L. G. A., Berggren, Å. (2020): Small field islands systems include a large proportion of the regional orthopteran species pool in arable landscapes. J. Insect Conserv. 24: 695–703.

Rockström, J., Steffen, W., Noone, K., Persson, Å., Chapin, F. S., Lambin, E. F., Lenton, T. M., Scheffer, M., Folke, C., Schellnhuber, H. J. (2009): A safe operating space for humanity. Nature 461: 472–475.

Roderus, D. (2020): Artensteckbrief Italienische Schönschrecke *Calliptamus italicus* (Linnaeus, 1758). Hessisches Landesamt für Naturschutz, Umwelt und Geologie.

Roe, R. M., Clifford, C. W., Woodring, J. P. (1980): The effect of temperature on feeding, growth, and metabolism during the last larval stadium of the female House Cricket, *Acheta domesticus*. J. Insect Physiol. 26: 639–644.

Röber, H. (1951): Die Dermapteren und Orthopteren Westfalens in ökologischer Betrachtung. Abh. Landesmus. Naturkde. Münster 14: 1–60.

Röller, O., Sturm, M. (2017): Erster Populationsnachweis der Atlantischen Bergschrecke, *Antaxius pedestris* (Fabricius, 1787) in Deutschland (Saltatoria, Tettigoniidae). Entomol. Nachr. Ber. 61: 55–58.

Rösch, V., Hoffmann, M., Diehl, U., Entling, M. H. (2019): The value of newly created wood pastures for bird and grasshopper conservation. Biol. Conserv. 237: 493–503.

Rothhaupt, G. (1994): Die Situation der Wanstschrecke *Polysarcus denticauda* in Bayern und Thüringen. Articulata 9: 79–87.

Rudy, H. (1925): Die Wanderheuschrecke *Locusta migratoria* L. phasa *migratoria* L. et phasa *danica* L. Beiträge zu einer Monographie. Badische Bl. Schädlingsbekämpf., Sonderbeil.: 1–34.

Rühlke, M. (2021): Erstfund der Westlichen Sattelschrecke (*Ephippiger diurnus*, Dufour, 1841) im Niederrheinischen Tiefland. Articulata 36: 23–30.

Rutschmann, F., Roesti, C. (2024): Orthoptera.ch. https://www.orthoptera.ch. (Zugriff am 15.03.2024).

Sändig, S. (2024): Tiefes Mulchen als wirksame Artenschutzmaßnahme für den in Baden-Württemberg vom Aussterben bedrohten Schwarzfleckigen Heidegrashüpfer *Stenobothrus nigromaculatus* (Herrich-Schäffer 1840). Articulata 39: 101–116.

Sändig, S. (2025): Rotflügelige Ödlandschrecke (*Oedipoda germanica*) – Fördermaßnahmen auf einer ehemaligen Weinbergsfläche im Kochertal bei Ingelfingen (Baden-Württemberg). Artenschutz und Biodiversität 6 (2): 1–12.

Sändig, S., Fritze, M.-A. (2022): Erstnachweis der Gestreiften Sumpfgrille *Pteronemobius lineolatus* (Brullé, 1835) für Deutschland. Articulata 37: 47–58.

Sänger, K. (1977): Über die Beziehungen zwischen Heuschrecken (Orthoptera: Saltatoria) und der Raumstruktur ihrer Habitate. Zool. Jahrb., Abt. Syst. Oekol. Geogr. Tiere 104: 433–488.

Samietz, J. (1994): Verbreitung und Habitatbindung der Zweifarbigen Beißschrecke, *Metrioptera bicolor* (Phil.), in Thüringen (Insecta: Saltatoria: Tettigoniidae). Faun. Abk. Staatl. Mus. Tierkde. Dresden 19: 153–166.

Samietz, J. (1996): Zur Mikrohabitatnutzung einer Heuschreckenart in Halbtrockenrasen: *Stenobothrus lineatus* (Panzer) (Insecta: Caelifera). Verhandl. Ges. Ökol. 26: 569–573.

Samietz, J. (1998): Populationsgefährdungsanalyse an einer Heuschreckenart. Methoden, empirische Grundlagen und Modellbildung bei *Stenobothrus lineatus* (Panzer). Cuvillier, Göttingen.

Samu, S., Wischhof, S. (2000): Individuenreiche Population von *Leptophyes albovittata* (Kollar, 1833) in einer stillgelegten Mergelgrube in Mecklenburg-Vorpommern – Ein Beitrag zur Habitatpräferenz der Art an ihrer nördlichen Arealgrenze. Articulata 15: 179–191.

Samways, M. J. (2005): Insect Diversity Conservation. Cambridge University Press, Cambridge.

Samways, M. J., McGeoch, M. A., New, T. R. (2010): Insect Conservation. A Handbook of Approaches and Methods. Oxford University Press, Oxford.

Sander, U. (1992): Fund eines Weinhähnchens, *Oecanthus pellucens* (Scopoli 1763) (Insecta, Saltatoria), bei Bonn (Nordrhein-Westfalen). Articulata 7: 51–54.

Sander, U. (1995): Neue Erkenntnisse über Verbreitung und Bestandssituation des Weinhähnchens *Oecanthus pellucens* (Scopoli 1763) (Gryllidae, Oecanthinae) im nördlichen Rheinland-Pfalz und in Nordrhein-Westfalen. Articulata 10: 73–88.

Sardet, E., Roesti, C., Braud, Y. (2015): Cahier d'identification des Orthoptères de France, Belgique, Luxembourg et Suisse. Biotope Éditions, Mèze.

Schädler, M. (2009): Ein neues Vorkommen des Zwerggrashüpfers, *Stenobothrus crassipes* (Charpentier, 1825) (Caelifera, Acrididae), in Deutschland. Entomol. Nachr. Ber. 53 (3–4): 203–206.

Schädler, M., Stadler, J. (2000): Verbreitung und Lebensraum des Kiesbank-Grashüpfers, *Chorthippus pullus* (Phillipi, 1830) (Acrididae: Gomphocerinae), in Sachsen. Articulata 15: 7–15.

Schädler, M., Nicolai, B., Schäfer, B., Schulze, M. (2019): Aktuelle Funde südlich verbreiteter Insektenarten (Blattodea, Mantodea, Orthoptera, Odonata) in Mittel- und Ostdeutschland (Sachsen-Anhalt, Sachsen, Brandenburg, Thüringen). Entomol. Nachr. Ber. 63 (3): 269–279.

Schäfer, B. (2013): Nachweis der Heideschrecke *Gampsocleis glabra* (Herbst 1786) (Ensifera) in der Colbitz-Letzlinger Heide (Sachsen-Anhalt). Articulata 28: 115–126.

Schäfer, B., Hennigs, S. (2020): Nachweise der Heideschrecke *Gampsocleis glabra* (Herbst, 1786) (Ensifera) in der Altengrabower sowie in der Klietzer Heide (Brandenburg/Sachsen-Anhalt). Articulata 35: 117–127.

Schielzeth, H. (2000): Mobilität von *Tetrix subulata* und *T. undulata* als Anpassung an die Dynamik der Aue. NUA-Seminarber. 6: 88–91.

Schirmel, J. (2011): Response of the grasshopper *Myrmeleotettix maculatus* (Orthoptera: Acrididae) to invasion by the exotic moss *Campylopus introflexus* in acidic coastal dunes. J. Coast. Conserv. 15: 159–162.

Schirmel, J., Fartmann, T. (2007a): Larvalökologie des Grünen Heupferds *Tettigonia viridissima* (Orthoptera: Tettigonidae. Articulata 22: 173–190.

Schirmel, J., Fartmann, T. (2007b): Gibt es Unterschiede in der tages- und jahreszeitlichen Aktivität von *Tettigonia caudata* und *T. viridissima* (Orthoptera: Tettigoniidae)? Articulata 22: 191–203.

Schirmel, J., Fartmann, T. (2013): Coexistence of two related bush-cricket species (Orthoptera: *Tettigonia caudata*, *T. viridissima*) in an agricultural landscape. Biologia 68: 510–516.

Schirmel, J., Blindow, I., Fartmann, T. (2010): The importance of habitat mosaics for Orthoptera (Caelifera and Ensifera) in dry heathlands. Eur. J. Entomol. 107: 129–132.

Schirmel, J., Mantilla-Contreras, J., Blindow, I., Fartmann, T. (2011): Impacts of succession and grass encroachment on Orthoptera in heathlands. J. Insect Conserv. 15: 633–642.

Schlick-Steiner, B. C., Steiner, F. M., Buschinger, A. (2003): Ameisengrillen (*Myrmecophilus* spp.) im wissenschaftlichen Niemandsland Mitteleuropas (Ensifera: Myrmecophilidae). Beitr. Entomofaunist. 4: 67–73.

Schlumprecht, H., Waeber, G. (2003): Heuschrecken in Bayern. Ulmer, Stuttgart.

Schlumprecht, H., Waeber, G. (2022): Artenhilfsprogramm. Schwarzfleckiger Heidegrashüpfer *Stenobothrus nigromaculatus*. Überprüfung der Vorkommen in Nordbayern und Empfehlungen zur Erhaltung und Förderung. Bayer. Landesamt Umwelt.

Schmidt, C., Schmitt, T. (2011): Vorkommen der Ameisengrille (*Myrmecophilus acervorum* (Panzer, 1799)) im Bochumer Raum. Online-Veröff. Bochumer Bot. Ver. 3 (3): 25–32.

Schmidt, R., Dufresnes, C., Krištín, A., Künzel, S.,Vences, M., Hawlitschek, O. (2024): Phylogenetic insights into Central European *Chorthippus* and *Pseudochorthippus* (Orthoptera: Acrididae) species using ddRADseq data. Mol. Phylogenet. Evol. 193: 108012.

Schoof, N., Luick, R., Zehm, A., Morhard, J., Nickel, H., Renk, J., Schaefer, L., Fartmann, T. (2024): Naturverträgliche Mahd von Grünland und Pflege von Straßenbegleitgrün – Technik, Verfahren, Auswirkungen und Empfehlungen für die Praxis. Naturschutz-Prax. Landschaftspfl. 4: 1–82.

Schuch, S., Bock, J., Leuschner, C., Schaefer, M., Wesche, K. (2011): Minor changes in orthopteran assemblages of Central European protected dry grasslands during the last 40 years. J. Insect Conserv. 15: 811–822.

Schuch, S., Bock, J., Krause, B., Wesche, K., Schaefer, M. (2012): Long-term population trends in three grassland insect groups: a comparative analysis of 1951 and 2009. J. Appl. Entomol. 136: 321–331.

Schuhmacher, O., Fartmann, T. (2003a): Offene Bodenstellen und eine heterogene Raumstruktur – Schlüsselrequisiten im Lebensraum des Warzenbeißers (*Decticus verrucivorus*). Articulata 18: 71–93.

Schuhmacher, O., Fartmann, T. (2003b): Wie mobil ist der Warzenbeißer? Eine populationsökologische Studie zum Ausbreitungsverhalten von *Decticus verrucivorus*. Naturschutz Landschaftspl. 35 (1): 20–28.

Schuhmacher, O., Kelm, H.-J. (2021): Expansion der Feldgrille (*Gryllus campestris*) in den Dürre-Jahren 2018–2020 im Nordosten Niedersachsens (LK Lüchow-Dannenberg). Articulata 36: 105–112.

Schüle, M., Heinken, T., Fartmann, T. (2023): Long-term effects of environmental alterations in protected grasslands – Land-use history determines changes in plant species composition. Ecol. Eng. 188: 106878.

Schulte, A. M. (2003): Taxonomie, Verbreitung und Ökologie von *Tetrix bipunctata* (Linnaeus 1758) und *Tetrix tenuicornis* (Sahlberg 1893) (Saltatoria: Tetrigidae). Articulata, Beih. 10: 1–226.

Schulz, B. (2003): Zur Bedeutung von Beweidung und Störstellen für Tierarten am Beispiel der Verteilung von Feldheuschreckengelegen im Grünland. Articulata 18: 151–178.

Schulze, M. (1998): Neue Funde von *Isophya kraussii* Brunner von Wattenwyl, 1878 und weiteren seltenen Heuschreckenarten in Sachsen-Anhalt. Articulata 13: 47–51.

Schwarz, C., Fartmann, T. (2022): Traditional grazing management creates heterogeneous swards and fosters grasshopper densities. Insect Sci. 29: 1805–1818.

Schwarz, C., Fumy, F., Drung, M., Fartmann, T. (2023): Insect-friendly harvest in hay meadows – Uncut refuges are of vital importance for conservation management. Glob. Ecol. Conserv. 48: e02731.

Schwarz, C. J., Keller, M., Berger, D. (2017): Neues zur Gottesanbeterin, *Mantis religiosa* Linnaeus, 1758, (Mantodea, Mantidae), dem Insekt des Jahres 2017. Entomol. Nachr. Ber. 61 (1): 1–18.

Sczepanski, S. (2005): Die Heuschreckenfauna des NSG „Holzplatz" bei Bönen (Kreis Unna) unter besonderer Berücksichtigung der Gestreiften Zartschrecke (*Leptophyes albovittata* Koll.) und der Gemeinen Sichelschrecke (*Phaneroptera falcata* Poda) (Insecta: Saltatoria). Nat. Heim. 65: 65–76.

Sczepanski, S., Jacobi, B. (2005): Notizen zur Ausbreitung der Südlichen Eichenschrecke (*Meconema meridionale* Costa) in Nordrhein-Westfalen (Insecta: Saltatoria). Nat. Heim. 65: 1–6.

Seehausen, M. (2016): Ein neuer Fundort von *Aiolopus thalassinus* (Fabricius, 1781) an der nördlichen Verbreitungsgrenze in Hessen (Orthoptera: Acrididae). Articulata 31: 45–48.

Smettan, H. (1991): Erstnachweis der Wanstschrecke (Saltatoria: *Polysarcus denticauda* Charp.) am Alpennordrand. Nachrichtenbl. Bayer. Entomol. 40 (1): 30–32.

Sobczyk, T., Köhler, G. (2024): Ein außergewöhnliches Massenvorkommen der Italienischen Schönschrecke, *Calliptamus italicus* (Linnaeus, 1758), in Sachsen (Orthoptera, Acrididae, Calliptaminae). Entomol. Nachr. Ber. 68: 492–499.

Sobczyk, T., Trampenau, M. (2011): Zum aktuellen Auftreten der Italienischen Schönschrecke (*Calliptamus italicus*) in der Oberlausitz, Sachsen (Caelifera: Acrididae). Sächs. Entomol. Z. 6: 90–95.

Squires, V.R., Dengler, J., Feng, H., Hua, L. (2018): Grasslands of the World. Diversity, Management and Conservation. CRC Press, Boca Raton.

Stadler, J. (1997): Zu Verbreitung, Verhalten und Ökologie der Nadelholz-Säbelschrecke (*Barbitistes constrictus*) in der Nationalparkregion Sächsische Schweiz. Artenschutzrep. 7: 23–25.

Stärz, C., Buchweitz, M., Fartmann, T. (2010): Feuer – (k)eine Chance für die Gottesanbeterin? Populations- und Larvalökologie von *Mantis religiosa*? Fire – a chance for the praying mantis? Population and larval ecology of *Mantis religiosa* in vineyard slopes. In: Fartmann, T. (Hrsg.): Brandmanagement in Rebböschungen des Kaiserstuhls: Auswirkungen auf thermophile Insekten als Zielarten des Naturschutzes – Fire management of vineyard slopes in the Kaiserstuhl region: impacts on thermophilous insects as flagship species for conservation. Arb. Inst. Landschaftsökol. 19: 21–76.

Stapf, J., Röller, O., Schirmel, J. (2021): Neue Nachweise der Grünen Strandschrecke (*Aiolopus thalassinus*) im Pfälzerwald. Pollichia-Kur. 37 (3): 31–32.

Statistisches Bundesamt (2023): Tiere und tierische Erzeugung – Schafbestände leicht gestiegen. https://www.destatis.de. (Zugriff am 21.11.2023).

Stelter, C., Reich, M., Grimm, V., Wissel, C. (1997): Modelling persistence in dynamic landscapes: lessons from a metapopulation of the grasshopper *Bryodema tuberculata*. J. Animal Ecol. 66: 508–518.

Stenger, M. (2020): Schrecken ohne Ende im Rheingraben – Zur aktuellen Verbreitung der Europäischen Gottesanbeterin, *Mantis religiosa religiosa* Linnaeus, 1758 (Mantodea, Mantidae), in Baden-Württemberg. Articulata 35: 105–116.

Strätz, Ch., Schlumprecht, H. (1999): Verbreitung ausgewählter „Kryptischer Heuschrecken" in Oberfranken – Plumpschrecke, Laubholz- und Nadelholz-Säbelschrecke. 73. Ber. Naturforsch. Ges. Bamberg: 107–124.

Straube, S. (2013): Zur Biologie und Ökologie der Ödlandschrecken *Sphingonotus caerulans* (L.) und *Oedipoda caerulescens* (L.) (Caelifera, Acrididae) unter Berücksichtigung verschiedener Bedingungen in einer mitteldeutschen Flusslandschaft. Shaker, Aachen.

Streitberger, M., Ackermann, W., Fartmann, T., Kriegel, G., Ruff, A., Balzer, S., Nehring, S. (2016a): Artenschutz unter Klimawandel: Perspektiven für ein zukunftsfähiges Handlungskonzept. Naturschutz Biol. Vielf. 147: 1–367.

Streitberger, M., Jedicke, E., Fartmann, T. (2016b): Auswirkungen des rezenten Klimawandels auf die Biodiversität in Mittelgebirgen – eine Literaturstudie zu Arten und Lebensräumen. Naturschutz Landschaftspl. 48 (2): 37–45.

Streitberger, M., Stuhldreher, G., Fartmann, T., Ackermann, W., Ludwig, H., Pütz, S., Züghart, W. (2024): The German insect monitoring scheme: establishment of a nationwide long-term recording of arthropods. Basic Appl. Ecol. 80: 81–91.

Stübing, S., Hundertmark, I., Reiners, T. E. (2019a): Beobachtungen zur Ausbreitung von Weinhähnchen (*Oecanthus pellucens*), Vierpunktiger Sichelschrecke (*Phaneroptera nana*) und Südlicher Grille (*Eumodicogryllus bordigalensis*) in Hessen. Articulata 34: 127–138

Stübing, S., Krummen, N., Hundertmark, I., Geske, C. (2019b): Vorkommen und Ausbreitung der Gottesanbeterin (*Mantis religiosa*) in Hessen. Articulata 34: 139–145.

Stuhldreher, G., Fartmann, T. (2018): Threatened grassland butterflies as indicators of microclimatic niches along an elevational gradient – Implications for conservation in times of climate change. Ecol. Indic. 94: 83–98.

Sturm, P., Zehm, A., Baumbach, H., von Brackel, W., Verbücheln, G., Stock, M., Zimmermann F. (2018): Grünlandtypen. Erkennen – Nutzen – Schützen. Quelle, Meyer, Wiebelsheim.

Sudfeldt, C., Dröschmeister, R., Wahl, J., Berlin, K., Gottschalk, T., Grüneberg, C., Mitschke, A., Trautmann, S. (2012). Vogelmonitoring in Deutschland – Programme und Anwendungen. Naturschutz Biol. Vielfalt 119: 1–266.

Sutton, P. G., Beckmann, B. C., Nelson, B. (2017): The current status of orthopteroid insects in Britain and Ireland. Atropos 59: 6–35.

Sutton, P. G., Beckmann, B., Roy, H. (2024): UK Orthoptera. https://orthoptera.org.uk. (Accessed: 15/03/2024).

Svennig, J.-C. (2002): A review of natural vegetation openness in north-western Europe. Biol. Conserv. 104: 133–148.

Termaat, T., Van Strien, A. J., Van Grunsven, R. H. A., Knijf, G. de, Bjelke, U., Burbach, K., Conze, K.-J., Goffart, P., Hepper, D., Kalkman, V. J., Motte, G., Prins, M. D., Prunier, F., Sparrow, D., Van den Top, G. G., Vanappelghem, C., Winterholler, M., WallisDeVries, M. F. (2019): Distribution trends of European dragonflies under climate change. Div. Distrib. 25 (6): 936–950.

Thorens, P. (1989): Descroptions comparee des ootheques et des oeufs de Chorthippus mollis (Charp.) et de *Chorthippus biguttulus* (L.) (Orthoptera, Acrididae). Mitt. Schweiz. Entomol. Ges. 62: 87–106.

Thorn, S., König, S., Fischer-Leipold, O., Gombert, J., Griese, J., Thein, J. (2022): Temperature preferences drive additive biotic homogenization of Orthoptera assemblages. Biol. Lett. 18: 20220055.

Tillmanns, O. (2007): Ein neuer Fund von *Myrmecophilus acervorum* (Panzer, 1799) an der nordwestlichen Verbreitungsgrenze. Articulata 22: 99–101.

Trampenau, M., Sobczyk, T. (2011): Wiederfund der Rotflügeligen Schnarrschrecke (*Psophus stridulus*) in der Oberlausitz, Sachsen (Caelifera: Acrididae). Sächs. Entomol. Z. 6: 96–98.

Trautner, J. (2024): Wanstschrecke (*Polysarcus denticauda*) – Zum Erstnachweis in Deutschland im 19. Jahrhundert im Neckartal bei Tübingen und ihrem dortigen Erlöschen. Artenschutz Biodiv. 5 (6): 1–8.

Trautner, J., Buchweitz, M. (2002): Kein Etikettenschwindel – die „Grille" aus dem Jagsttal. Articulata 17: 89–92.

Trautner, J., Hermann, G. (2008): Die Sumpfschrecke (*Stethophyma grossum* L., 1758) im Aufwind – Erkenntnisse aus dem zentralen Baden-Württemberg. Articulata 23: 37–52.

Trautner, J., Rietze, J. (2020): 10.14 Sicherung der Östlichen Grille in einem Weinberggebiet. In: Trautner, J. (Hrsg.): Artenschutz. Rechtliche Pflichten, fachliche Konzepte, Umsetzung in der Praxis. Ulmer, Stuttgart: 250–257.

Treiber, R. (2016): Klimabedingte Ausbreitung der Großen Schiefkopfschrecke in Baden-Württemberg. Naturschutz Landschaftspfl. Bad.-Württ. 78: 307–323.

Treiber, R., Stephan, U. (1995): Nachweis von *Pholidoptera aptera* (Fabricius, 1793) im östlichen Elbsandstein-Gebirge (Sachsen). Articulata 10: 93–95.

Unruh, M., Klaus, D. (2005): Nachweise der Nadelholz-Säbelschrecke (*Barbitistes constrictus* Br. v. W. 1878) im Süden des Burgenlandkreises/Sachsen-Anhalt (Orthoptera, Ensifera, Phanopteridae). Entomol. Mitt. Sachsen-Anhalt 13: 112–125.

van Elst, A., Schulte, T. (1995): Freilandfunde der Südlichen Grille, *Tartarogryllus burdigalensis* (Latr., 1804) und der „Exotischen Grille", *Gryllodes sigillatus* (Walk., 1869) (Orthoptera: Gryllidae) im südlichen Rheinland-Pfalz. Articulata 10: 185–191.

Van Strien, A. J., Van Swaay, C. A. M., Van Strien-van Liempt, W. T., Poot, M. J., WallisDeVries, M. F. (2019): Over a century of data reveal more than 80% decline in butterflies in the Netherlands. Biol. Conserv. 234: 116–122.

Veen, P., Jefferson, R., de Smidt, J., van der Straaten, J. (2009): Grasslands in Europe of High Nature Value. KNNV Publishing, Zeist.

Venne, C. (2008): Beitrag zur Verbreitung der Dornschrecken (Saltatoria: Tetrigidae) im Landschaftsraum Senne mit Erstnachweisen der Westlichen Dornschrecke (*Tetrix ceperoi* [Bolívar, 1887]) für Ostwestfalen-Lippe. Mitt. Arbeitsgem. Westf. Entomol. 24 (1): 1–13.

Vogels, J. J., Van de Waal, D. B., WallisDeVries, M. F., Van den Burg, A. B., Nijssen, M., Bobbink, R., Berg, M. P., Olde Venterink, H., Siepel, H. (2023): Towards a mechanistic understanding of the impacts of nitrogen deposition on producer-consumer interactions. Biol. Rev. 98: 1712–1731.

Voith, J., Beckmann, A., Sachteleben, J., Schlumprecht, H., Waeber, G. (2016): Rote Liste und Gesamtartenliste der Heuschrecken (Saltatoria) Bayerns. Stand 2016. Bayer. Landesamt Umwelt.

Vossen, B., Piper, W. (1996): Wiederfund der Rotflügeligen Schnarrschrecke *Psophus stridulus* (L., 1758) für Brandenburg. Articulata 11: 103–108.

Wagner, G. (2000): Eine Populationsgefährdungsanalyse der Rotflügeligen Ödlandschrecke, *Oedipoda germanica* (Latr. 1804) (Caelifera: Acrididae). Articulata, Beih. 9: 1–126.

Wallaschek, W. (1995): Untersuchungen zur Zoozönologie und Zönotopbindung von Heuschrecken (Saltatoria) im Naturraum „Östliches Vorland". Articulata, Beih. 5: 1–153.

Wallaschek, M. (2005): Zur Heuschreckenfauna (Ensifera et Caelifera) des Landes Schollene unter besonderer Berücksichtigung der Heideschrecke *Gampsocleis glabra* (Herbst, 1786).

Entomologische Mitteilungen Sachsen-Anhalt, Sonderheft: 1–121.

Wallaschek, M. (2013): Die Geradflügler des Landes Sachsen-Anhalt (Insecta: Dermaptera, Mantodea, Blattoptera, Ensifera, Caelifera): Aktualisierung der Verbreitungskarten. Ent. Mitt. Sachsen-Anhalt, Sonderh.: 1–100.

Wallaschek, M. (2014): Zur Zoogeographie und Zoozönologie der Geradflügler (Dermaptera, Blattoptera, Orthoptera) des Huy (Sachsen-Anhalt). Entomol. Mitt. Sachsen-Anhalt 22 (2): 97–106.

Wallaschek, M. (2020): Rote Liste Sachsen-Anhalt. 31. Heuschrecken. Ber. Landesamt Umweltschutz Sachsen-Anhalt 1/2020: 505–511.

Wallaschek, M., Langner, T. J., Richter, K. (2004): Die Geradflügler des Landes Sachsen-Anhalt. Heuschrecken, Ohrwürmer, Fangschrecken und Schaben. Ber. Landesamt. Umweltschutz Sachsen-Anhalt, Sonderh. 5: 1–288.

WallisDeVries, M. F., Bobbink, R. (2017): Nitrogen deposition impacts on biodiversity in terrestrial ecosystems: Mechanisms and perspectives for restoration. Biol. Conserv. 2012: 387–389.

Weibart, M., Fischer, K. (2006): Populationsstruktur, Dispersionsverhalten und Habitatpräferenzen der Rotflügeligen Schnarrschrecke (*Psophus stridulus* L.) in der Fränkischen Schweiz. Articulata 21: 169–182.

Weidner, H. (1953): Die Wanderheuschrecken. Die Neue Brehm-Bücherei 96: 1–48.

Weidner, H. (1972): Das Heimchen, oder die Hausgrille, *Acheta domesticus* (Linnaeus, 1758). Prakt. Schädlingsbekämpf. 24: 72–76.

Weipert, J., Köhler, G. (2022): Beiträge zur Fauna des Kyffhäusergebirges. Teil XI: Die Heuschrecken (Insecta: Ensifera, Caelifera) aus Effizienzkontrollen im Naturschutzgroßprojekt „Kyffhäuser" (2004–2006). Thüring. Faunist. Abh. 27: 129–143.

Wells, C. E., Hodgkinson, D., Huckerby, E. (2000): Evidence for the possible role of beaver (*Castor fiber*) in the prehistoric ontogenesis of a mire in northwest England, UK. Hewiit 10: 503–508.

Wiesener, C., Wanner, M., Xylander, W. E. R. (2004): *Chorthippus pullus* (Philippi, 1830) (Saltatoria) auf dem Truppenübungsplatz „Oberlausitz". Entomol. Nachr. Ber. 48 (1): 47–50.

Wiesner, K., Habel, J. C., Gossner, M. M., Loxdale, H. D. Köhler, G., Schneider, A. R. R., Tiedemann, R., Weisser, W. W. (2014): Effects of habitat performance and land-use intensity on the genetic structure of the grasshopper species *Chorthippus parallelus*. R. Soc. Open Sci. 1: 140133.

Willemse, L. P. M., Kleukers, R. M. J. C., Odé B. (2018): The grasshoppers of Greece. EIS Kenniscentrum Insecten, Naturalis Biodiversity Center, Leiden.

Winkler, C., Haacks, M. (2019): Die Heuschrecken Schleswig-Holsteins – Rote Liste, 4. Fassung. Landesamt für Landwirtschaft Umwelt und ländliche Räume des Landes Schleswig-Holstein, Flintbek.

Worschech, K. (2007): Zum Vorkommen der Westlichen Dornschrecke *Tetrix ceperoi* (Bolívar, 1887) im Landkreis Altenburger Land. Landschaftspfl. Naturschutz Thüring. 44 (3): 97–103.

Worschech, K. (2017): Vorkommen des Feldgrashüpfers *Chorthippus apricarius* (Linnaeus, 1758) (Saltatoria: Acrididae) in der intensiv genutzten Agrarlandschaft des Altenburger Landes (Thüringen). Mauritiana 32: 360–386.

Wöss, G., Denner, M., Forsthuber, L., Kropf, M., Panrok, A., Reitmeier, W., Zuna-Kratky, T. (2020): Insekten in Wien – Heuschrecken. Österreichische Gesellschaft für Entomofaunistik, Wien.

Wranik, W. (2024): Zum Vorkommen der Italienischen Schönschrecke *Calliptamus italicus* (Linnaeus, 1758) in Mecklenburg-Vorpommern. Articulata 39: 25–36.

Wranik, W., Kleeberg, A. (2011): Interessante Neu- und Wiederfunde von Heuschrecken in Mecklenburg-Vorpommern. Articulata 26: 151–162.

Wranik, W., Lange, L. (2007): Nachweis der Blauflügeligen Sandschrecke (*Sphingonotus caerulans*) für Mecklenburg-Vorpommern. Arch. Freunde Naturgesch. Mecklenburg 46: 203–207.

Wranik, W., Königstedt, D. G., Röbbelen, F. (1996): Rote Liste der gefährdeten Heuschrecken Mecklenburg-Vorpommerns. Ministerium für Landwirtschaft und Naturschutz des Landes Mecklenburg-Vorpommerns, Güstrow.

Wranik, W., Meitzner, V., Martschei, T. (2008): Verbreitungsatlas der Heuschrecken Mecklenburg-Vorpommerns. Beiträge zur floristischen und faunistischen Erforschung des Landes Mecklenburg-Vorpommerns. Landesamt für Umwelt, Naturschutz und Geologie Mecklenburg-Vorpommerns, Güstrow.

Wünsch, Y., Schirmel, J., Fartmann, T. (2010): Habitatnutzung juveniler *Myrmeleotettix maculatus* (Caelifera: Gomphocerinae) und *Platycleis albopunctata* (Ensifera: Tettigoniinae) in Küstendünenheiden. Articulata 25: 167–183.

Wünsch, Y., Schirmel, J., Fartmann, T. (2012): Conservation management of coastal dunes for Orthoptera has to consider oviposition and nymphal preferences. J. Insect Cons. 16: 501–510.

Zacher, F. (1917): Die Geradflügler Deutschlands und ihre Verbreitung. Gustav Fischer, Jena.

Zahn, A. (2020): Ausbreitung der Blauflügeligen Sandschrecke in Südostbayern? ANLiegen Natur 42 (1): 144–145.

Zahn, A., Englmaier, I., Drobny, M. (2010): Food availability for insectivores in grasslands – arthropod abundance in pastures, meadows and fallow land. Appl. Ecol. Environ. Res. 8: 87–100.

Zechner, L., Klapf, H. (2005): Aktuelle Nachweise des Kiesbank-Grashüpfers *Chorthippus pullus* (Philippi, 1830) in der Steiermark (Saltatoria). Joannea Zool. 7: 171–178.

Zechner, L., Koschuh, A. (2000): Zur Verbreitung der Italienischen Schönschrecke *Calliptamus italicus* (Linné, 1758) in der Steiermark (Saltatoria, Acrididae). Joannea Zool. 2: 83–97.

Zechner, L., Fachbach, G., Lazar, R. (2000): Verbreitung und Habitatansprüche der Östlichen Grille *Modicogryllus frontalis* (Fieber, 1844) in der Steiermark, Österreich (Saltatoria, Gryllidae). Joannea Zool. 2: 37–69.

Zehm, A. (1997): Zur Koinzidenz von Sandvegetation, ihrer Struktur und Heuschrecken-Zönosen (Orthoptera) in der hessischen Oberrheinebene. Tuexenia, N. F. 17: 193–222.

Zehm, A. (2003): Analyse der vertikalen Vegetationsstruktur: Methodenentwicklung und Beziehungen zu Pflanzengesellschaften, Mikroklima, Beweidung und Heuschrecken-Gemeinschaften. Diss., TU Darmstadt.

Zimmermann, P., Hafner, A. (2011): Eine Ödlandschrecke erobert den Hochschwarzwald – Neufunde der Lauchschrecke (*Mecostethus parapleurus*) in Baden-Württemberg. Carolinea 69: 127–132.

Zinke, J. (2000) Nachweis der Höhlenschrecke *Troglophilus neglectus* Krauss, 1879 in Deutschland (Ensifera, Rhaphidophoridae, Troglophilinae). Entomol. Nachr. Ber. 44 (3): 161–163.

Zinner, F., Schlegel, C., Richter, K. (2000): Zum Vorkommen von *Podisma pedestris* (L. 1758) in der Königsbrücker Heide, Sachsen (Caelifera: Acrididae, Melanoplinae). Articulata 15: 245–249.

Zöller, S. (1995): Untersuchungen zur Ökologie von *Oedipoda germanica* (Latreille, 1804) unter besonderer Berücksichtigung der Populationsstruktur, der Habitatbindung und der Mobilität. Articulata 19: 21–59.

Zuna-Kratky, T., Karner-Ranner, E., Lederer, E., Braun, B., Berg, H.-M., Denner, M., Bieringer, G., Ranner, A., Zechner, L. (2009): Verbreitungsatlas der Heuschrecken und Fangschrecken Ostösterreichs. Naturhistorisches Museum Wien, Wien.

Zuna-Kratky, T., Landmann, A., Illich, I., Zechner, L., Essl, F., Lechner, K., Ortner, A., Weißmair, W., Wöss, G. (2017): Die Heuschrecken Österreichs. Denisia 39: 1–880.

Index

Scientific genus and species names are highlighted in *italics*; the first page of each species profile is printed in **bold** type.

Contributor profiles

Prof. Dr. Thomas Fartmann is an ecologist and biogeographer. He has been the head of the Department of Biodiversity and Landscape Ecology at Osnabrück University since 2016. His main research and teaching topics are biodiversity monitoring, disturbance ecology, global change ecology and restoration ecology. To date, he has published more than 270 scientific publications – among them 125 articles in international journals and 15 books – on animal ecology, vegetation ecology and nature conservation. Since the beginning of his scientific career in the early 1990s, Mantodea and Orthoptera have been an important model group in his research. He has dealt with these insects in more than 70 publications.

> t.fartmann@uos.de

Dr. Dominik Poniatowski is a landscape ecologist with a special interest in animal ecology and vegetation science. He has been a Research Associate in the Department of Biodiversity and Landscape Ecology at Osnabrück University since 2016. His fields of research cover different aspects of conservation biology and community ecology. Currently, he is analysing the effects of land-use change and global warming on species diversity. Grasshoppers are among his most important research objects. He has been studying this group of insects for more than 20 years. During this time, he has published over 40 papers on grasshoppers in national and international journals and books. The new red list of the German grasshopper and mantis species, which he supervised, has just been published.

> dominik.poniatowski@uni-osnabrueck.de

Jonas Brüggeshemke is a landscape ecologist currently doing his PhD at Osnabrück University. His research particularly focuses on disturbance ecology and impacts of land-use change on biodiversity in Central Europe. A part of his recent work has been devoted to the impacts of military and civil heathland management on grasshoppers.

Prof. Dr. Peter Detzel has a degree in biology and worked for many years as a self-employed biologist/landscape ecologist in Baden-Württemberg. His scientific focus was always on grasshoppers. He has written several reference books on this subject. During his active career, he supervised numerous theses in the field of zoology, also with a focus on grasshoppers, and published numerous scientific papers. He retired in 2020 and has since maintained his passion for grasshoppers as a hobby.

Arne Drews works in the Department of Biodiversity at the Schleswig-Holstein State Authority for Environment on the technical requirements for faunistic species protection, in particular on the compilation of red lists of various animal groups. The coordination of species surveys for monitoring and protection projects for endangered species is also a focal point of his work.

Dr. Daniel Elias works at the Anhalt University of Applied Sciences in Bernburg. He has been involved in a large number of applied research projects to enhance biodiversity in agricultural landscapes. One of his research areas is restoration and management of grassland habitats. He is the author of many scientific papers on this subject. In addition, he appears as a speaker at national and international conferences.

Felix Maximilian Freienstein is a landscape ecologist working at Osnabrück University. His research focuses on biodiversity in the context of global change, habitat management and restoration. Current key activities include the conceptualisation and realisation of the grasshopper monitoring scheme in Germany as well as participation in the state monitoring programmes in North Rhine-Westphalia and Lower Saxony.

Dr. Florian Fumy is a landscape ecologist working at Osnabrück University. In his research, he focuses on the impact of climate and land-use change on grassland ecosystems and their biodiversity in Central Europe. A major aspect of his recent work involves studying grasshopper communities in semi-natural pastures and meadows. He regularly publishes his research in international scientific journals.

Felix Helbing is a landscape ecologist working as a research assistant in the Department of Biodiversity and Landscape Ecology at Osnabrück University. His research interests include various aspects of restoration ecology and the impact of land-use and climate change on biodiversity in Central Europe. In his work, grasshoppers play an important role as indicators in various research and monitoring projects.

Dr. Lisa Holtmann is a landscape ecologist working as a research assistant at Osnabrück University. During her doctoral thesis, she investigated the biodiversity of insects and plants in urban stormwater ponds. In current projects, she primarily focuses on the monitoring of grasshoppers and butterflies as well as on the effects of different types of grassland management on biodiversity. To date, she has published ten scientific papers.

Inga Hundertmark is a biologist and works at the Hessian Society for Ornithology and Nature Conservation (HGON). She is a member of the 'Grasshopper Network' of the Hessian state working group for Orthoptera and contributes to the creation of distribution maps for this state through her own surveys and by collecting and processing further survey data.

Dr. Sigfrid Ingrisch is a retired entomologist who formerly worked at the universities in Giessen, Aachen, and Zurich and at the Zoological Research Museum in Bonn. His research focused on distribution, ecology and embryonic development of bush-crickets, as well as on taxonomy and systematics of that group. To date, he has authored more than 160 scientific publications, including textbooks, and is still active in research and as a reviewer of manuscripts for international journals.

Dr. Steffen Kämpfer is a landscape ecologist and research associate at the Department of Biodiversity and Landscape Ecology at Osnabrück University. His research focuses on the influence of climate and land-use change on biodiversity and the development of conservation measures for endangered species. His current research is particularly concerned with the protection of open-grassland species.

Dr. Marcel Kettermann is a conservation biologist and research associate at Osnabrück University. A focus of his research is on studies of biodiversity in human-influenced habitats such as quarries and sand pits. He also investigates environmental factors that affect the habitat selection and distribution of grasshoppers. Currently, he is involved in various conservation projects, including grasshopper monitoring in Lower Saxony and North Rhine-Westphalia.

Dr. Günter Köhler is an ecologist and entomologist at the Friedrich Schiller University of Jena, Germany. Currently, his focus is on regional ecofaunistics, including alien insects. Many of his 370 papers deal with Orthoptera. He is the author of several checklists and red lists, co-author of the *Stresemann – Exkursionsfauna von Deutschland* and editor of the well-selling *Bestimmung wirbelloser Tiere*.

Dr. Sebastian König is an ecologist working at the Technical University of Munich. In his research and teaching, he particularly focuses on the patterns, dynamics and protection of biodiversity in the context of global change. Furthermore, he studies trophic interactions between organisms and publishes in international journals. Orthoptera, which he has been fascinated by since childhood, constitute a common element of his research.

Dr. Andreas Kronshage is a geographer and landscape ecologist. He worked for several years at the Metelen Biological Institute. He currently works for the Education and Research Center Heiliges Meer at the LWL Museum of Natural History. In North Rhine-Westphalia, Andreas Kronshage heads the grasshopper working group. His scientific publications cover a wide range with a focus on faunistics as well as on animal and water ecology.

Dr. Franz Löffler is a landscape ecologist based at Osnabrück University. In his research, he especially focuses on changes in Central European biodiversity and their drivers. Grasshoppers form an important part of his studies. He is currently coordinating the grasshopper monitoring in the German Federal States of Lower Saxony and North Rhine-Westphalia. To date, he has authored more than 30 scientific publications, 22 of which have been publsihed in international journals.

Thorsten Münsch is a landscape ecologist working in the Department of Biodiversity and Landscape Ecology at Osnabrück University. Current research projects include the monitoring of butterflies and grasshoppers in North Rhine-Westphalia (Tagfalter- und Heuschreckenmonitoring) and the research project 'Insect-friendly Management of Roadside Verges' (Insektenfreundliche Gestaltung von Straßenseitenflächen).

Manfred Alban Pfeifer studied biology and forestry. As a student, he came across grasshoppers during a conservation project. Today, he works – among other things – as an entomologist. He is the lead author of the *Atlas of the Mantids, Crickets and Grasshoppers of Rhineland-Palatinate* and the red list of the orthopteroid insects of this federal state. He was also the coordinator for Rhineland-Palatinate of the German red list of grasshoppers.

Dr. Michael Reich is a Professor of Nature Conservation and Landscape Ecology (retired) at Leibniz University Hannover. Since the mid-1980s, he has been researching dynamic processes in Alpine rivers and their significance for grasshoppers and their habitats. In several research projects for nature conservation authorities, he has investigated the effects of hydropower utilisation on the terrestrial communities of the floodplain and their biodiversity.

Dr. Carsten Renker is a biologist and works as the head of the zoological collections at the Mainz Natural History Museum. He is also the managing director of ConchBooks Publishing House and a freelance expert in nature and species conservation. Since 1993, he has been working on the distribution and ecology of grasshoppers and molluscs. To date, he has authored more than 130 scientific publications.

Jörg Rietze is a biologist and scientific staff member at the Working Group for Animal Ecology and Planning Company in Filderstadt. He has worked on the dispersal of grasshoppers at roadside verges and green bridges and has conducted surveys of grasshopper communities in many regions of southwestern Germany. Moreover, he has contributed to several red lists, is involved in species-specific monitoring projects and is the author of several scientific publications.

Dr. Jens Sachteleben is a biologist and managing partner at PAN GmbH, Munich – a planning agency for applied nature conservation. His fields of work are faunistic studies, nature conservation planning, nature conservation-related research and development projects (e. g. on the Habitats Directive Monitoring Scheme) and the implementation of nature conservation projects. He has been working on grasshoppers – with a focus on southern Germany – since the 1980s.

Sebastian Sändig is a biologist and scientific staff member at the Working Group for Animal Ecology and Planning Company in Filderstadt. In his current work, he focuses on monitoring as well as on development and application of species conservation measures, among others for threatened grasshopper species in Baden-Württemberg. He has contributed to several red lists and is author of some scientific publications.

Gwydion Scherer works as a conservation biologist at Osnabrück University. The main focus of his research is the EU-wide protected Marsh Fritillary (*Euphydryas aurinia*). In addition to autecological studies, he analyses restoration measures for the threatened butterfly and its habitats. Another research focus is on the effects of global change on butterflies, grasshoppers, wild bees and plants as model organisms.

Dr. Cinja Schwarz works as landscape ecologist at Osnabrück University. In her current work, she studies the effects of grassland management on insect diversity. She dealt with the topic of land-use change in her PhD thesis: among other things, she compared the diversity of grasshopper communities in traditionally managed pastures with those of the surrounding grasslands in the foothills of the Alps.

Dr. Merle Streitberger is a landscape ecologist and works at Osnabrück University. She has been working on the conceptualisation of the German Insect Monitoring Scheme, which includes the monitoring of grasshoppers in grasslands, since 2018. Beyond that, her research focuses on land-use effects on biodiversity. In addition to 15 publications in international journals, she is the co-author of a book that outlines the drivers of insect decline in Central Europe.

Stefan Stübing is a biologist and has been fascinated since childhood by all animals that can fly. As a co-owner of the planning agency for faunistics and scientific instructor of the Hessian Society for Ornithology and Nature Conservation (HGON), he is involved in the recording, monitoring and protection of birds, grasshoppers and dragonflies. He initiated the 'Grasshopper Network' of the Hessian state working group for Orthoptera, which coordinates the recording of this animal group in Hesse.

Dr. Gregor Stuhldreher is a landscape ecologist and post-doctoral researcher at Osnabrück University. His work covers various topics of community and restoration ecology, with a special interest in the effects of land-use and climate change on butterflies, grasshoppers and vegetation and their implications for nature conservation. Furthermore, he is involved in the development of monitoring schemes for insect populations throughout Germany.

Jürgen Trautner is a landscape ecologist and the managing director of the Working Group for Animal Ecology and Planning Company in Filderstadt. Moreover, he founded a non-profit company that supports the conservation of highly threatened species. Methodological issues and the interface between technical and legal aspects of the protection of species and biotopes are among his work priorities. He has authored many scientific publications.

Johannes Voith is a landscape ecologist that works at the Bavarian Centre for Species Protection within the Bavarian Environmental Agency in Augsburg. There, he is responsible for basic research, species action plans and red lists of invertebrates. He was significantly involved in the drafting of the Bavarian distribution atlas of grasshoppers in 2003. His main interests focus on different groups of insects in the Bavarian Alps, including grasshoppers.

Christian Winkler has been studying grasshoppers for more than 20 years as part of his freelance work and as a member of the Schleswig-Holstein grasshopper working group. He is the author of the red list of grasshoppers in Schleswig-Holstein and is coordinating the work for a new distribution atlas. At the federal level, he was involved in the preparation of endangerment analyses for grasshoppers and the current version of the red list.

Dr. Wolfgang Wranik worked as a marine biologist for several decades at the University of Rostock (macrozoobenthos, biodiversity of Yemen). As part of entomological leisure research, he has been working on the grasshopper fauna in Mecklenburg-Western Pomerania since the 1980s. The resulting publications include the red list and a distribution atlas of the grasshoppers of this state.

Dr. Andreas Zehm is a biologist/ecologist. After positions at the Bavarian Environmental Agency, the Government of Swabia, the Bavarian Agency for Geoinformatics and Land Development and the Academy for Nature Conservation and Landscape Management, he has been at the Bavarian State Ministry of the Environment and Consumer Protection since 2020. Topics of interest are: biodiversity, biotope connectivity, botany and community ecology.

Image Credits

Cover photos: Front cover: Male Rattle Grasshopper (*Psophus stridulus*): Dominik Poniatowski; back cover left: Male Meadow Grasshopper (*Pseudochorthippus parallelus*): Thomas Fartmann; right: Female Praying Mantis (*Mantis religiosa*): Dominik Poniatowski.
Photos inside the book: p. 54/55: Female Stripe-winged Grasshopper (*Stenobothrus lineatus*): Dominik Poniatowski. The copyright information for all other photos can be found in the captions. Drawings, graphics and maps: inside cover, map of major geographical features in Germany: Felix Helbing; p. 2: Praying Mantis (*Mantis religiosa*): Rainer Poniatowski. Unless otherwise stated, all other maps and graphics were created by Felix Helbing, with the assistance of Thomas Fartmann, Hannah Kalthoff, Marcel Kettermann, Franz Löffler, Dominik Poniatowski, Merle Streitberger and Gregor Stuhldreher.

Imprint

Bibliographic information of the German National Library
The German National Library lists this publication in the German National Bibliography; detailed bibliographic data is available on the Internet at http://dnb.d-nb.de.

Wollgrasweg 41, 70599 Stuttgart (Hohenheim)
E-Mail: info@ulmer.de
Internet: www.ulmer-verlag.de
Project management: Ulf Müller, Wuppertal
Manufactoring and Cover design: Verlag Eugen Ulmer
Typesetting: Fotosatz Buck, Kumhausen
Printing and binding: Livonia Print, Lettland
Printed in Latvia

ISBN: 978-3-8186-25696